SABKHA ECOSYSTEMS

Tasks for vegetation science 36

SERIES EDITORS

Sabkha Ecosystems

Volume I: The Arabian Peninsula and Adjacent Countries

Edited by

HANS-JÖRG BARTH

University of Regensburg,
Department of Physical Geography,
Regensburg, Germany

and

BENNO BÖER

UNESCO Regional Office in the Arab States of the Gulf,
Doha, Qatar

KLUWER ACADEMIC PUBLISHERS

DORDRECHT/BOSTON/LONDON

A C.I.P. Catalogue record for this book is available from the Library of Congress.

ISBN 978-90-481-5972-7

Published by Kluwer Academic Publishers,
P.O. Box 17, 3300 AA Dordrecht, The Netherlands.

Sold and distributed in North, Central and South America
by Kluwer Academic Publishers,
101 Philip Drive, Norwell, MA 02061, U.S.A.

In all other countries, sold and distributed
by Kluwer Academic Publishers,
P.O. Box 322, 3300 AH Dordrecht, The Netherlands.

Printed on acid-free paper

The designations employed and the presentation of material throughout this publication do not imply the expression of any opinion whatsoever on the part of UNESCO, the University of Regensburg, and ERWDA concerning legal status of any country, territory, city or area or of its authorities, or concerning the delimination of its frontiers or boundaries.

Printed in the Netherlands

Contents

List of Authors

Abbas, Jameel: Department of Biology, University of Bahrain, P.O. BOX 32038, Isa Town, Bahrain.

Abed, Abdulkader M: Department of Geology, University of Jordan, 11943 Jordan.

Al-Farraj, Asma: Geography Department. UAE University, P.O. Box 17771, Al Ain, UAE.

Al-Gifri, Abdul Nasser: Department of Biology, University of Aden, P.O. Box 6014, Khormaksar, Aden, Yemen.

Al-Hajiri, Saif: The Friends of the Environment Centre, Qatar

Alsharhan, Abdul Rahman. S.: Faculty of Sciences, UAE University, P.O. Box 17551 Al Ain, UAE.

Aspinall, Simon: Abu Dhabi Islands Archaeological Survey, P.O. Box 45553, Abu Dhabi, UAE.

Barth, Hans K.: University of Paderborn, Hohefeld 6, 33100 Paderborn Dahl, Germany.

Barth, Hans-Jörg: Department of Physical Geography, Universitätsstr. 31, 93040 Regensburg, University of Regensburg, Germany.

Beech, Mark: Department of Archaeology, The King's Manor, University of York, York Y01 7EP, UK.

Belnap, Jayne: US Geological Survey, Biological Resources Division, 2282 S. Resource Blvd., Moab UT 84532.

Böer, Benno: UNESCO Regional Office in the Arab States of the Gulf, P.O. Box 3945, Doha, Qatar.

Breckle, Siegmar-Walter: Department of Ecology, Faculty of Biology, University of Bielefeld, P.O. Box 100131, D-33619 Bielefeld, Germany.

Clüsener-Godt, Miguel: UNESCO - Division of Ecological Sciences, Man and the Biosphere Programme (MAB) 1, rue Miollis, 75732 Paris, Cedex 15, France.

Cohen, A.: Department of Geological Sciences, University of South Carolina, Colombia, SC 29208, USA.

Cunningham, Peter L.: Polytechnic of Namibia, School of Natural Resources and Tourism, Dept. Nature Conservation, Private Bag 13388, Windhoek 9000, Namibia.

de Soyza, Amrita G.: TERC/ERWDA, P.O Box 45553, Abu Dhabi, UAE.

Evans, Graham: School of Earth and Ocean Sciences, Southampton Oceanography Centre, University of Southampton, European Way, Southampton, England, UK.

Fahmy, Ahmed G.: Dept. of Botany, Faculty of Science, University of Helwan, Cairo, Egypt.

Gabali, Saeed A.: Department of Biology, University of Aden, P.O. Box 6014, Khormaksar, Aden, Yemen.

Ghazanfar, Shahina A.: Dept. of Plant Sciences, Downing Site, Cambridge CB2 3EA, U.K.

Gill, Anthony C.: Fish Research Group, Department of Zoology, The Natural History Museum, Cromwell Road, London SW7 5BD, UK.

Gul, Bilquees: Department of Botany, University of Karachi, Karachi-75270, Pakistan.

Hellyer, Peter: Abu Dhabi Islands Archaeological Survey, P.O. Box 45553, Abu Dhabi, United Arab Emirates.

Hogarth, Peter, J.: Department of Biology, University of York, P.O. Box 373, York YO10 5YW, UK.

John, David M.: The Natural History Museum, Department of Botany, Cromwell Road, London SW7 5BD, UK.

Kendall, Christopher G. St. C.: Department of Geological Sciences & Marine Sciences, University of South Carolina, Columbia, South Carolina 29208, USA.

Khan, Ajmal: Department of Botany, University of Karachi, Karachi-75270, Pakistan.

Kirkham, Anthony: Reservoir Characterisation, Research & Consulting (UK) Limited, 5 Greys Hollow, Rickling Green, Saffron Walden, Essex, CB11 3YB, England, UK.

Lieth, Helmut: c/o Institut für Umweltforschung, Artilleriestr. 34, D-49076 Osnabrück, Germany.

Loughland, Ronald A.: Commission of Environmental Research Emirates Heritage Club Abu Dhabi, P.O. Box 41464, Abu Dhabi, UAE.

Menzel, Uwe: University of Osnabrück, Artilleriestr. 34, D-49069 Osnabrück, Germany.

Misak, Raafat F.: Environmental & Earth Sciences Division, Remote Sensing Group, P.O. Box 24885, 13109 Safat, Kuwait.

Omar, Samira: Kuwait Institute for Scientific Research, P.O. Box 24885, 13109 Safat, Kuwait.

Osman, Omyma: University of Juba, P.O. Box 12311, Khartoum, Sudan.

Phillips, Ron: 3100 South Kinney Road 77, Tucson, Arizona 85713, USA.

Round, Frank: School of Biological Sciences, Woodland Road, Bristol BS8 1UG, England, UK.

Sanford, Ward, E.: U. S. Geological Survey, 431 National Center, Reston, VA 20192, USA.

Shahid, Shabbir, A.: Kuwait Institute for Scientific Research, P.O. Box 24885, 13109 Safat, Kuwait.

Tigar, Barbara, J.: Department of Biological Sciences, University of Paisley, High Street, Paisley, A12BE, Scotland, UK.

Vistro, Najamuddin, B.: Environmental Research and Wildlife Development Agency, P.O. Box 45553, Abu Dhabi, United Arab Emirates.

Wood, Warren: Research Hydrologist, MS 430, National Center, US Geological Survey, 12201 Sunrise Valley Drive, Reston, VA 20192, USA.

Zajonz, Uwe; Section Ichthyology, Senckenberg Research Institute and Natural History Museum, Senckenberganlage 25, 60325 Frankfurt am Main, Germany.

Sabkhas can potentially be converted into man-made, seawater-irrigated ecosystems.

Foreword

Sabkhas are an integral part of the natural heritage of the Arabian Peninsula Region, including Abu Dhabi Emirate. Extensive inter-tidal and inland saline flats are locally known as sabkhas, a term which describes saline desert flats devoid of vegetation; the term has been adopted by scientists. Research on sabkha dynamics commenced in the early 1960s.

On first sight sabkhas seem to be without any life at all, and of little value for humans, because of the lack of freshwater and rangeland for livestock. A closer look however, reveals abundant evidence of plant and animal life in the sabkhas. Additionally, sabkhas can potentially be converted into man-made, seawater-irrigated ecosystems, either for the benefit of wildlife, and for the generation of jobs and income.

In order to manage these unique systems to our nations' and region's best benefit, it is necessary to study both, the biotic and abiotic factors of sabkha ecosystems in detail, as well as the history and the potential development prospects. Once these factors are understood, we can generate recommendations on how best to develop and conserve the vast sabkhas of the Arabian Peninsula Region.

In this book the editors and their colleagues have collated information on regional sabkhas. Throughout they have shown great sensitivity towardc the sabkha systems, towards the present and future well-being of the region as a whole. This scholarly work is an essential multi-disciplinary contribution towards the sustainable utilization and the conservation of the sabkhas as real ecosystems, as well as towards the understanding of our natural heritage, and I warmly welcome it.

Hamdan Bin Zayed Al Nahyan

Minister of State for Foreign Affairs
Deputy Chairman
Environmental Research and Wildlife Development Agency

مقدمـــة

من الضروري دراسة العوامل الحيوية وغير الحيوية للنظام البيئي للسبخة بهدف فهم هذه العوامل ووضع توصيات لتطوير وحماية السبخات في المنطقة.

من النظرة الأولى يعتقد الإنسان بأن السبخة خالية من مظاهر الحياة . ولا قيمة لها للإنسان، بسبب افتقارها للمياه العذبة والنباتات الضرورية لرعي الماشية. ولكن من خلال التعرف على السبخات عن قرب يتم الكشف عن أدلة تؤكد وجود حياة للنباتات والحيوانات. بالإضافة إلى إمكانية تحويلها إلى نظم بيئية اصطناعية يمكن زراعتها بمياه البحر لتنمية الحياة الفطرية. بالإضافة إلى أنها توفر فرصا للعمل كأحد مصادر الدخل.

وتعتبر السبخات جزءاً هاما من التراث الطبيعي لمنطقة شبه الجزيرة العربية. ويشمل ذلك إمارة أبو ظبي. "والسبخة" تعبير يستخدم محليا لوصف مناطق المد والجزر والمسطحات الملحية الصحراوية الخالية من النباتات في المناطق الداخلية؛ كما أستخدم العلماء هذا التعبير. وعلى الرغم من انتشار السبخات في المنطقة منذ زمن طويل الا أن الأبحاث العلمية لدارسة نظم السبخات لم تبدأ بشكل جدي الا مع بداية الستينات.

وللاستفادة من هذه النظم الفريدة بما يخدم أمتنا ومنطقتنا كان من الضروري إجراء دراسة تفصيلية عن العوامل الحيوية وغير الحيوية للنظام البيئي للسبخة. ومن خلال فهم هذه العوامل. يمكن وضع توصيات لتطوير وحماية السبخات المنتشرة في منطقة شبه الجزيرة العربية.

وفي هذا الكتاب جمع المؤلفون وزملاؤهم معلومات عن السبخات في المنطقة. عبروا خلالها عن اهتمام واضح بنظم السبخات من اجل حاضر ومستقبل أفضل للمنطقة ككل. ويمثل هذا العمل العلمي مساهمة فعالة من قبل متخصصين في عدة مجالات تهدف إلى الاستفادة المستدامة من السبخات كنظام بيئي حقيقي لفهم تراثنا الطبيعي بشكل أفضل.

حمدان بن زايد آل نهيان

وزير الدولة للشؤون الخارجية

نائب رئيس مجلس إدارة

هيئة أبحاث البيئة والحياة الفطرية وتنميتها

Preface

As a very distinctive geomorphological and geo-ecological phenomenon sabkhat are to be found in semi-arid and arid regions all over the world. Where ever those parameters defining aridity determine a desert environment, sabkhat are developed due to specific morphodynamics in combination with certain hydrologic preconditions. In general their genesis and appearance depend on geo-structural and environmental elements which are best developed within the tropical and subtropical regions around the Arabian Peninsula. Therefore, the Arab term "sabkha" is justified to describe the wide variety of these extensive and barren saltflats which are intermittendly inundated in continental regions or periodically flooded in coastal plains.

In spite of the global distribution and the large variety in appearance, sabkhat did not attract scientific interest for a long time. Due to the hostile environment with high salinity of soils and absence of higher plants, they were regarded as wastelands and unsuitable for development or any kind of economic exploitation. Scientific attention as well as the perception of economic and ecologic importance were changing in the last decades.

In fact, the here presented collection of contributions from experts of various disciplines demonstrate that their research led already to a far-reaching understanding of the sabkha-phenomenon. This volume with 30 papers summarises earlier research and considerable detailed work of recent studies in a regional context of the Arabian Peninsula and the surrounding areas. It is a successful first attempt to present the existing knowledge not only with regard to the physical features of sabkhat, but equally to describe and to understand the sabkha environment as a complex physical and biotic ecosystem.

It is unquestionable that the here published research results of authors and editors can not be estimated high enough. They will be inspiring and initiate further research and they will be also a basic source of information for decision-makers with respect to ecologic value and potential development and conservation requirements.

Hans Karl Barth
University of Paderborn, Germany

Acknowledgements

This multi-disciplinary volume on sabkhat could not have been done, without the enthusiastic participation of the article contributors. We therefore offer our sincere thanks to all the ladies and gentlemen, listed as authors of this book.

Furthermore, most of the time we spent on the organisation of the volume, while being employed at the Environmental Research and Wildlife Development Agency (ERWDA), in Abu Dhabi, and at the Faculty of Physical Geography of the University of Regensburg, Germany.

We wish to express our gratitude to the following persons for their interest, encouragement, and support: HH Sheikh Khalifa bin Zayed al Nahyan (Chairman of ERWDA), HH Sheikh Hamdan bin Zayed al Nahyan (Vice Chairman of ERWDA), the ERWDA Executive Committee members, Mr. Mohammed al Bowardi (Managing Director), Mr. Ahmad al Sayagh, and Mr. Majed al Mansouri, Prof. Dr. Klaus Heine (Head of Department, Faculty of Physical Geography, University of Regensburg) and Ernst Ardelean (Cartographer, University of Regensburg). We thank the referees who, with their expertise helped to improve the manuscripts.

We also thank the Kluwer Academic Publishers, for their support and flexibility, and professional way of handling the publishing process. Special thanks are due to Dr. Ursula Hertling, Dr. Mary Kelly, Prof. Dr. Helmut Lieth and Prof. Dr. Hans Karl Barth.

INTRODUCTION

HANS-JÖRG BARTH & BENNO BÖER

Sabkha is an Arabic term for flat salt-crusted desert. The local terminology of the Arabian Gulf region describes the extensive, barren, salt encrusted and periodically flooded coastal flats as well as inland salt flats.

Besides *sabkha* in other languages there are numerous terms for salt flats, which is not surprising, considering the wide spread occurrence of these features throughout the arid environments. Some examples include: *pan* (southern Africa), *salt pan* (Australia), *nor* (Mongolia), *takyr* (Russia), *kavir* (Iran), *salar* (Chile) *dry lake* and *playa* (USA). However, these terms do not necessarily represent geomorphologically and geologically identical features. A substantial literature exists on the subject of salt flats, yet they remain little known geomorphic features. Quite some material on playas has been brought together by Neal (1975) and in vol. 54 (1986) of Palaeogeography, Palaeoclimatology, Palaeoecology. But the *sabkha* as a significant feature of arid coastal lowlands and some inland areas has not received much attention yet. Even in most recent works (Briere 2000), where the attempt was made to classify the different salt flats, the continental or inland sabkha is considered to be equivalent to playa, ignoring the significant differences in sedimentology, hydrology and genesis.

With this volume we would like to emphasise out, that *sabkha* is in a geomorphic sense not an equivalent term for playa, which is the widespread opinion though. In the scientific literature the term *sabkha* is applied in different contexts basically dealing with salt flats in one or the other form. There is conformity as long as *sabkha* stands for flats having a surface of high salt content which is the result of an equilibrium between sedimentation and deflation processes usually controlled by the shallow groundwater level. Often *coastal* and *inland sabkha* are distinguished. In the latter case *sabkha* is used equally to playa which is the general term for arid internal drainage basins that periodically flood and accumulate sediment. Because sabkhat are geologically quite different features a universal use of *sabkha* or *playa* respectively is confusing.

Therefore some authors restrict the use of the term *sabkha* only to coastal salt flats like the ones described at the UAE Gulf coast. Kinsman (1969), Glennie (1970) and Whitten & Brooks (1972) differentiate between *coastal sabkha* meaning a supratidal surface, which is periodically flooded by sea water, and *inland sabkha* located in the interior. Warren (1991) also differentiates *coastal sabkha* and *inland sabkha* and additionally *interdune sabkha* depending on the dominant physical

Barth & Böer (eds.), SABKHA ECOSYSTEMS, 1-5

processes (apart from evaporation), which can be a) marin-coastal b) fluvio-lacustrine and c) aeolian.

These examples of which there are numerous demonstrate that there has been and still is quite some discussion going on about the terminology. In this context we see the necessity to define the scientific use of the term *sabkha* in a geomorphic as well as ecological view. It is essential for good science, as well as for science based environmental management, that there is a uniform use throughout the disciplines of the geo-ecological sciences.

What makes the sabkha ecosystem distinguishable from other ecosystems is the total absence of higher plants, except for the sabkha edges, and areas with the influence of low salinity water. Opposite the term sabkha in Arabic *seih* may be used, which means a flat area with vegetation. This is normally an interdune valley with solonchak related soils and some halophytic shrubs. So far ecologists and naturalists alike use the term sabkha for all kinds of saline deserts, with and without vegetation, which is basically incorrect, and causes unnecessary confusion to what a sabkha actually is.

There is also the widespread believe that sabkhat are wastelands. Therefore many sabkhat in the Arabian Peninsula region undergo currently a rapid process of uncoordinated development, including land filling, top-soiling and dredging. Possibly sabkhat are of inferior ecological value when compared to other ecosystems, however, we do not know yet. The ecological importance of sabkhat has so far not been defined. There may be factors which are frequently being overlooked. For example is there very little known about primary productivity, cyanobacterial mats, diatoms, birds and benthic organisms possibly depending on the sabkha ecosystem. This volume of TVS will present the existing knowledge about sabkhat and discuss their ecological value as well as potential development techniques and conservation requirements.

Since *sabkha* is an Arabic term used to describe the abundant and extensive salt flats around the Arabian Peninsula we would like to apply it scientifically to geomorphic features similar to those. In this context *sabkha* is a flat and barren area basically consisting of sand or finer unconsolidated substrate showing surficial evaporite accumulations. *Coastal sabkhat* are periodically flooded by sea water. The ground water of *coastal sabkhat* is of marine origin as well as most sabkha sediments. *Inland sabkhat* are only flooded episodically by precipitation. They are not catchments of internal drainage and mostly aeolian in nature. Their surface as well as the surface *of coastal sabkhat* is an equilibrium of deflation and aeolian sedimentation, controlled by the local shallow ground water level which is the lower limit of deflation.

Sabkhat occur in arid environments where evaporation by far exceeds precipitation which lies usually below 250 mm/year. Therefore they are most common within the tropical and subtropical arid zones. Besides climatic aspects geomorphic and geological as well as hydrological conditions are important parameters determining sabkha formation. Sabkhat can only develop where unconsolidated sandy or finer sediments are present. All described sabkhat show a shallow ground water level between a few cm and two meters below the surface. In the case of coastal sabkhat low lying coastlines and tide is necessary to provide an extensive flooding area for sedimentation.

Within the limits set by climatic conditions sabkhat are quite common geomorphic features around the world. Our objectives are threefold:

- to define and classify sabkha terminology (see table 1 below).

  Beside the land form and sub-classes, based on the water inundation regime, or based on the moisture regime, there are a number of common anthropogenic factors influencing the sabkhat so, that they form man-made systems. Among others one of the most important factors is, of course, the presence of low to medium salinity water, which changes the biology, chemistry, as well as the physical environment of the sabkha ecosystem drastically. We therefore decided to add anthropogenic classes.

- to generate the most comprehensive multidisciplinary review on sabkhat.

- to generate recommendations on sabkha-management for land-owners, land users and decision makers.

It turned out that there will be too much material if we do not confine the subject to specific regions. Therefore we intend to publish a compilation of recent scientific studies focused on the sabkhat of the Arabian Peninsula and adjacent regions as volume I. This includes ecological as well as geomorphic studies. Other regions will follow in separate volumes. The 31 papers included are grouped into three principal parts.

Part I, Occurrence of sabkhat within the Arabian Peninsula and the adjacent countries. The 13 contributions of part one deal with sabkha occurrence mostly on country scale. According to the discipline and interest of the authors the papers focus on a variety of different aspects.

Part II, Sabkha ecology. The 14 papers of this section describe various subjects regarding the ecology of sabkha environments.

Part III, Sabkha land use and development. The 4 contributions of the last section cover some historical information and they provide prospects on future sabkha development and conservation.

Table 1. Sabkha classification

LANDFORM

1. Coastal sabkha: If the sabkha is frequently inundated by marine tides, then it is a coastal sabkha.

2. Supralittoral sabkha: If the sabkha is rarely inundated by marine tides, then it is a supralittoral sabkha. The supralittoral sabkha is an intermediate type between types 1 and 3, and it is often recognisable by the presence of the most landward fringe of old marine driftwood.

3. Inland sabkha: If the sabkha is never inundated by marine tides, then it is strictly speaking a terrestrial sabkha. For the ease of understanding we decided to continue the term inland sabkha, since it was frequently used for this type of sabkha-landform.

SUBCLASSES

Permanently wet sabkha: Permanently wet sabkha systems such as lakes and ponds. After complete evaporation of all surface water or retreat of tidal water, the groundwater level never drops more than 10 cm below surface level. This implies that the sabkha sediment is permanently wet due to capillary rise of groundwater. This sub class is often related to anthropogenic action.

Periodically flooded sabkha: Periodically flooded sabkha systems can be either permanently wet or periodically dry. Characteristic is that they are flooded on a regular basis, by high tides in the case of coastal sabkhat, or once ore twice a year due to precipitation during the winter season.

Permanently dry sabkha: Permanently dry sabkha systems are characterised by a relatively dry surface sediment (depending on the hygroscopic effects of salts). The ground water level is between 10 and 200 cm below the surface level. Temporarily wet surface can occur during sparse rain events. Episodically heavy rains may cause a flooding which usually does not persist.

ANTHROPOGENIC CLASSES

Low salinity: The presence of low to medium salinity water.
Landfill: Any evidence of landfill
Excavation: Any evidence of excavation.
Plantation: Any evidence of plantation.
Other: Any evidence of other anthropogenic factors: needs specification.

Section I of this book "Distribution of sabkhat" deals with the occurrence of sabkhat mostly on the country scale. Authors present reviews of sabkha distribution in the country of interest as well as descriptions of special examples they have been working on. Since the authors are a team of interdisciplinary scientists, several different aspects will be covered. This presents the reader with a comprehensive picture of the sabkhat of the Arabian Peninsula and the adjacent regions.

We started with the sabkhat of the UAE because they are the largest and best documented sabkhat in the region. Evans & Kirkham were among the first scientists who published important research results of the sabkhat of Abu Dhabi between 1960 and 1970. In this volume they present an overview of the sabkhat around Abu Dhabi with special emphasise on evolution, sedimentology and geology. Kendall et al. focus on sedimentary and diagenetic facies zonation of the Abu Dhabi sabkhat. Barth continues with an introduction into the sabkhat of Saudi Arabia and differences between coastal and inland sabkhat. Abbas describes the plant communities bordering the sabkhat of Bahrain. Böer & Al Hajiri present a paper on sabkhat distribution and halophytes in Qatar. Omar et al. review the sabkhat of Kuwait with special interest in soil taxonomy and halophytes. Abed presents the sabkhat of Jordan with results of his work on the Taba Sabkha and Ghazanfar reports about the sabkha vegetation of Oman. Breckle reviews sabkha environments in Iran and Afghanistan whereas Khan & Gul focus on the vegetation of the coastal sabkhat in Pakistan. Al-Gifri provides a description of sabkha ecosystems in Yemen. Böer et al. review Djibouti, Eritrea, Ethiopia, Somalia and Sudan regarding sabkha occurrence and associated vegetation. Finally Fahmy & Böer present an introduction on the vegetation of the Read Sea sabkhat in Egypt.

References

Briere, P.R. 2000. Playa, playa lake, sabkha: proposed definitions for old terms. *Journal of Arid Environments*, 45: 1-7.

Glennie, K.W. 1970. Desert sedimentary environments. *Developments in Sedimentologie* 14: 222.

Kinsman, D.J. 1969. Modes of formation, sedimentary associations and diagnostic features of shallow water and supratidal evaporites. *Amer. Assoc. Petrol. Geologists, Bull.* 53: 830-840.

Neal, J.T. (ed.) 1975. Playas and dried lakes. Stroudsburgh, Pennsylvania: Dowden, Hutchinson and Ross.

Warren, J.K. (1991). Sulfate dominated sea-marginal platform evaporative settings. In: Melvin, J.L. (ed.), Evaporites, Petrolium and Mineral Resources, *Developments in Sedimentology* 50: 69-187.

Whitten, D.G. & Brooks, J.R. 1972. Penguin Dictionary of Geology, Penguin books Ltd.

PART I

DISTRIBUTION OF SABKHAT WITHIN THE ARABIAN PENINSULA AND THE ADJACENT COUNTRIES

THE ABU DHABI SABKHA

GRAHAM EVANS[1] & ANTHONY KIRKHAM[2]

[1] *School of Earth and Ocean Sciences, Southampton Oceanography Centre, University of Southampton, European Way, Southampton, England, UK.*
[2] *Reservoir Characterization, Research & Consulting (UK) Limited, 5 Greys Hollow, Rickling Green, Saffron Walden, Essex, CB11 3YB, England, UK.*

Abstract

The coastal plain -the sabkha- of Abu Dhabi developed by drowning of the outer parts of the deserts of Arabia during the Flandrian transgression. Subsequent deposition and progradation of an inner lagoonal shoreline protected by a chain of barrier islands, together with deflation of the adjacent desert dunefields down to the level of the water table has produced a relatively flat coastal plain. Evaporation and complex reaction between the sediment and the highly saline groundwater has produced a complex association of evaporitic minerals. Differences in elevation of microbial deposits developed and buried during the coastal progradation and the present day contemporary microbial mat suggest modern changes in the contemporary sea level are possibly recorded. This suggests that the low altitude of the present sabkha plain makes this coastal area very prone to increased inundation caused by contemporary accelerated rise in sea level and such changes need to be carefully considered during any future economic development of the coastal zone.

Barth & Böer (eds.), SABKHA ECOSYSTEMS, 7-20

INTRODUCTION

Although the term 'sabkha' appears often in early travelers' accounts of Arabia, this morphological feature attracted very little scientific attention until several decades ago when even then the only scientific interest that seems to have been aroused by this feature was that of some French scientists working in North Africa (Tricart 1954). However, during the 1960's, work by Shell geologists in Qatar and geologists from Imperial College, London, in Abu Dhabi led to considerable interest in this environment (see Evans 1995). The early research especially along the 'Evans line' (Fig. 1, Evans et al. 1969) led to the understanding of the evolution of the sabkha. This was followed by considerable detailed work by Kinsman, Patterson, Bush, Butler, McKenzie and Hsu and others (see reviews by Evans 1995, and Kirkham 1998b) on the mineralogy and groundwater chemistry of the sabkha. With the aid of satellite imagery, Kirkham (1998b) provided an updated description of the sabkha sedimentary sequence exposed along a newly excavated canal together with a comprehensive reference list of the early work on the mineralogy, diagenesis and geochemistry of the sabkha.

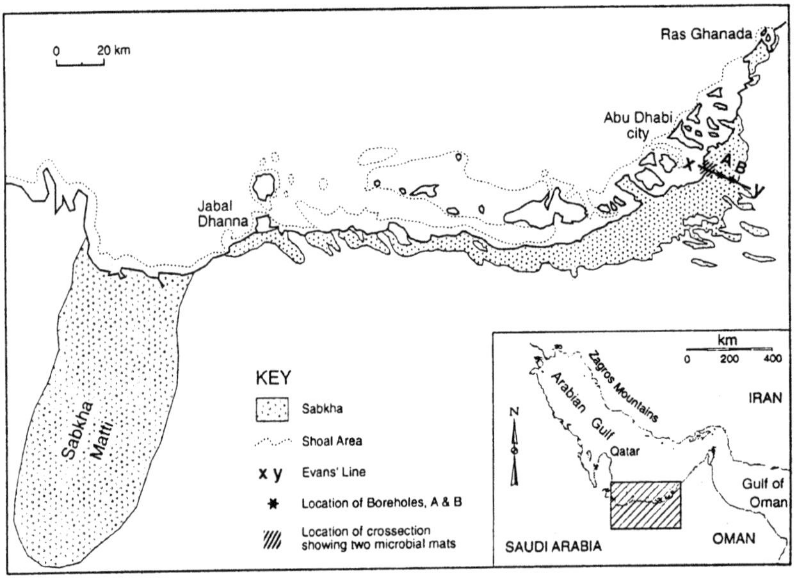

Figure 1. Location of sabkha in the UAE (showing locations of two boreholes illustrated in Figure 4 and the location of the cross-section shown in Figure 9).

Today, the term 'sabkha' appears in almost all accounts of nearshore peritidal carbonate–evaporite sediments and certainly in all textbooks dealing with this subject. Interest was first aroused by the discovery of modern dolomite associated with gypsum forming in the sediments of the sabkha and adjacent intertidal flats in Qatar (Wells 1962); and secondly, and possibly a discovery that led to even greater interest, was that anhydrite ($CaSO_4$) was forming today in the modern sabkha sediments of Abu Dhabi (Curtis et al. 1963). Although rare occurrences of this mineral had been reported from contemporary sediments of areas of central Asia, these were only isolated crystals. In contrast, the anhydrite was widespread in the modern sediments of the Abu Dhabi sabkha. Furthermore, of particular significance was that it was identical in form to the extensive anhydrite deposits found in the rocks of the geological column and, importantly, so often form the cap rock which seals petroleum reservoirs in the subsurface, particularly in many of the prolific oilfields of the Middle East (Shearman 1966, Kinsman 1969). Previously, it had been assumed that anhydrite had been deposited in a submarine environment and in most areas was produced by a secondary change from primary gypsum ($CaSO_4.H_2O$) during and after burial. However, the work in Abu Dhabi revealed that this mineral formed in coastal plain sediment over wide areas, above the water table. These studies, together with an enormous amount of work elsewhere on shallow water carbonates, led to the re-interpretation of the geological history and environment of deposition of many carbonate–evaporite successions. Many deposits were attributed incorrectly to coastal and nearshore environments of genesis and the term 'sabkha' was applied rather indiscriminately to many rock sequences in a spate of enthusiasm after this discovery.

GENERAL SETTING

Much of the coastline of Abu Dhabi is formed by a chain of barrier islands which enclose an extensive complex of lagoons (Fig. 1, Kendall & Skipwith 1969, Purser & Evans 1973, Evans et al. 1973). On the mainland coast, bordering the lagoons, is a rather flat, featureless plain with a salt encrusted surface, the sabkha, with its surface just above high water mark of normal astronomical tides and with a monotony which is only broken occasionally by scattered jebals (Fig. 2, Evans et al. 1964). The plain is bordered to landward by a scarp of mainly Miocene rocks capped by dune sands, although the latter are often absent on the crest of the scarps due to deflation and scour. Pleistocene rocks form the scarp east of Abu Dhabi Island. Occasionally, small alluvial fans encroach onto the sabkha's inner edge. Seawards, the salt encrusted flats of the sabkha pass gradually into intertidal flats where extensive microbial mats pass lagoonwards into rippled carbonate sands and muddy carbonate sands colonised by swarms of gastropods and occasional bivalves. In places, mangroves stands occur on the intertidal flats but they are rarely extensive, except where they have been recently artificially planted.

Figure 2. Photo of jebal on "Evans line" looking seawards across typical polygonal salt encrusted sabkha.

MORPHOLOGY

The sabkha is a Holocene coastal plain that is widest in central Abu Dhabi, where the scarp of Tertiary rocks and its small alluvial fans are furthest from the coast. To the northeast and west it becomes narrower and, to the west especially, is split up into isolated segments by long seaward extending ridges and inliers of Tertiary rock (Fig. 3). However, west of Jebal Dhanna it extends inland as a long tongue bordered on the west by the southward extension of the Qatar Peninsula to form the sabkha of Sabkha Matti which passes southwards from a marine to a continental sabkha (Goodall 1995).

Isolated Tertiary inliers together with outcrops of cross–stratified Pleistocene aeolianite, often capped by skeletal limestones or caliche, form small jebals (zeugen) which break the monotony of the sabkha surface (Kirkham 1998a). The latter are the remnants of a sheet of carbonate-rich, quartzose sand which has been removed by deflation. The skeletal capping is the only remains of a deposit formed at a sea level several meters above that of today and is presumed to be of last interglacial age. Whereas the composition of the aeolianite is very similar to that of the uncemented aeolian sand which underlies the sabkha surface (Fig. 4), it does appear in places to be rather lighter in colour and may represent a younger, more carbonate-rich facies of the latter.

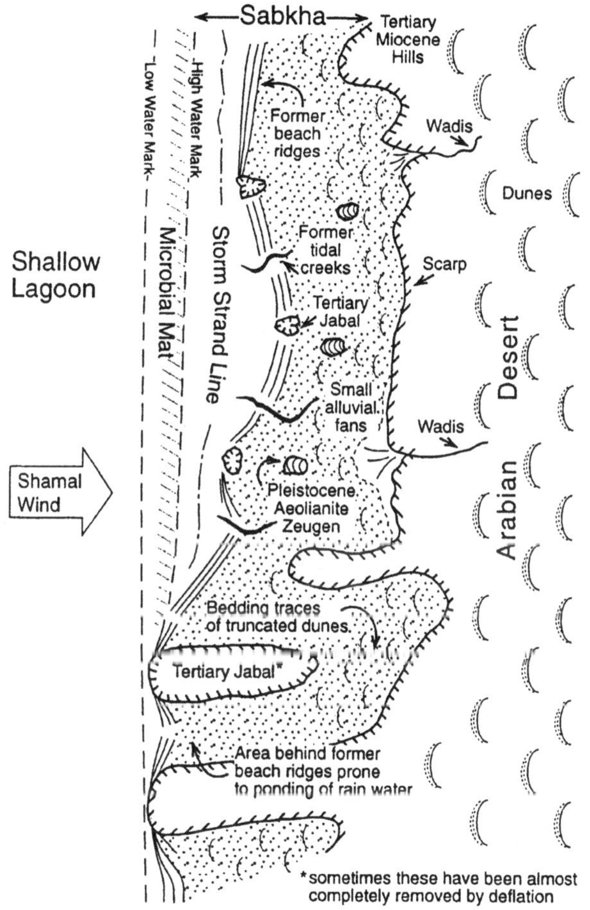

Figure 3. Schematic sketch showing the main morphological features of the sabkha.

Superficially, the sabkha appears to be a rather flat featureless plain with a salt encrusted surface moulded in many cases into large polygonal patterns with upturned edges. These have been produced by the lateral expansion of the surface sediment due to salt precipitation from water drawn to the surface by capillary action. However, more detailed examination reveals a variable microtopography which shows up particularly well on aerial photographs and satellite images (Fig. 5, Kirkham 1997).

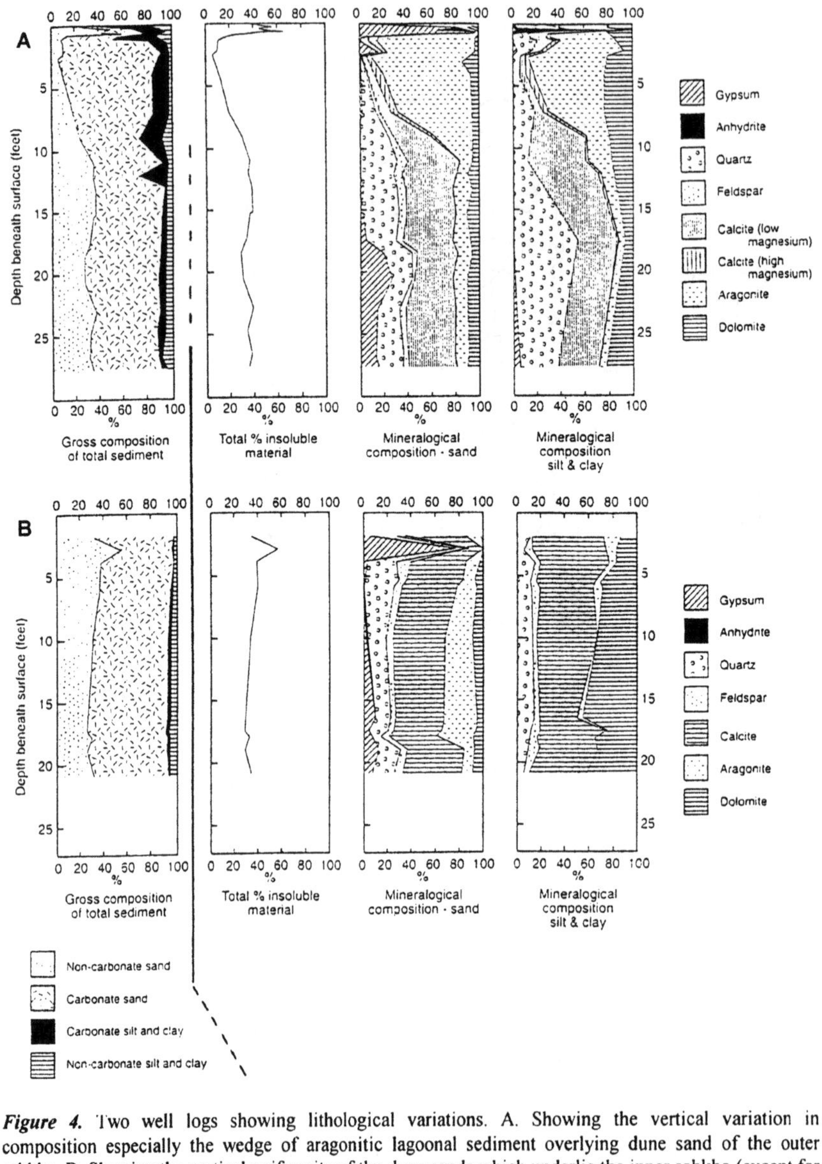

Figure 4. Two well logs showing lithological variations. A. Showing the vertical variation in composition especially the wedge of aragonitic lagoonal sediment overlying dune sand of the outer sabkha. B. Showing the vertical uniformity of the dune sands which underlie the inner sabkha (except for a small increase in evaporitic minerals near the surface).

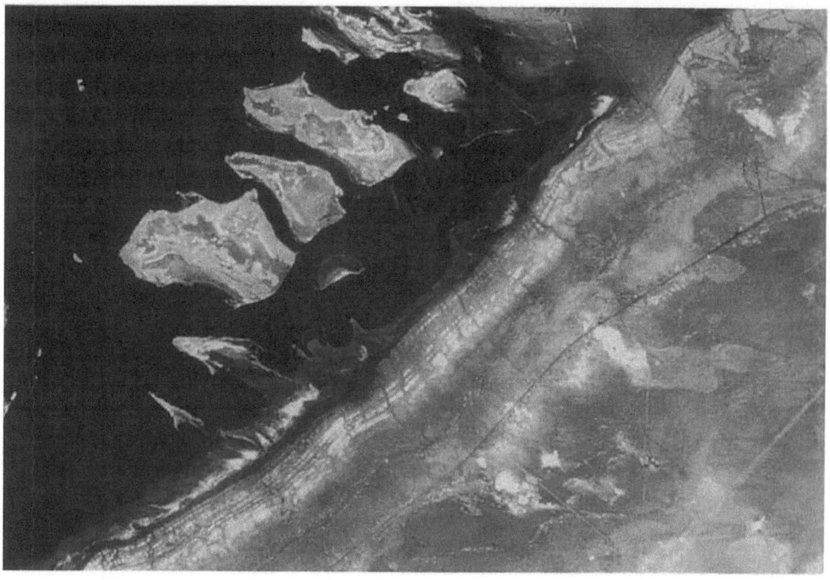

Figure 5. Satellite image of a typical part of the sabkha plain showing the numerous parallel storm beaches, traversed by ephemeral tidal channels, along the seaward edge of the coastal sabkha. Black microbial mat is developed seaward of the storm beaches at the eastern end of the Khor Al Bazam Lagoon where it abutts against the south eastern end of Al Dabb'iya.

The most prominent of the features are long linear former storm beach ridges which rise only a few cm or dm above the surrounding surface. They are lighter in colour than the orange brown surface of the sabkha and are often colonised by scattered halophytic plants. They fan out from the Tertiary rocky jebals and headlands, splaying out from their extremities into a series of finger-like ridges or 'winged spits'. They are formed of shelly sand, which often has been extensively winnowed, and are capped by a small gravel lag composed mainly of gastropods with occasional bivalves and fragments of beachrock. In places, the ridges are cut by low depressions, aligned at right-angles to them. These depressions are the remnants of infilled former tidal creeks which die out to landward and seaward. Seaward of the ridges, there are often shore-parallel strips of white salt with lines of debris which mark the approximate maximum landward limits of storm events. Outcrops of giant foresets of deflated, barchanoid aeolian sands often cover huge areas along the inner margin of the sabkha, beyond the limits of the marine sediments deposited by the last, Flandrian transgression.

For much of the year the sabkha plain is dry. However, during exceptional meteorological events, which may occur several times a year, strong onshore winds (Shamals) drive seawater from the adjacent lagoons over the outer parts of the sabkha. Also, during periods of heavy rains the surface is flooded by water draining from the upland areas at its landward margin. This water sometimes remains on the surface for weeks or months before being completely evaporated to form extensive sheets of glistening white salt.

Thus the seaward part of the sabkha, which is not flooded during normal astronomical tides becomes inundated, in part, by marine waters during exceptional meteorological events and is best described as a supratidal environment, i.e. although it is not covered by the seawater during normal tides, it is affected by occasional marine flooding and so, in this way, it differs from the adjacent intertidal flat environment which is covered either daily or in its most landward parts at least several times a lunar month.

Although the sabkha has water contributed to it from the adjacent marine areas by marine flooding and from rain and surface run-off from the adjacent higher ground, these are not the only sources of its groundwater. Subsurface seaward seepage from the adjacent uplands and subsurface rocks into the sabkha sediments is likely to be another major source of its groundwater (see paper by Warren Wood, this volume). Landward seepage from the adjacent lagoons is unlikely to be important.

EVOLUTION

The sabkha plain was initiated by the drowning of the outer parts of the aeolian dunefields and the underlying Pleistocene and Miocene rocks of the Arabian Peninsula by the waters of the Flandrian transgression (Fig. 6). Although not proven offshore in Abu Dhabi, the sands of these dune fields are known from borings to lie beneath the surficial marine sediment and waters of the Gulf of Salwa (Darwish & Conley, 1989).

Unsubmerged remnants of Tertiary and Pleistocene rocks formed small islands around some of which reefs grew and carbonate sediment produced by chemical and biochemical processes accumulated to form a string of barrier islands. Between the islands and the mainland coastline, carbonate sands and muds were deposited in what are now the lagoons and, as these islands grew in area, the average temperatures and salinities rose in this increasingly sheltered environment.

Along the mainland shoreline of the lagoons, the rising sea truncated the aeolian dune sands and extended landwards to approximately 4 km from the present coastline where it appears sea level reached its maximum landward extent and its maximum Holocene level at approximate 4000 yr. BP (Evans et al. 1969). Thereafter, the sea level fell slightly and the mainland shoreline advanced seawards and the numerous parallel beach ridges and intertidal flats formed as the sea regressed. Sedimentation and the progressive infilling of the sheltered lagoons and colonisation of the emergent areas led to progradation (probably with minor offlap) of the mainland shoreline.

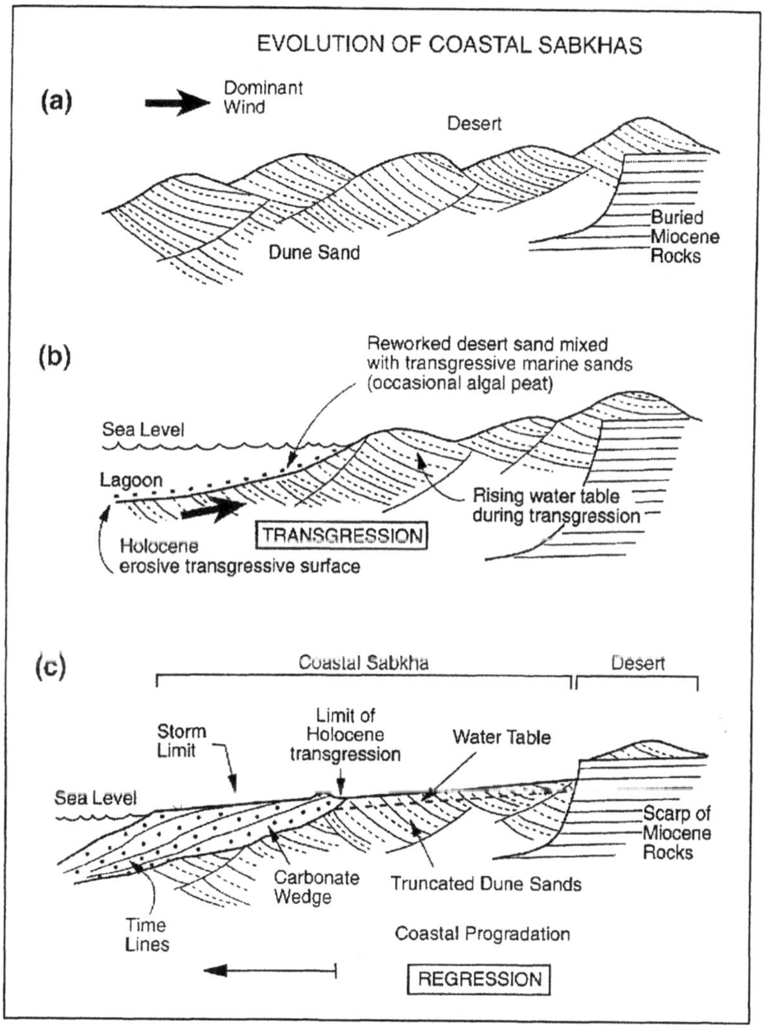

Figure 6. Evolution of the coastal sabkha of Abu Dhabi (modified after Evans 1995).

On the intertidal flats and adjacent sabkha plain, evaporation led to precipitation of gypsum and anhydrite and the higher salinities of the interstitial waters led to the conversion of the original calcium carbonate ($CaCO_3$) to dolomite ($CaMg(CO_3)_2$)

within the sediments. Other minerals developed within the sediment as the diagenetic dolomite sometimes reacted further with the highly saline interstitial waters to produce magnesite ($MgCO_3$), and the replacement of calcium carbonate by dolomite led to precipitation of minor celesite ($SrSO_4$) (Evans, 1995). Precipitation of sulphate from the high salinity interstitial waters within the sediment together with deposition of wind blown dust raised the surface above the level of astronomical tides. Hence the sediment surface was raised mainly by wind deposited dust from the hinterland, storm supply of sediment from adjacent lagoon during unusual meteorological events and "jacking up" of the surface by continued sulphate deposition from interstitial waters. Due to the elevated temperatures and salinities in the upper parts of the sediment previously deposited, gypsum was dehydrated to anhydrite and primary anhydrite was deposited; first starting in isolated centres and then gradually expanding and spreading – the impingement of adjacent nodules producing the so called "chickenwire" structure and due to crystallisation pressure layers buckled and produced contorted horizons (ptygmatic structures, Fig. 7).

Figure 7. A cutting in the bank of the Mussafah Channel exposing banded anhydrite. Some of the bands are enterolithic with complex ptygmatic folding About thirty centimetres of gypsum mush underlie the anhydritic sequence. Vertical scale about one metre.

Two main phases of anhydrite formation can be recognised and they form shore-parallel bands either side of the former beach ridges about half a kilometre wide across the largest area of coastal sabkha. Although the evaporites are generally regarded as having formed during the late Holocene regressive phase, those associated with the innermost anhydrite band, which forms an elevated area or ridge may be transgressive as they appear to be partly related to the final stages of the Flandrian transgression (Kirkham, 1997). This latter anhydrite is the arid zone equivalent of the transgressive peat which underlies transgressive lagoonal deposits in more humid areas (Evans et al. 1970). The Abu Dhabi sabkha sedimentary succession produced by the coastal progradation gave rise to the "idealised" sabkha succession (Fig . 8).

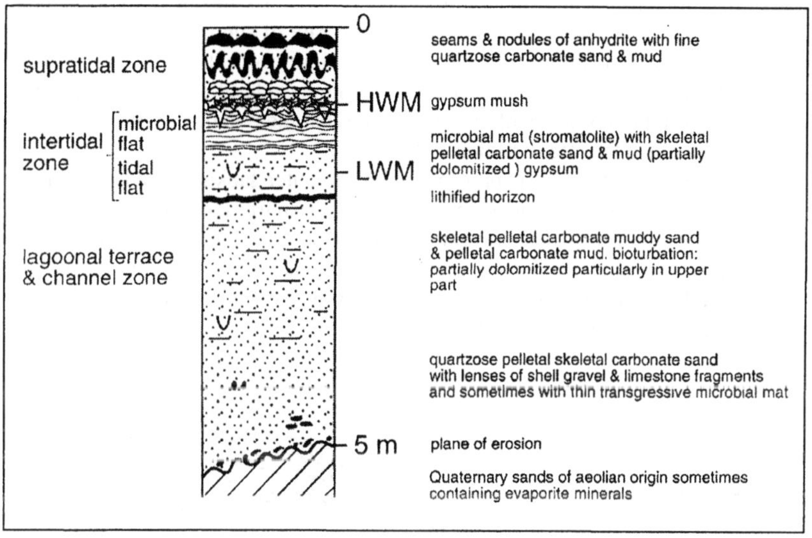

Figure 8. Idealised sabkha sequence produced by coastal progradation.

Although the innermost former storm beach ridges on the sabkha of Abu Dhabi appear to have formed when sea level was above present day normal high water level, the evidence is not conclusive as the enhanced heights of the ridges may be due to the fact that before the islands, which shelter the lagoon, had grown to their present day size and the lagoons were deeper and relatively unfilled with sediment, the waves reaching the mainland shoreline would have been larger and hence may have thrown up higher ridges. However, there is evidence elsewhere, in the form of wave-cut platforms in the UAE and beach ridges in Bahrain and Qatar, which also indicate that sea level was slightly higher than that of the present at approximately 4000 BP. Further evidence is seen in the fact that older Holocene lagoonal sediments and mangrove rhizoliths occur just above present day mean sea level along the Abu Dhabi coastline. Hence, it appears that progradation, took place under conditions of falling sea level, as is indicated by the gradual fall in level of the buried microbial mat as it is traced seawards in the subsurface (Evans et al. 1969). Iron oxide bands which mark former water tables within the sabkha sediments also appear to testify to a progressively lowering water table associated with a progressively falling sea level (Kirkham 1998b).

A further interesting and intriguing point is that the buried microbial mat produced during the regression does not pass seaward into the modern mat of the contemporary lagoonal shoreline but lies below it and is separated from the latter by a wedge of muddy carbonate sand and mush of gypsum crystals (Fig. 9). Thus the modern microbial mat is at a slightly higher level. This may be due to local tectonics or to some change in coastal morphology and the rate of supply of sediment within the lagoon, but there is the interesting possibility that it reflects a sea level rise and that it is the manifestation of the beginning of a new cycle of sedimentation (a new transgression) and of coastal evolution indicating that the rate of rise of sea level has become greater than the rate of sedimentation.

Whilst the progradation was proceeding on the outer sabkha the adjacent desert areas just inland of the maximum extent of the Flandrian transgression was still being affected by the dominantly onshore winds. Due to the drowning of the desert upwind by marine waters, they no longer received significant sand and thus have been gradually deflated down to a level determined by that of the ground water table to expose the cross-bedding of the former dunes.

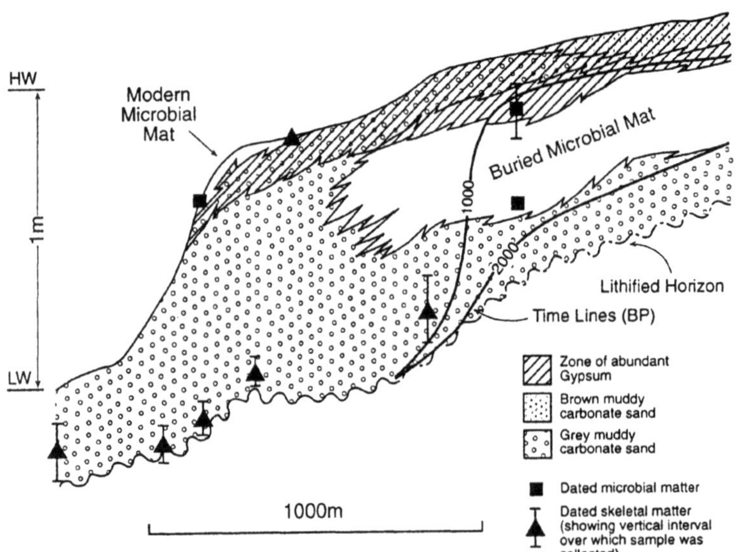

Figure 9. Cross section showing the relationship of the two microbial mats at the seaward end of the Evans line (after Evans et al. 1969).

HUMAN ACTIVITY AND GLOBAL WARMING

The sabkha plain of Abu Dhabi is a complex physiographic feature that is partly depositional and partly erosional. The outer sabkha plain reflects dominantly marine deposition and evaporite precipitation above the water table, whereas the inner sabkha reflects dominantly erosion by deflation of aeolian deposits.

The abundant construction of roads, bunds and pipelines on the adjacent intertidal flats and parts of the sabkha is leading to very fundamental changes. Increasingly, large areas are becoming artificially protected from natural marine flooding at high tides and/or storm conditions. Although surface-ponding of rain water is a natural (almost annual) event on the deflated aeolian sands landward of the innermost beach ridges and the innermost anhydrite ridge, the civil engineering projects are interfering extensively with the original ponding pattern. In addition, the planting of trees along the roads which cross the sabkha has necessitated the continual introduction of large quantities of fresh - brackish water to the sabkha for their survival and this must have a profound effect on the ideal water salinities (see Wood, this volume).

The future evolution of the sabkha and of the natural geochemistry of its sediments and interstitial waters is uncertain due to this continued and extensive interference by man. However, it is likely, with the predicted rise in sea level due to global warming, that the coastline will retreat landward and most or all of the low-lying sabkha will be flooded by the advancing waters of the marine transgression relatively quickly. Consequently, the coastal settlements of Abu Dhabi may have a relatively limited lifespan. They could receive limited protection in the short-term by the construction of bunds, or even by artificially raising the ground level by extensive dumping of material dredged from offshore or trucked from inland as is done today. However, even such fill is likely to be affected by the rising levels of saline groundwater which are likely to produce engineering problems. The disparity between sea level and the level of the reclaimed enclosed sabkha surface will gradually increase with the inevitable result that without extensive and expensive engineering works population migration to higher topographic levels will be necessary.

ACKNOWLEDGEMENTS

The authors wish to thank P. Bush for carrying out X-ray diffraction analyses; K. Davis for drafting the diagrams; L. Kirkham for typing the manuscripts; and B. Marsh for producing the photographs.

References

Curtis, R., Evans, G., Kinsman, D.J.J. & Shearman, D.J. 1963. Association of dolomite and anhydrite in the recent sediments of the Persian Gulf. *Nature*, 197: 679-680.

Darwish, A.H. & Conley, C.D 1989. Pleistocene-Holocene sedimentation and diagenesis along the King Fahd Causeway between Saudi Arabia and Bahrain. JKAH Earth Science, Vol. 3, Special Issue: 1st Saudi Symposium on Earth Sci. Jeddah 1989, 63-79.

Evans, G. 1970. Coastal and nearshore sedimentation: a comparison of clastic and carbonate deposition. Proc. Geol. Assoc., London, Vol. 81, 493-508.

Evans, G. 1995. The Arabian Gulf: a modern carbonate-evaporite factory; a review. Cuadernaos de Geologia Iberica, 19: 61-96. Servicio de Publicaciones. Universidad Complutense, Madrid.

Evans, G., Kendall, C.G.St.C. & Skipwith, Sir P. 1964. Origin of the coastal flats, the sabkha, of the Trucial Coast, Persian Gulf. *Nature*, Vol. 2. No. 4934: 759-761.

Evans, G., Murray, J.W., Biggs, H.E., Bate, R. & Bush, P.R. 1973. The oceanography, ecology, sedimentology and geomorphology of parts of the Trucial Coast barrier island complex, Persian Gulf. In: Purser, B.H. (ed.), The Persian Gulf, Holocene Carbonate Sedimentation in a Shallow Epi-continental Sea. Springer-Verlag, New York, 233-278.

Evans, G., Schmidt, V., Bush, P & Nelson, H. 1969. Stratigraphy and geological history of the sabkha, Abu Dhabi, Persian Gulf. *Sedimentology*, 12: 145-159.

Goodall, T.M. 1995. The geology and geomorphology of the Sabkhat Matti region (United Arab Emirates): a modern analogue for the ancient desert sediments from North-West Europe. Ph. D. Thesis, Aberdeen University.

Kendall, C.G.St.C. & Skipwith, Sir Patrick A.d'E.Bt. 1969. Geomorphology of a recent shallow water carbonate province: Khor al Bazam, Trucial Coast, southwest Persian Gulf. Geological Society of America Bulletin, 8: 865-891.

Kinsman, D.J.J 1969. Modes on formation, sedimentary associations and diagenetic features of shallow water and supratidal evaporates. *Bull. Amer. Ass. Petrol. Geol.*, 53: 830-840.

Kirkham, A. 1997. Shoreline evolution, aeolian deflation and anhydrite distribution of the Holocene, Abu Dhabi. *GeoArabia*, 2: 403-416.

Kirkham, A. 1998a. Pleistocene carbonate seif dunes and their role in the development of complex past and present coastlines of the U.A.E. *GeoArabia*, 3: 19-32.

Kirkham, A. 1998b. A Quaternary proximal foreland ramp and its continental fringe, Arabian Gulf, U.A.E.. In: Wright, V.P. & Burchette, T.P. (eds.), Carbonate ramps. Geological Society, London, Special Publications, 149: 15-41.

Purser, B.H. & Evans, G. 1973. Regional sedimentation along the Trucial Coast, S.E. Persian Gulf. In: Purser, B.H. (ed.), The Persian Gulf, Holocene Carbonate Sedimentation in a Shallow Epi-continental Sea. Springer-Verlag, New York, 211-232.

Shearman, D.J. 1966. Origin of evaporites by diagenesis. Trans. Inst. Min. Metal. (Sect. B: Appl. Earth Science), 75, B208-15.

Tricart, J. 1954. Un forme de relief climatique: les sebkhas. *Géomorphologie Dynamique*, 5: 97-101.

Wells, A.J. 1962. Recent dolomite in the Persian Gulf. *Nature* 194: 274 -275.

THE HOLOCENE TIDAL FLAT COMPLEX OF THE ARABIAN GULF COAST OF ABU DHABI

C. G. St. C. KENDALL[1], A. S. ALSHARHAN[2] & A. COHEN[1]

[1]Department of Geological Sciences, University of South Carolina, Colombia, SC 29208, USA
[2]Faculty of Science, UAE University, P.O. Box: 17551, Al-Ain, United Arab Emirates

Abstract

Carbonate sand and mud of the offshore bank flanking the Khor Al Bazam lagoon, the lagoons of the western Abu Dhabi and the mainland coast are accreting and prograding. Seaward of the Khor Al Bazam lagoon the offshore bank is progressively extending its shoal and channel area, coral banks, tidal deltas, and nearshore coastal terraces. Where wave energy is minimal cyano-bacterial mats colonise protected intertidal sediments, building seaward, binding any sediment washed onto them and raising the sediment surface to the high water mark. here the cyano-bacterial mats are entrapped beneath prograding supratidal carbonate/evaporite coastal sabkhat. Mangroves flourish in tide-dominated areas protected from all but the smallest waves, and aid the entrapment of sediment later colonised by algal mats. Local beach ridges develop from spits that acquire beach faces, berms and dunes. These dunes are deflated as the beach line is stranded by coastal accretion. Landward of the beach ridges windblown and wind tide floodwater sediment has accumulated.
Traced landward from the middle intertidal zone of protected coasts the following are found: lime muds and sand flats in association with carbonate cemented hardgrounds passing to cyano-bacterial mat which exhibit a cindery mammillated character, and then polygons and algal crinkles. These latter are associated with a gypsum mush that thickens landward and is replaced by 20 cm thick anhydrite layer. Landward these sediments lie beneath a salt dominated crumpled polygonal surface and near surface interlayers of thin anhydrite and nodules that are flanked by stranded beach ridges of an earlier coastline. The cyano-bacterial materials are preserved as peats beneath the supratidal sabkhat. They have a high potential for being preserved in the geologic record.

Barth & Böer (eds.), SABKHA ECOSYSTEMS, 21-35

HOLOCENE FACIES TIDAL FLAT

Wide, long, cyano-bacterial tidal flats and coastal sabkha line the coastal lagoons of Abu Dhabi (Fig. 1) (Kendall & Skipwith 1968). At the east end of the Khor Al Bazam lagoon, the largest cyano-bacterial flat and the associated lower tidal flat sands and muds parallels the coast for 42 km and a smaller one parallels the coast for some 9 km. These flats, part of the seaward edge of the prograding coastal plain, have an average width of about two km.

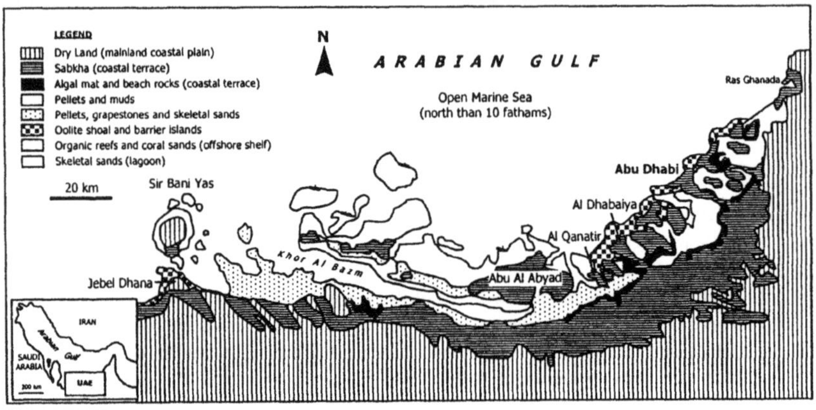

Figure 1. Map of the general marine facies of the coast of Abu Dhabi in the UAE.

The cyano-bacterial flats are underlain by compacted peat-like remains of living cyano-bacteria mat that landward reaches some 5-30 cm in thickness. This mat is prograding seawards, and commonly it extends landward in the subsurface beneath a layer of displacive evaporites and windblown and storm-wash-over sediment. Carbon dating suggests that locally the cyano-bacterial mat has prograded as much as 7 km in the last 4000 years (Kinsman 1964). Seaward of the mats are subtidal and mid intertidal accumulations of carbonate sediment. Landward of the cyano-bacterial mats are stranded beach ridges of an earlier shoreline subdividing the topography of the lower and upper sabkha. Traced from the present shoreline to the line of ancient beaches the character of the sediment surface reflects the topography and tidal range of the Holocene facies tidal flat (Butler 1969), (Fig. 2).

It should also be noted that similar wide tidal flats are located in the protected lee of islands, but here the middle tidal flat is marked by the growth of the mangroves (*Avicennia marina*) while landward lime muds, highly bioturbated by crabs, are accumulating. As with the more protected coasts with cyano-bacteria mats, a crinkled and crenulated mat extends over an accumulation of the gypsum mush of the upper tidal flat. In some places, this zonation of mangroves, crab-burrowed lime mud and gypsum mush is foreshortened so that mangroves abut against beach ridges of cerithids, or mangroves and lime muds adjoin the ridges.

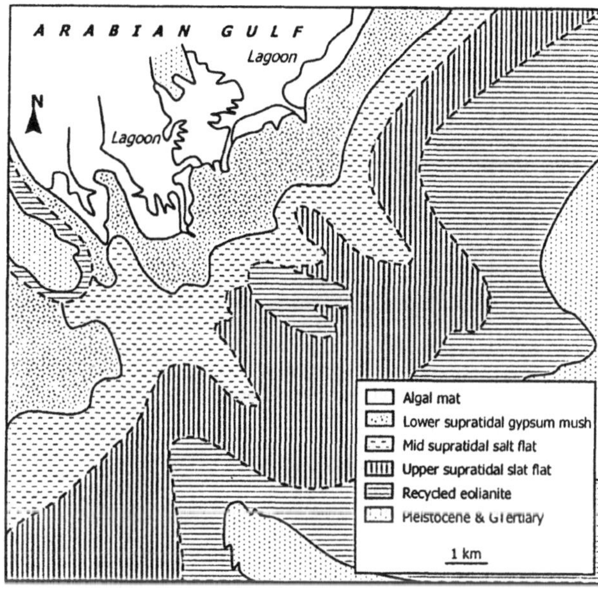

Figure 2. Sabkha sedimentary facies distribution along the southern Arabian Gulf coast (modified from Butler et al., 1982). Location of this section to the southwest of Abu Dhabi island (see Figure 3).

Lower intertidal carbonate flats

Traced seaward to landward across the protected tidal flats, particularly those flanked by cyano-bacterial or "algal" mats, the seaward subtidal to intertidal sediments are composed of foraminiferal-gastropod sands and/or muds and are often overlain by intertidal hardgrounds (Tab. 1). These hardgrounds are discontinuous and are capped by either "peat-like" layers of cyano-bacterial flat sediment adjacent to open lagoons or are capped by intertidal burrowed muds adjacent to the protected lagoons. Both these peats and burrowed muds are capped landward by supratidal evaporites. The hardground is cemented by a calcium carbonate and/or locally by later gypsum. This cemented layer tends to form a seal between the underlying more marine sediments with their marine waters and the overlying supratidal salt flats with their saline brines. This hardground is not present everywhere and seldom overlies the cyano-bacterial-peat layer. This latter occurs where the stranded shoals and coastal spits lie almost perpendicular to the shore and are elevated above the adjacent tidal flats. Here the hardground extends above the peat layer to be on-lapped by a layer of gypsum mush with a surface of cyano-bacterial mat.

Table 1. Sedimentary facies and diagenetic facies in Holocene tidal flat of UAE.

Sedimentary facies	Thick-ness (cm)	Main components	Diagenetic facies
Lower intertidal carbonate flats	30 - 120	Fragments of corals, skeletal sands, ooid and stabilised by the grass Halodule sp., bioclastic and peloidal carbonate sand and mud, cerithid and small shelled bivalves.	Isopach calcite cement, miniscus cement, leaching, skeletal and bioclastic carbonate sand
Middle to upper intertidal algal flats	10 - 55	Laminated cyano-bacterial mats intercalated with aragonite mud, this may be replaced by mangroves swamp and crab-burrowed lime mud. In upper intertidal gypsum crystals and protodolomite.	Gypsum mush, celestite, protodolomite, calcite
Lower supratidal salt flats	10 - 30	Gypsum mush forms a layer which is composed of a mass of loose but separate grain-supported gypsum crystals, with interspersed aragonite or dolomitised aragonite mud.	Gypsum mush, bassonite, dolomite ($MgCO_3{\cdot}3H_2O$)
Mid-supratidal salt flats	20 - 60	Crumpled surface of halite, capped polygonal anhydrite layer, and carbonate wash-over sands.	Pockets of gypsum crystals and anhydrite. The anhydrite partially replaces the gypsum mush and is accompanied by the partial replacement of aragonite muds with protodolomite
Upper supratidal salt flats	5 - 65	Near-surface modular anhydrite in a matrix of carbonate mud. Layer of chicken-wire anhydrite, represent the complete replacement of the gypsum mush	Anhydrite nodules, large gypsum disks, proto-dolomite, magnesite, hunite [$Mg_3Ca(CO_3)_2$], halite, polyhalite and the filling of intertidal molds in gastropods shells by anhydrite
Continental salt flat and recycled aeolian sands	15 - 45	Pre-Holocene eolianites, re-cycled Holocene eolianite, wash-over Tertiary sediments and bedded storm wash-over skeletal debris.	Anhydrite layer, secondary gypsum, dolomite and anhydrite are secondary. Anhydrite can rehydrate to gypsum

The cyano-bacterial facies of the mid intertidal to lower supratidal

Cyano-bacterial mats accrete and prograde on intertidal and periodically inundated supratidal zones where they are protected by islands, sand bodies, wide shoals, and wide intertidal sand flats (Fig. 1). The seaward edge of the cyano-bacterial flats, and where they line ebb channels, is limited by wave and/or tidal scour that erode the surface mats. These cyano-bacteria tolerate temperature variations of over 43 °C and tolerate salinities from 0 to 196 PPT (Kinsman 1964).

The cyano-bacterial mats of the upper intertidal to low supratidal zones of the restricted and protected tidal flat areas of the Abu Dhabi coast are dominated by *Microcoleus* and *Schizothrix*. These microbial mats of Golubic, (1992) are characterised by surface morphologies related to frequency of exposure to the atmosphere above the tidal waters, the salinity of these waters, and the resulting cyano-bacterial community (Fig. 3, 4 and 5), (Kendall & Skipwith 1968, Golubic 1992). The surface of the cyano-bacterial flats can be informally divided from seaward to landward on the basis of surface morphology into: 1) cinder zone of the low to mid-intertidal zone; 2) polygonal zone where there are shallow pools and channel areas with a high flood frequency in the higher intertidal zone; 3) pinnacle mat in the more elevated well-drained tidal flat areas; 4) crinkle zone from the highest intertidal area; and 5) flat mat of the low supratidal zone (Kendall & Skipwith 1968), (Fig. 4).

The most seaward cyano-bacterial mat forms the cinder mat of Kendall & Skipwith (1968). Here at the seaward edge of the algal flats, the surface is pustular or mamillated (Golubic 1992), and has the appearance of an irregular surface of small black cinders. As the cinder zone of cyano-bacteria is traced landward, the surface becomes smoother and is covered by desiccation polygons from 15-30 cm across which are coated by leather-like, dark-green layers of cyano-bacteria. The edges of the polygons upturn, and provide a small scale niche for pinnacled mat algae. The pinnacles of this mat are abundant in areas with low tide exposure, and occur with several environmental modifications nearer the shore. Where tidal creeks drain the algal flat, these meandering bodies of water are obstructed by further growths of cyano-bacteria forming dammed ponds. The ponds may be up to 10 m across, but are usually no more than 2-3 m in diameter. Here, pink surfaced layers of peat collect, broken into huge lily pad-like bodies measuring several meters across, which themselves may be desiccated into large polygons 0.5 to 1 m in diameter. These polygons are preserved and can be identified well landward of the present algal flat beneath the sabkha surface in trenches cut in the supratidal salt flats.

The algal-peat of the polygonal zone varies in composition from nearly 100% organic material to a very sandy organic material (Kendall et al. 1991, Schreiber et al. 2001). The peat richest in organics is associated with the ponds formed where the tidal creeks are dammed. These peat-filled creeks extend from the middle of the intertidal cinder zone to well back into the upper intertidal crinkle zone. The most common occurrence of the algal mat peat is associated with the polygonal zone and is represented by the stacked saucer-like structures of this zone. The cracked, unturned margins of the polygons are filled by carbonate sand. The peats of the ponds and polygonal zone are 30-45 cm thick at their seaward edge, but compress beneath the sabkha surface landward to some 25 cm.

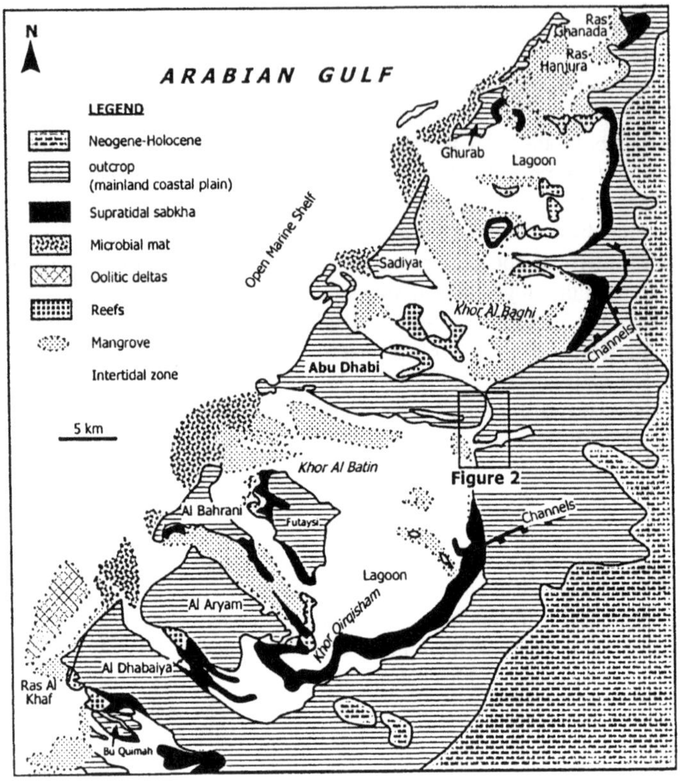

Figure 3. Holocene tidal flat facies distribution of the Abu Dhabi coastal carbonate environment (modified from Kenig et al., 1991).

The upper intertidal zone, which is flooded during middle-to high tide, is commonly covered with a crenulated or crinkled green to black cyano-bacterial mat (Kendall & Skipwith 1968). This is the convoluted mat of Golubic (1992) and the crenulations are probably formed by the rapid lateral growth of its surface. A mush of gypsum crystals begins to accumulate beneath the surface of this mat in the upper intertidal zone. The crenulated mat is a leathery, wrinkled organic layer, blackened on the upper surfaces of the folds in response to air-exposure, but commonly retaining the pinkish-beige colour on lower less exposed surfaces. The crenulated folds trap air and gases (probably methane) that are expelled when walked upon. A common morphological variation of these mats includes the formation of "tufts" on the upper more aerated surface of the crenulated mat, usually in the mid-intertidal zone at transitions between the pinnacle and polygonal mats with the crenulated mat. Occasionally in the upper-most intertidal zone, the crenulated mat can desiccate and shrivel to form a dried crust at low tide. In the lower supratidal zone there is a thin layer of microbial material (less than a centimeter) at the surface in this zone, and the sediment surface becomes horizontal, probably because the cyano-bacteria do not grow vigorously and the pink cyano-bacterial surface no longer expands laterally. Gypsum continues to actively precipitate beneath a centimeter or less of the surface sediment. The sediment is moist, but dewaters where footsteps and vehicle tracks have compressed it and anhydrite can form in the resulting depressions. A surface of carbonate wash-over sediment may contain halite, but it does not form a crust at the surface.

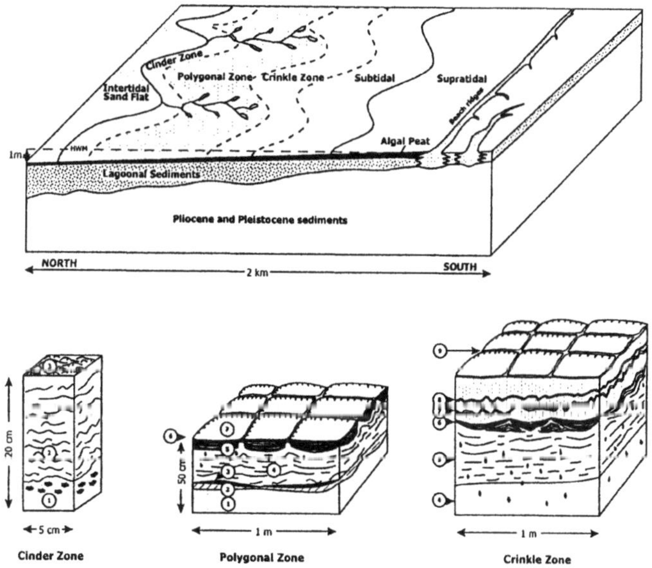

Figure 4. Composite idealized model showing distribution of sabkha sediments based on the barrier island – lagoon sabkhat of Abu Dhabi region (modified from Butler et al. 1982). (1) Lagoonal carbonate sands and/or muds with carbonate hardgrounds, (2) Vaguely laminated lower tidal-flat cyano-bacterial peat, (3) Upper tidal-flat algal mat; commonly laminated and cross-cut by desiccation polygons, (4) Large gypsum crystals (prismatic or lenticular), (5) Cemented carbonate layer, (6) Hightidal flat to supratidal mush of gypsarenite and carbonate, (7) Supratidal anhydrite polygons with windblown carbonate and quartz, (8) Anhydrite layer replacing gypsum mush and forming diapiric structures, (9) Halite crust formed into compressional polygons.

The crystals of gypsum vary in size from a few millimeters to a centimeter in diameter and form a layer that ranges in thickness from a few centimeters to as much as ten centimeters. Lime muds overlain by and caught up in this crystal mush show evidence of dolomitisation. The crystals overlie a cyano-bacterial·peat, commonly forming a transition zone of organic-rich gypsum layers. This facies overlies discontinuous hardgrounds of carbonate sand that are often cemented by both carbonate and/or gypsum.

The preservation of the peat-like remains of the cyanobacteria probably can be linked to the "pickling" effects from the high salinities of the area and the reducing environment in which the cyano-bacteria are accumulating. The presence of stranded beach ridges landward of the cyano-bacterial flats suggests that the cyano-bacterial-peats are not the first Holocene sediments to accumulate here. In fact, during the early stages of the Holocene, the western Abu Dhabi (Khor Al Bazam area) and portions of the lagoons behind the Abu Dhabi island were undoubtedly open bodies of water, subject to higher wave energies than today. The result is that chenier-like beach ridges, formed of cerithid gastropod debris, line the coasts. Subsequently, the lagoon margins filled, shallowed, became more restricted, and the beach ridges ceased to develop. On the coasts of the Khor Al Bazam lagoon, intertidal cyano-bacterial mats began to form while offshore carbonate sands or muds accumulated. These cyano-bacterial stromatolites have since prograded five to six kilometres seaward, and close to Abu Dhabi City reach seven kilometres.

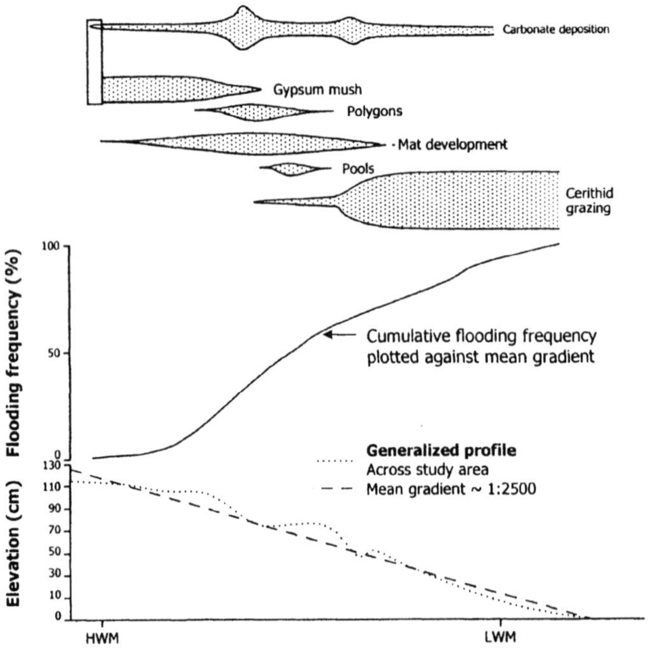

Figure 5. Distribution of microbial mat types and sedimentary features across the Abu Dhabi tidal flats (modified after Park, 1977).

Mid to upper supratidal evaporites

As the buried layer of gypsum mush is traced landward from flat zone of the cyano-bacterial flats or from the crab-burrowed muds of the mangrove swamps, the surface of this gypsum mush becomes oxidised and is replaced by an initially thin layer of anhydrite (1-2 cm thick) which is interlayered with storm wash-over carbonate sediment. Over a distance of tens of meters the layer of gypsum mush becomes completely replaced by a 20 cm thick upper sequence of anhydrite that lies beneath a polygonally to convoluted and cracked surface crust of halite. The underlying anhydrite is finely crystalline and has the consistency of cream cheese. It is thixotropic and contains pseudomorphs after discoidal gypsum, which ranges from 0.05 mm to 30 cm in diameter (Butler 1969, Butler et al. 1982, Warren & Kendall 1985). This anhydrite is a predominantly secondary mineral that replaces gypsum. Locally it replaces carbonates or occurs as a direct precipitate into voids. This anhydrite-rich layer and the underlying peat disappear close to the line of stranded beach ridges. The beach ridges overlie laminated and burrowed lime muds. At some localities, these burrows remain open. The cyano-bacterial-peat usually occurs seaward beneath the surface evaporite layer, although in local areas protected by islands, a lime mud replaces this peat as it does close to the beach ridges.

Thus in the mid to upper supratidal zone halite polygons commonly overly, in descending order, a sequence of storm wash-over carbonates, anhydrite, minor gypsum, cyano-bacterial peat, and medium-to-very coarse carbonate sand. Large (5 to 10 cm across) gypsum crystals are common within or immediately below the lowest cyano-bacterial peat layer, often preserving original cyano-bacterial laminations. The buried cyano-bacterial peat commonly displays the desiccation polygons seen on the surface of the tidal flats in the upper intertidal zone and overlies discontinuous hardgrounds of lower tidal flat carbonate muds and sands, cemented by carbonates or gypsum. This vertical sabkha sequence, which extends from the cyano-bacterial flats to the beach ridges, reflects the previously described surface relationships for lateral sedimentary settings that can be traced from the land to the sea.

The major difference landward of the cerithid beach ridges is that the sediments overlying the gypsum and anhydrite being precipitated here are covered by an accumulation of thicker storm wash-over carbonates and clastics, which are derived from the tidal flat or are blown from inland. This wash-over sediment includes gypsum crystals that were eroded from the upper cyano-bacterial flats. These crystals now form the nucleus to the "cumulus cloud-like" nodules of anhydrite found just within and beneath the sabkha surface. Trenches dug landward of the beach ridges may expose shallow lagoonal sediments dominated by carbonate mud. Large gypsum rosettes are found in this mud. The waters associated with the ridges and adjacent landward are stained with iron which leaves a vivid red stain of hematite marking the top of the water table. Early Holocene aeolian carbonates and sediments washed out as fans from Tertiary outcrops have accumulated on the most landward side of the sabkha. Here, particularly within the sabkhat close to the Abu Dhabi Island, the anhydrite layers and nodules are replaced by gypsum. This replacement is a response to the influx of the fresher continental waters from the Arabian interior entering the coastal system (Patterson & Kinsman 1981). Meanwhile, to the landward side of the buried lagoon, the water takes on a milky appearance, probably a response to the presence of suspended carbonate mud as described by Swart et al. (1987) and Muller et al. (1991) from the Holocene sediments of Qatar and Abu Dhabi respectively.

ORGANIC MATTER IN A RECENT CARBONATE ENVIRONMENT

Cyano-bacterial material and associated carbonate sediments collected along the coast of western Abu Dhabi were investigated in terms of their: 1) surface conditions at deposition; 2) geometry and relationship to surrounding facies; and 3) organic chemical composition from whole rock, bitumen, and kerogen analysis.

Microscopic evidence shows that the solid organic matter is mostly highly fluorescent algal material that is relatively free of microbial maceration. This suggests that the organic matter is well enough preserved to become a source rock.

Total organic carbon (TOC) values in some samples are as high as 8.4%, with Rock-Eval pyrolysis yields as high as 564 mg of hydrocarbons per gram of TOC. Although TOC and Rock-Eval techniques were developed to assess petroleum source-rock potential for ancient rocks that have at least undergone organic diagenesis (a major loss of organic functional groups and condensation to a functionally more stable geopolymer), analyses of these recent cyano-bacterial sediments indicate significant organic enrichment, with potential to generate hydrocarbons upon thermal degradation. These results represent significant evidence that rocks of this type can be potential petroleum sources.

Laboratory studies indicate that concentrations of organics in these cyano-bacterial deposits meet productivity requirements based on TOC analyses (Fig. 6). Hydrocarbon generation potentials based on Rock-Eval pyrolysis (HI values) suggest that the organic matter is capable of generating oil and gas upon thermal maturation; however, these results also indicate that the organic matter may be more gas prone than oil prone. On the other hand, elemental analysis, which is usually more accurate than Rock-Eval pyrolysis, indicates that the organic matter is more oil prone than gas prone. Both Rock-Eval pyrolysis and elemental analysis show that the organic matter follows the Type II kerogen evolution pathway.

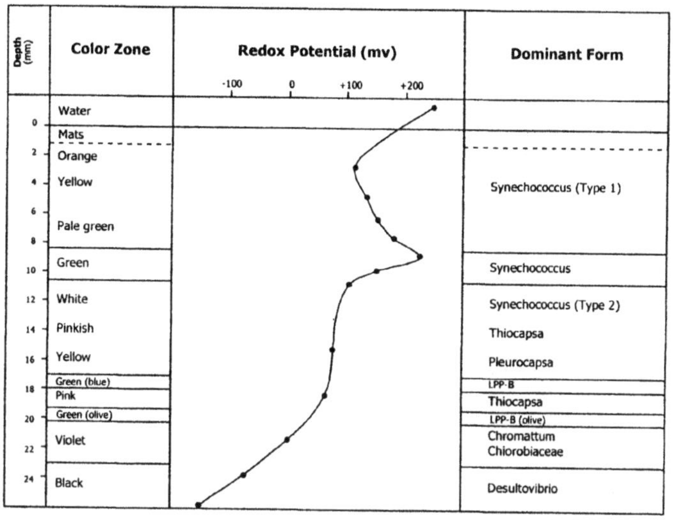

Figure 6. Chemical and physical details through a typical cyano-bacterial mat on a sabkha (composite). Thickness of mat (mm) is shown on the left, with color of the various 'sublayers'. Redox potential shows anoxic conditions below about 2 cm deep. Main microorganism groups at each depth are indicated on the right. (modified from Sheppard et al., 1992).

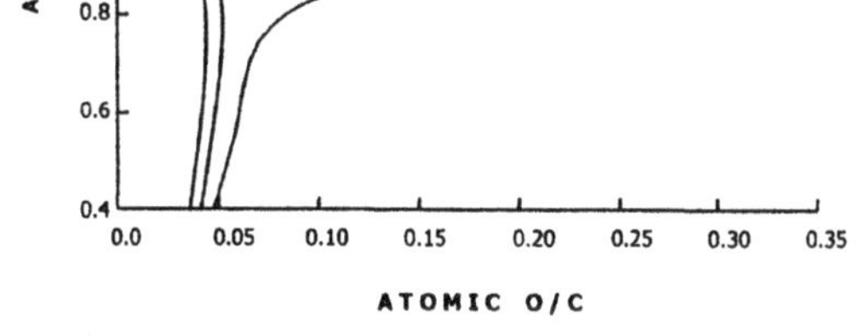

Figure 7. Modified van Krevelen diagram showing H/C to O/C relationships from elemental analyses of cyano-bacterial mat samples from Abu Dhabi, UAE.

Analyses of mats collected from western Abu Dhabi are compared with those collected by Cardoso *et al.* (1978) and Kenig *et al.* (1990) from other areas of Abu Dhabi. TOC of the sediment in the field area has a wide range of organic enrichment (0.46-8.40% TOC). Whole rock pyrolysis yields moderately high Hydrogen Indices (HI) of 389-597, typical of marine Type II kerogens. These values (0.5-2.7% TOC and HI – 510 to 675) were published by Kenig *et al.* (1990). Elemental composition (C, H, N, O, S) of isolated solid organic material (Kerogen precursors) showed atomic hydrogen to carbon (H/C) ratios (1.20-1.54) that fall directly on the Type II evolutionary pathway, as shown in a modified van Krevelen diagram (Tissot & Welte, 1984). Stable carbon isotope ratios ranged from –8.41 to –10.78% (Fig. 7).

PRESERVATION POTENTIAL OF CYANO-BACTERIAL MAT

As the cyano-bacterial mat becomes buried more deeply, the gross morphologic structures are less well preserved (Fig. 8). For example, below the gypsum mush facies polygonal and mamillate mats are easily distinguishable in buried sections. Beneath the lower supratidal facies (further inland) only the polygonal mat was identified by field inspection. The crenulated cyano-bacterial mat subfacies was not evident in the subsurface at all. This mat may not be preserved because: 1) it forms more inland in a higher and a more oxidizing depositional setting and/or 2) it forms in a setting conducive to precipitation of post-depositional and displacive gypsum.

Previous studies of Abu Dhabi cyano-bacterial mats assessed relative "preservation potential" of mat types based on the preservation of gross structures (Park, 1977). Preservation of the mat is a function of the subsurface extent (thickness and distribution) of the mat and the assumption that the mat will not be altered physically beyond easy recognition of the gross structures. The first part of this assessment may not be correct since the distribution and thickness of buried mat types cannot be unequivocally assumed based on the present extent of living equivalents. For example, if conditions 500 years ago were not conducive to mamillate mat growth in this area, it cannot be assumed that lack of mamillate mat in sediment 500 year old means that the mat was not preserved.

The second assumption is that the buried mat should "look like" its living counterpart on the surface. Geometric associations of the mat with inorganic sediment, such as well-formed laminae and desiccation polygons of the flat mat are easily recognised. Therefore, the flat mat can, without a doubt, be recognised in the subsurface. The mamillate mat, on the other hand, does not have such distinctive features. Additionally, it is prone to compression. Consequently, the "occurrence", or rather recognition, of the mamillate mat in the subsurface may not be as common as that of flat mat. However, this is a function of the initial extent, condition, and composition of the mat, as well as the criteria used for identification, not a result of poor preservation potential.

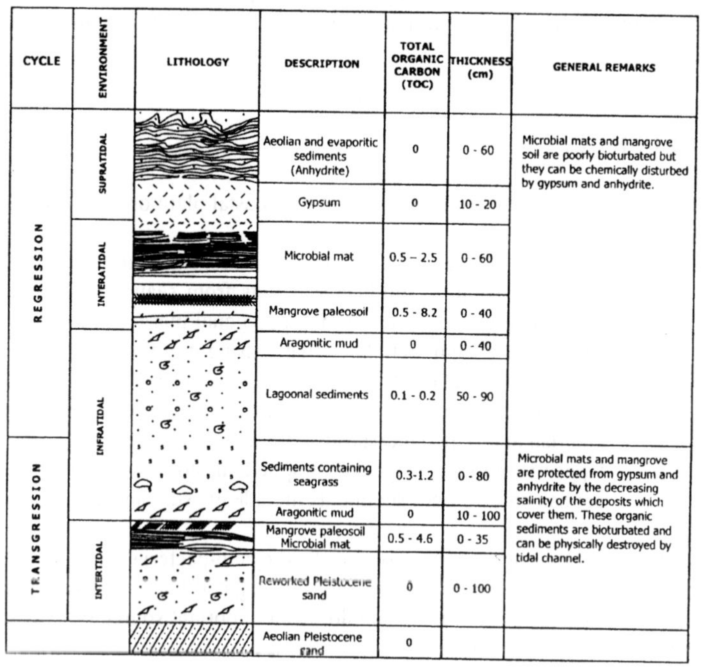

Figure 8. Holocene organo-sedimentary sequence of the Abu Dhabi coastal carbonate environment (modified from Kenig et al., 1991).

Petrographic analysis provides an alternate means of recognising mat types in the subsurface. Additionally, study of microstructures can delineate subtle diagenetic changes in the mat as well as the degree of preservation of the cyano-bacterial constituents of the mat. Thus, the following observations of cellular and amorphous organic components of the mats are offered:

Coccoid species (or at least the cell walls of these organisms) are relatively well preserved, regardless of the mat type in which they are found. Literature citing fossil examples of cyano-bacteria emphasises this point, since many of the examples are coccoid species similar (or perhaps identical) to those collected from western Abu Dhabi (Golubic & Barghoorn, 1977; Awramik, 1984; Golubic & Yun, 1985).

1. The amorphous organic material (sheath material and probably cell fluids) may degrade in a manner that destroys depositional microstructure. If or when this material is preserved, it will probably be preserved in an amorphous form, the origins of which are difficult or impossible to decipher.

2. The dissolution of inorganic components (e.g., pods of micritic carbonates) and subsequent compaction of the mats may alter the depositional structure of the mat.

CONCLUSIONS

The tidal flats with the seaward margin of the coastal plain or sabkha can divided into:

(a) Lower subtidal carbonate flats with muds, sands and hardgrounds
(b) Upper intertidal cyano-bacterial flats from seaward to landward of cinder like cyano-bacterial mat, polygonal mat (with local pinnacles and channels) and crenulated mat passing into flat mat
(c) Lower supratidal evaporites with gypsum mush
(d) Mid supratidal evaporites with gypsum being replaced by anhydrite and flanked to the landward by stranded beach ridges
(e) Upper supratidal evaporites trapped in storm wash-over, recycled aeolian and Tertiary sediments include gypsum and anhydrite

The potential for the preservation of this cyano-bacterial material in these intertidal and supratidal settings is still being investigated but it looks as if tidal flat settings similar to those of Abu Dhabi could be an important source of hydrocarbons in the future and have been in the past.

References

Awramik, S.M. 1984. Ancient stromatolites and microbial mats. In: Microbial Mats Stromatolites: Alan R. Liss, Inc., New York, pp. 1-22.

Butler, G.P. 1969. Modern evaporite deposition and geochemistry of co-existing brines, the sabkha, Trucial Coast, Arabian Gulf. *Journal of Sedimentary Petrology* 39: 70-89.

Butler, G.P., Harris, P.M. & Kendall, C.G.St.C. 1982. Recent evaporites from Abu Dhabi coastal flats. In: Handford, C.R., Loucks, R.G. & Davies, G.R. (eds.), Depositional and diagenetic spectra of evaporites: SEPM Core Workshop 3, pp. 33-64.

Cardoso, J.N., Watts, C.D., Maxwell, J.R., Goodfellow, R., Eglinton, G. & Golubic, S. 1978. A biogeochemical study of the Abu Dhabi algal mats: a simplified ecosystem. *Chem. Geol.* 23: 273-291.

Golubic, S. 1992. Microbial mats of Abu Dhabi. In: Margulis, L. & Olendzenski, L. (eds.), Environmental evolution; effects of the origin and evolution of life on planet Earth, pp. 103-130.

Golubic, S. & Barghoorn, E.S. 1977. Interpretation of microbial fossils with special reference to the Precambrian. In: Flugel, E. (ed.), Fossil Algae. Springer-Verlag, Berlin, pp. 1-14.

Golubic, S., & Yun, Z. 1985. Phycological expertise in geological applications. *Hydrobiologia* 123: 193-198.

Kendall, C.G.St.C. & Skipwith, P.A.d'E. 1968. Recent algal mats of a Persian Gulf lagoon. *Journal of Sedimentary Petrology* 38: 1040-1058.

Kendall, C.G.St.C., Cohen, D.A., Birdwell, B., Alsharhan, A.S., Colling, E.L., Rollins, M.S. & Durig, J.R. 1991. Cyano-bacterial mats in strandline evaporite/carbonate deposits along the coast of Abu Dhabi, U.A.E.: potential sources of petroleum. In: Bosellini, A., Brander, R., Flugel, E., Purser, B.H., Schalger, W., Tucker, M., & Zenger, D. (eds.), Dolomieu Conference on Carbonate Platforms and Dolomitization, Abstracts, pp. 130-131.

Kenig, F., Huc, A.Y., Purser, B.H. & Oudin, J.L. 1990. Sedimentation, distribution and diagenesis of organic matter in a recent carbonate environment, Abu Dhabi, UAE. *Organ. Geochem.* 16: 735-745.

Kenig, F., Boichard, R. & Purser, B.H. 1991. Holocene organo-sedimentary sequence of the Abu Dhabi coastal carbonate environment, United Arab Emirates. In: Bosellini, A., Brander, R., Flügel, E., Purser, B.H., Schalger, W., Tucker, M. & Zenger, D. (eds.), Dolomieu conference on carbonate platforms and dolomitization, Abstracts pp. 134-140.

Kinsman, D.J.J. 1964. The recent carbonate sediments near Halal al Bahraini, Trucial Coast, Persian Gulf. In: van Straaten, L.M.J.U. (ed.), Deltaic and Shallow Marine Deposits. *Developments in Sedimentology* 1: 189-192, Amsterdam, Elsevier.

Muller, D.W., McKenzie, J.A., Kendall, C. G. St. C. & Alsharhan, A.S. 1991. Application of strontium isotopes to the study of brine genesis during sabkha dolomitization. In: Bosellini, A., Brander, R., Flügel, E., Purser, B.H., Schalger, W., Tucker, M. & Zenger, D. (eds.), Dolomieu conference on carbonate platforms and dolomitization, Abstracts, p. 181.

Park, R.K. 1977. The preservation potential of some recent stromatolites. *Sedimentology* 24: 485-506.

Patterson, R.J. & Kinsman, D.J.J. 1981. Hydrologic framework of a sabkha along Arabian Gulf. *American Association of Petroleum Geologists Bulletin* 65: 1457-1476.

Schreiber, B. C., R. P. Philip, S. Benali, M. L. Helman, J. A. de la Peña, R. Marfil, P. Landais, A.D. Cohen, & C. G. St. C. Kendall. 2001. Characterization of Organic matter Formed in Hypersaline carbonate/Evaporite Environments: Hydrocarbon Potential and Biomarkers Obtained through Artificial Maturation Studies. *Journal of Petroleum Geology* 24:309-338.

Sheppard, C., Price, A. & Roberts, C. (eds.) 1992. Marine Ecology of the Arabian Region: Patterns and Processes in Extreme Tropical Environments. Academic Press, Harcourt Brace Jovanovich Publishers, 359 pp.

Swart, P.K., Shinn, E.A., McKenzie, J., Kendall, C.G.St.C. & Hajari, S.F. 1987. Spontaneous dolomite precipitation in brines from Umm Said sabkha, Qatar: Geological Society of America Annual Meeting Abstracts, p. 862.

Tissot, B.P. & Welte, D.H. 1984. Petroleum Formation and Occurrence, 699 pp. Springer-Verlag, Berlin.

Warren, J.K., & Kendall, C.G.St.C. 1985. Comparison of sequences formed in marine sabkha (subaerial) and salina (subaqueous) settings-modern and ancient. AAPG Bulletin 69: 1013-1023.

THE SABKHAT OF SAUDI ARABIA
AN INTRODUCTION

HANS-JÖRG BARTH

University of Regensburg, Department of Physical Geography, Universtiätsstr. 31, D-93040 Regensburg, Germany

Abstract
In Saudi Arabia there are four regions where sabkhat occur as a distinctive geomorphological land form. In Eastern Saudi Arabia sabkhat are common along the shoreline from Kuwait to the southern end of the Arabian Gulf. Coastal sabkhat are dominant but somewhat more inland, continental or inland sabkhat are abundant too, but smaller and more widely dispersed. The coastal lowland bordering the Red Sea is the second region where sabkhat occur. Here the sabkhat are restricted to flat and sheltered depositional environments. The third and fourth region are the large inland basins of the Rub'al Khali embayment in the south and the much smaller Al Jawf-Sakakah basin of the Great Nefud in the north. Between the aeolian sands of these basins inland sabkhat are common at several places, often interspersed among the huge sand dunes. The sabkhat of the central coastal lowlands in the Eastern Province are studied in detail. The results are compared with the findings of other authors who have been working on the Saudi Arabian sabkhat.

INTRODUCTION

Saudi Arabia can be divided in two geo-structural provinces. The western part, known as the Arabian Shield, is part of the Precambrian crustal plate, which is slightly tilted toward the eastern Arabian Gulf region. The eastern part, called the Arabian Shelf, consists of a thick sedimentary sequence covering the plate (Chapman 1978). This geologic situation provides a narrow western coastal plain between the Red Sea and the scarp mountains and a much wider plain at the Arabian Gulf coast. It is here where the abundant presence of unconsolidated sediments combined with shallow ground water levels and poor relief created sabkhat as a distinctive

Barth & Böer (eds.), SABKHA ECOSYSTEMS, 37-49

geomorphological feature. The extensive coastal flats at the southern Gulf coast display the largest and most impressive sabkhat of the world. Consequently, most research has been carried out in this region. But compared to numerous studies in the United Arab Emirates (UAE), the Saudi Arabian part generally has been neglected. The following descriptions focus on the sabkhat of the central coastal lowlands of the Eastern Province.

OCCURRENCE

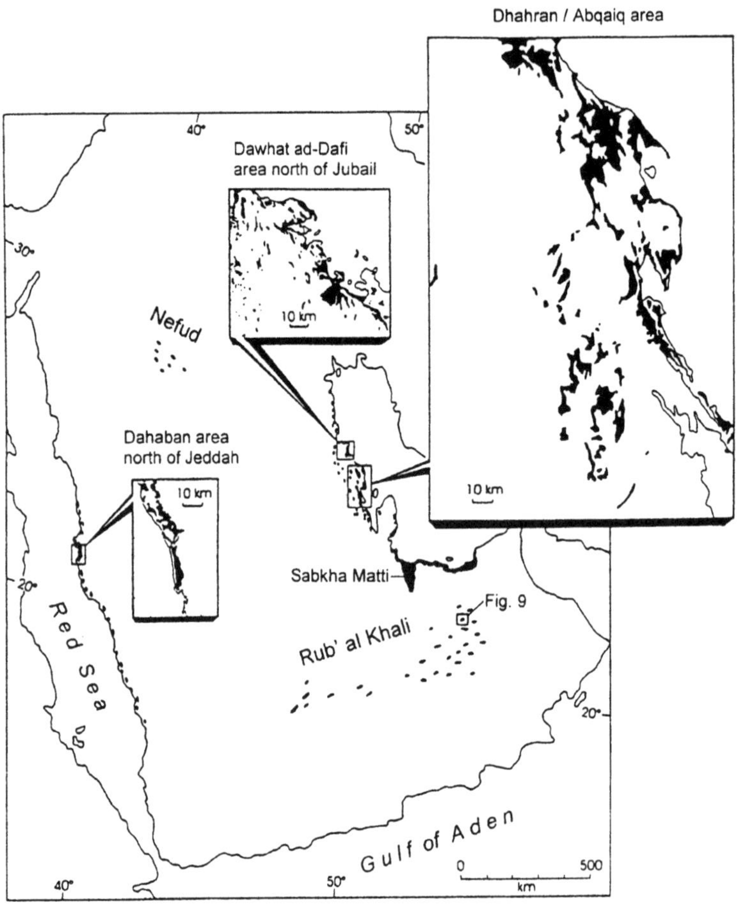

Figure 1. Location of sabkhat in Saudi Arabia (see black spots). The regions where significant research has been done in the past are highlighted.

The sabkhat of eastern Saudi Arabia are common along the shoreline between Kuwait and the southern end of the Arabian Gulf. Sabkhat Matti reaching far into the UAE and Saudi Arabia is the largest with an area of about 6000 square kilometres (Fig. 1). Coastal sabkhat alternate with flat and extensive sand sheets and inactive dune systems from Kuwait to the area north of Dhahran. Further south sabkhat also occur between barchanoid dune systems. In the same area somewhat more inland, inland sabkhat are abundant but they are often much smaller and more widely dispersed than their coastal counterpart.

The Red Sea on the western side of the Arabian Shield is also bordered by a coastal lowland which is known as the *Tihamah.* It is a narrow band of alluvial gravel and sands covering pediments cut on the crystalline and Tertiary rocks of the scarp mountains to the east. The western part of the Tihamah is basically a depositional surface on coralline rocks. Here sabkhat are common, mostly of the coastal type. They occur along the Saudi Arabian coast of the Red Sea, but are restricted to flat and sheltered depositional environments like the larger sabkhat between Dahaban and Jeddah (Fig. 1).

The third and fourth region where sabkhat can be found in Saudi Arabia are the large inland basins of low relief which are generally occupied by aeolian sands. There is the huge tectonic depression of the Rub' al Khali (*Empty Quarter*) in the south and the much smaller northern counterpart, the Al Jawf-Sakakah Basin of the Great Nefud. Inland sabkhat are located at several different spots within the large sand deserts, but they are most prominent in the eastern part of the Rub' al Khali Embayment, where an enormous group of sabkhat is interspersed among the sand dunes.

COASTAL SABKHAT

The slightly dipping sedimentary rocks of the Arabian Shelf are not only responsible for the shallowness of the Arabian Gulf, but also for the low relief of the whole coastal region, the width of the intertidal zone, the far reaching inlets, and the forward leaping peninsulas. A particular feature in this type of environment is the wide spreading coastal sabkha which occurs along the Gulf coast between Kuwait and the UAE. These supratidal surfaces are the continuation of the intertidal zone, sometimes divided by flat lying sand sheets. They consist of marine sediments sometimes covered or intercalated by aeolian terrestrial deposits.

The origin of coastal sabkhat is primarily in the flatness of the area combined with the erosive force of strong unidirectional winds. The unconsolidated sediments of the supratidal area are deflated until the damp zone of the capillary fringe is reached, which is the downward limit of aeolian erosion. Continuous evaporation results in the precipitation of evaporites in the sediments near the sand-air interface and directly at the surface.

Characteristic for the present sabkha environment in the Gulf region is a prograding depositional system. Holm (1960) recognised two types of sabkhat and sedimentation processes respectively at the Arabian Gulf coast. One type he called „arenaceous" (meaning „filled with sand") and the second he called „argillaceous" sabkhat which form from the manufacture of calcareous mud by marine organisms. Changes in sea level probably resulted in a change of the dominant sedimentation type. The model of marine sedimentation (argillaceous) during rising sea level and aeolian sedimentation (arenaceous) during standstill or regression was presented by Fryberger et al. (1983) for the Dhahran area (Fig. 2).

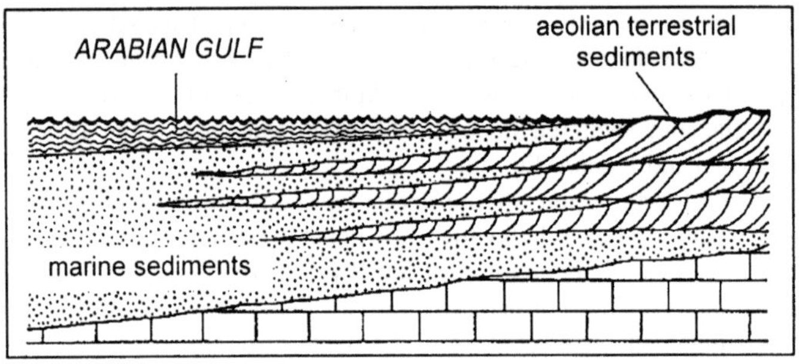

Figure 2. Schematic profile of the Saudi Arabian East coast at the Dhahran area (Fryberger et al. 1983).

Presently, for large parts of the Saudi Arabian coastal regions aeolian sedimentation of terrestrial quartz sand is the dominant process, driven by winds coming from north to north-west. Recent studies by the author (Barth 2001) concerning sedimentation rates reveal, that the average progradation rate at the Dawhat ad-Dafi region north of Jubail was about 3 meters per year for the last 700 years. In a distance of 2 km from the present intertidal zone in a depth of 70 cm the sediments of Sabkha Murayr (Fig. 3) provided excellent samples of laminated old cyanobacteria (Fig. 4). They are in the manner we find them in the intertidal zone today (Fig. 5). Radiocarbon dates of the old cyanobacteria indicated an age of about 700 years. This implies that 700 years ago the intertidal was located about 2 km south-west of the present intertidal zone. This means the shoreline moved at a mean rate of at least three meters per year in seaward direction.

Because the sediment above the laminated layers basically consists of terrestrial quartz sand (Fig. 6), sedimentation characteristics must have changed rapidly about 700 years ago. Whether a climatic change or human impact led to this change in the sedimentation process, is presently unknown. Cross bedding of terrestrial quartz sand which Fryberger et al. (1983) described for the area of Dhahran was not observed. But the reasonable explanation is provided by the fact that in the Dhahran area near the Jafurah sand sea whole dune bodies moved into the sea whereas in Jubail that much sand was never available. The same process still can be observed at the coast of Qatar where barchan dunes move into the Gulf (Shinn 1969, Al-Sheeb 1995).

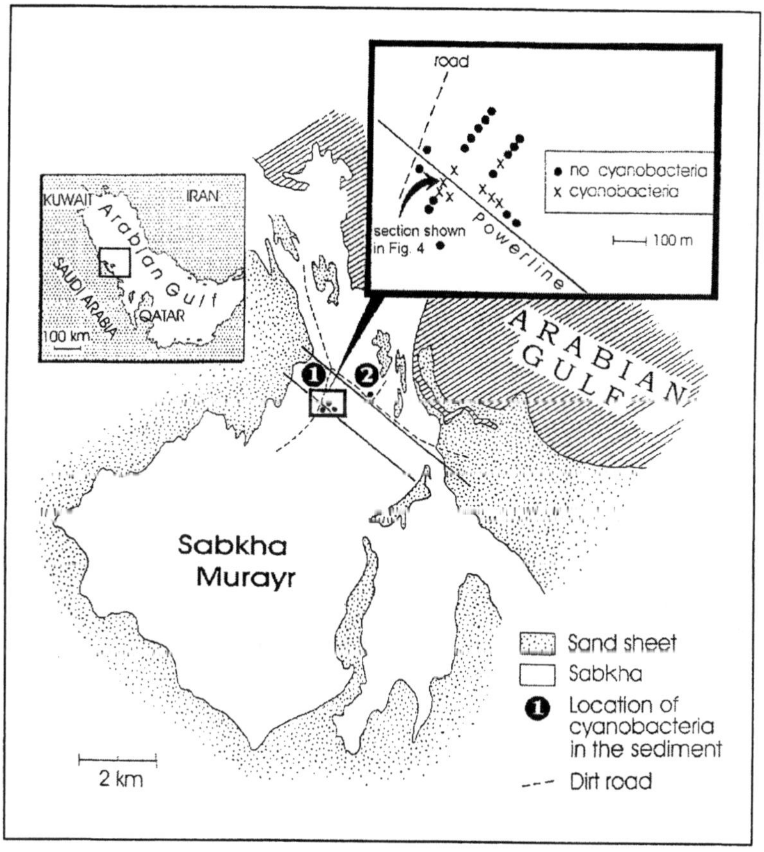

Figure 3. Location of Sabkha Murayr and the old cyanobacteria in its sediments.

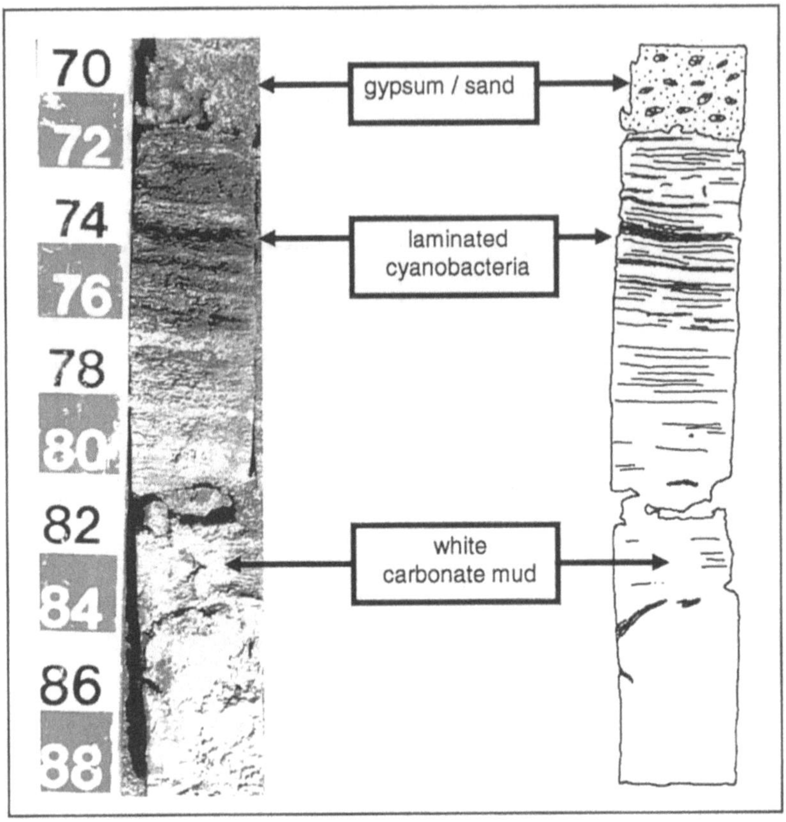

Figure 4. Laminated 700 years old cyanobacteria below 72 cm of sediment.

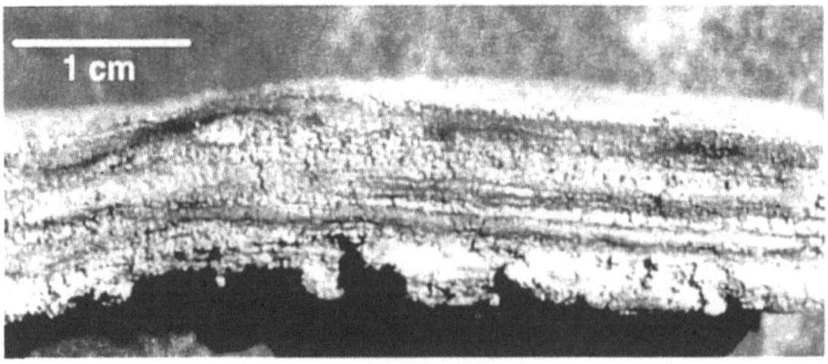

Figure 5. Laminated cyanobacteria collected in the present intertidal zone.

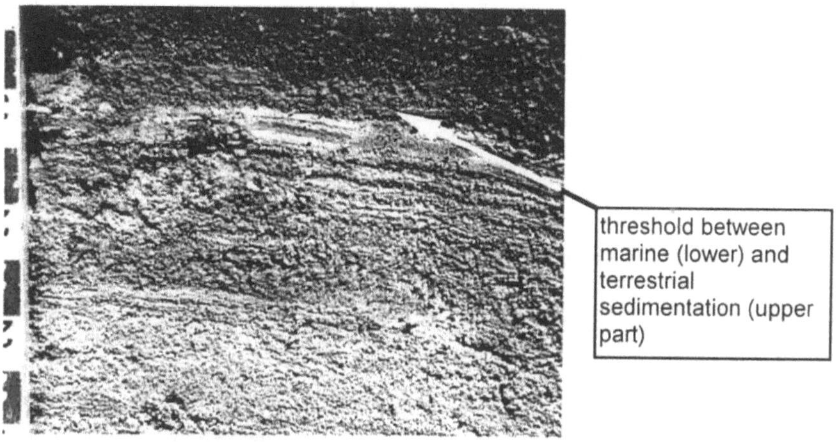

Figure 6. Section showing the transition between marine and terrestrial sedimentation.

Figure 7 shows a coastal sabkha divided from the lower supratidal and intertidal by a flat sand sheet at the north eastern end. Typical are various layers rich of carbonate sediments like *cerithid* gastropods and shells. Remarkable is the fact that gypsum only was found in a thin surface near layer where small (0.5-5 mm) clear monoklin flat crystals were distributed within the sand. This was a general observation for the whole area between Jubail and Ras az-Zaur (60 km north of Jubail). Gypsum crystals were only associated with organic material (cyanobacteria). This could be near the surface or quite rare in this area, deeper in the sediments. There, old laminated cyanobacteria form sort of an algal peat like the one described by Alsharhan et al. (1995, and this volume) for the sabkha of Al Qanatir (UAE). Massive gypsum crystallisation near the water table which is typical for the inland sabkhat in the Eastern Province was never found.

The ground water chemistry shows a general increase in concentration of dissolved solids towards the sabkha centre. Maximum conductivity values range between 190 and 215 mS/cm. During the course of a year the concentration of dissolved solids varies significantly (Fig. 7). This might be due to crystallisation and dissolution processes within the sediment. In figure 7 March and November values (before the first rainfall after summer drought) are displayed.

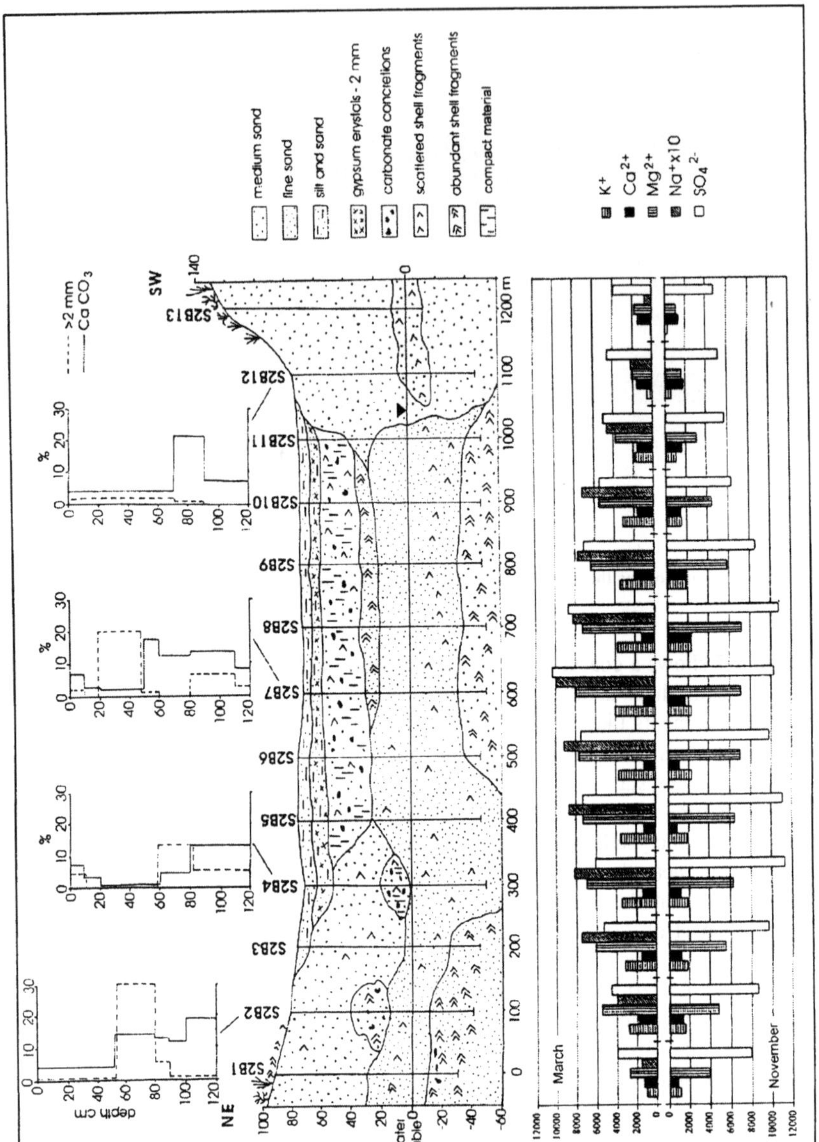

Figure 7. Transect in a coastal sabkha showing sediment characteristics, fractions larger than 2 mm (concretions and shell fragments), $CaCO_3$ fraction and ground water chemistry in March and November (before the first rain events).

The coastal sabkhat of the Red Sea coast are marine depositional environments where sand and marine carbonates accumulate in sheltered areas. Bahafzullah et al. (1993) presents a valuable description of the sabkhat in the vicinity of Jeddah, especially of the Dahaban sabkha. In contrast to the coastal sabkhat near Jubail, gypsum seems to be quite common in the saturated zone as well as in association with organic material. Further studies were carried out in the same region regarding sabkha classification by the means of satellite data by Al-Saifi & Quari (1996). Sabtan et al. (1995) published results about physical properties of sabkha soils also focusing on Dahaban sabkha near Jeddah. Theses studies indicate a growing interest in the sabkhat of Saudi Arabia and promise further results contributing to a better understanding of these remarkable geomorphological desert features.

INLAND SABKHAT

On the subject of inland sabkhat there are only few publications (Glennie et al. 1970, 1995, Goodall et al. 2000, Fookes et al. 1985, Fryberger et al. 1983, 1988, Hardie 1984, Holm 1960, Johnson et al. 1978, Kendall 1984, Warren 1989,). Additionally, confusion is caused by the synonymous use of the term „playa" for sabkha (see introduction of this book). Concerning sabkhat of the Eastern Province, even in early publications (Glennie 1970, Holm 1960, Johnson et al. 1978) inland sabkhat are distinguished from coastal sabkhat. Whereas Holm (1960) and Glennie (1970) base their differentiation on the geographical location, Johnson et al. (1978) describe inland sabkhat as sabkhat not influenced by marine groundwater. The following section again presents results of studies carried out on sabkhat near Jubail in the central coastal lowlands. They will be compared with findings of other authors who have been working at other locations in order to be able to present a general view of the subject.

Sabkhat al-Haidaruk is located between 49°12'10'' - 49°13'00''E and 27°04'00'' - 27°05'35''N in a distance of 15 km from the Gulf coast. The elevation is 22 m above sea level. Adjacent to the north-western fringe are some rocky outcrops of the Tertiary Hadrukh formation covered by sand at the baseline. To the south and east the sabkha is limited by extensive sand sheets. More detailed information about this study is given in Barth (1998). The salts within the surface crust consist basically of halite (80%). The remaining part is made of calcium, potassium (sylvite) and magnesium salts (epsomite). A similar surface crust chemistry is reported from Johnson et al. (1978) 50 km south of Jubail, Fryberger et al. (1983) south of Abqaiq and Goodall et al. (2000) in the Sabkhat Matti region. The crust morphology is pustular, vermiform or polygonal depending on the hydrological situation of each area.

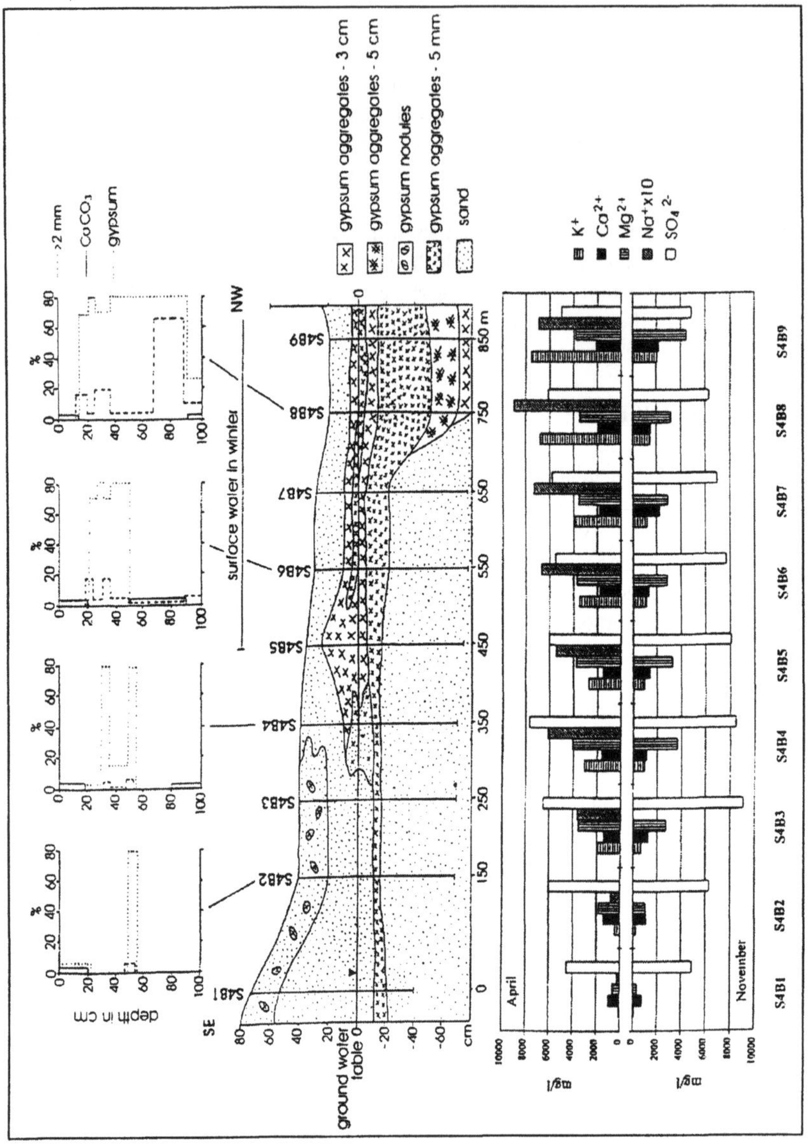

Figure 8. Transect in an inland sabkha showing sediment characteristics, fractions larger than 2 mm (concretions and gypsum), $CaCO_3$ fraction, gypsum fraction and ground water chemistry in April and November (before the first rain events).

The near surface substrate below the salt crust is mainly composed of quartz sand. Finer material like silt is usually below 15%. The portion of carbonates is less than 3%. 20-30 cm below the surface, horizontal layers of gypsum crystals (prismatic, 0.5-5 mm diameter) are intercalated with sand layers. The gypsum horizons increase in number and thickness towards the sabkha centre where highest ion concentrations were recorded (up to 230 mS/cm). Gypsum crystals of cm size are quite frequent until more than 50 cm below the water table. Again Johnson et al. (1978) make the same observations in the sabkhat north of Dhahran, where they found prismatic crystals up to 10 cm, as well as Fryberger et al. (1983, 1988) in the Abqaiq region. Studies of the brine chemistry reveal that the concentrations of typical ions follow the pattern of the overall concentration. The variations during the course of one year are similar to the results of coastal sabkhat. Brine studies carried out by Johnson et al. (1978) 50 km north of Dhahran in Sabkhat Ar Riyas show results within the same range. The vertical distance between the surface and the water table varies considerably during the course of the year, but was never observed to be more than 2 meters.

Typical sediments of inland sabkhat in the Eastern Province consist largely of quartz sand with a very low portion of carbonates (< 3%). The presence of various horizontal gypsum layers at different depths between the upper capillary fringe and the saturated zone is characteristic. Gypsum crystals of cm size are especially clustered just above, and extending 10 or more cm below the water table. The surface salt crusts are quite similar to the salt crusts in coastal sabkhat. Brine studies in general show the same characteristics like coastal sabkha brines except a lower potassium and magnesium concentration. But this can change dramatically as soon as influx of water from springs alter the brine chemistry.

Within the Rub' al Khali in southern Saudi Arabia inland sabkhat are widely dispersed. They often form interdunal spaces between large pyramidal, domal and megabarchan dunes for example in the north central Rub' al Khali (Fig. 9). An absolutely important feature are the sabkhat in the eastern Rub' al Khali, the 'Uruq al Mu'taridah, where giant crescentic dunes up to 230 meters high are developed on a widespread inland sabkha foundation (Edgell 1990). These sabkhat basically consist of a sandy substrate with a large gypsum content. Compared to the inland sabkhat closer to the Arabian Gulf coast, the portion of NaCl in the sediments is very low.

Figure 9. Sabkha in the Rub' al Khali.

CONCLUSION

There have been very few studies on the sabkhat of Saudi Arabia in the past. So we know almost nothing beyond their occurrence. The recent studies especially regarding the Red Sea coastal sabkhat in Saudi Arabia demonstrate that there is an increasing interest in these landforms. Especially coastal sabkhat offer a great potential of different kinds of uses (i.e. regarding agricultural systems combined with sea water irrigation). In order to quantify these potentials correctly further research on sabkha ecology should be encouraged. Primarily it should focus on:

- Sedimentology and soil characteristics (i.e. infiltration capacity)
- Ground water chemistry (regarding the formation of valuable salts as well as future sea water irrigation)
- Cyanobacteria and Halophytes occurring in sabkha environments (biotic resources)

ACKNOWLEDGEMENTS

I am grateful to NCWCD, Riyadh for the help they provided during different projects. Especially I would like to thank H. E. Prof. Dr. Abdulaziz H. Abuzinada for his support and encouragement. Assistance, continuous help and cooperation both in the field and the laboratory in Jubail were provided by numerous individuals.

References

Al-Saifi, M.M. & Quari, M.Y.H.T. 1996. Application of Landsat Thematic Mapper data in sabkha studies at the Red Sea coast. *International Journal of Remote Sensing* 17 (3): 527-536.

Alsharhan, A.S., Kendall, C.G.St.C. & Whittle, G.L. 1995. Holocene carbonate evaporites of Abu Dhabi, United Arab Emirates-field trip guidebook. IGCP-349, United Arab Emirates University, Al Ain.

Al-Sheeb, A.I. 1996. Sand dune movement in the state of Qatar. Paper presented at the International Conference on Desert Development in the Arab Gulf Countries, State of Kuwait 23-26 March, 1996, Abstracts.

Bahafzulla, A., Fayed, L.A., Kazi, A. & Al-Saify, M. 1993. Classification and distribution of the Red Sea coastal sabkhat near Jeddah - Saudi Arabia. *Carbonates and Evaporites* 8 (1): 23-38.

Barth, H.-J. 1998. Sebkhas als Ausdruck von Landschaftsdegradation in den zentralen Küstentiefländern der Ostprovinz Saudi Arabiens. Dissertation an der Ruhr-Universität-Bochum. *Regensburger Geographische Schriften* 29, 279 pp. Regensburg.

Barth, H.-J. 2001. Understanding coastal fluctuations at the Arabian Gulf leading to the „lost city of Gerrha". *Palaeoecology of Africa* 27: 291-303.

Chapman, R.W. 1978. Geology and geomorphology. In Al-Sayari, S.S. & Zötl, J. (eds.), Quaternary period in Saudi Arabia. Vol.1: 4-30, New York.

Edgell, H.S. 1990. Evolution of the Rub' al Khali Desert. J.K.A.U.: Earth Sciences, Vol. 3, Special Issue: 1st Saudi Symp. on Earth Sciences, Jeddah 1989, pp. 109-126.

Fookes, P.G., French, W.J. & Rice, S.M. 1985. The influence of ground and groundwater geochemistry on construction in the Middle East. *Quarterly Journal of Engineering Geology* 18: 101-128.

Fryberger, S.G., Al-Sari, A.M. & Clisham, T.J. 1983. Eolian dune, interdune, sand sheet, and siliciclastic sabkha sediments of an offshore prograding sand sea, Dhahran area, Saudi Arabia. *American Association of Petroleum Geologists Bulletin* 67-2: 280-312.

Fryberger, S.G., Schenk, C.J. & Krystinik, L. F. 1988. Stokes surfaces and the effects of near-surface groundwater-table on aeolian deposition. *Sedimentology* 35: 21-41.

Glennie, K.W. 1970. Desert sedimentary environments. *Developments in Sedimentologie* 14: 222 pp.

Glennie, K.W., Al-Belushi, J. & Al-Maskery, S. 1995. The inland sabkhat of the Huqf, Oman - A product of past extremes of humidity and aridity. Poster presentation on the international conference on Quaternary deserts and climatic change, Al Ain, UAE.

Goodall, T.M., North, C.P. & Glennie, K.W. 2000. Surface and subsurface sedimentary structures produced by salt crusts. *Sedimentology* 47: 99-118.

Hardie, L.A. 1984. Evaporites: marine or non-Marine. *American Journal of Science*, 284: 193-240.

Holm, D.A. 1960. Desert geomorphology in the Arabian Peninsula. *Science* 132: 1369-1379.

Johnson, D.H., Kamal, M.R., Pierson, G.O. & Ramsay, J.D. 1970. Sabkhahs of Eastern Saudi Arabia. In: Al-Sayari, S.S. & Zötl, J.G. (eds.), Quaternary period in Saudi Arabia. Vol. 1: 84-92, New York.

Kendall, A.C. 1984. Facies Model 11: Continental and supratidal (sabkha) evaporites. In: Walker, R.G. (ed.), Facies models. Second edition, Geoscience Canada Reprint Series 5: 66-78.

Sabtan, A., Al-Saify, M. & Kazi, A. 1995. Moisture retention characteristics of coastal sabkhat. *Quarterly Journal of Engineering Geology* 28: 37-46.

Shinn, E.A. 1969. Submarine lithification of Holocene carbonate sediments in the Persian Gulf. *Sedimentology* 12: 109-144.

Warren, J.K. 1989. Evaporite sedimentology: its importance in hydrocarbon accumulations. Prentice-Hall Scientific Publications, 285 pp. Englewood Cliffs, New Jersey.

PLANT COMMUNITIES BORDERING THE SABKHAT OF BAHRAIN ISLAND

JAMEEL A. ABBAS

Department of Biology, University of Bahrain, P.O. BOX 32038, Isa Town, Bahrain

Abstract

Sabkha and neighbouring saline habitats were recognised in four major areas in the coastal lowland of Bahrain Island which represents more than 40% of the total area of the Island. These areas are the western coastal plain, the southwest sabkha, the south sabkha, and the northeastern coastal area. Plant communities and plant species of the sabkha and saline habitats were identified. The main plant communities include *Avicennia marina* community, *Zygophyllum qatarense* community, *Arthrocnemum macro-stachyum* community, and *Sporobolus ioclados* community. The plant species composition and the distribution of sabkha communities in Baharin Island is similar to that in the rest of the Arabian Peninsula, particularly Eastern Saudi Arabia and Qatar.

INTRODUCTION

The term solonchak has been applied to salty soils in which capillary rise of solutions containing soluble salts occur. On the onset of evaporation, deposition of soluble salts takes place in the upper surface of the soil. Doornkamp et al. (1980) distinguished two types of solonchak soils in Bahrain. The first type is the natural solonchak that represents soils with no signs of previous cultivation, and possess high gypsum contents. The second type is the cultivated solonchak, which is found in areas with current or previous cultivation activities.

Historically, villages and date palm plantations in Bahrain were concentrated along the coastal areas integrated with the maritime activities of Bahraini people throughout ages. In the last few decades, urbanisation and mismanaged use of the limited freshwater has led to salinisation of soil in most of the traditional agricultural

Barth & Böer (eds.), SABKHA ECOSYSTEMS, 51-62

areas of Bahrain. These new saline soils (cultivated solonchak), in addition to the naturally occurring saline soils (natural solonchak), have increased the total area of saline soils in Bahrain tremendously. Recently, the area under cultivation decreased from 36.96% in 1995 to 33.89% in 1997 (GCC, 1999). Taking this into account, and based on the traditional use of halophytic plants in Bahrain as food, fodder, and in folk medicine (Abbas et al. 1992, Alsaleh et al. 1997), it is necessary to consider the introduction of halophyte crop agricultural systems and to use seawater for irrigation (Rozema et al. 1995, Böer & Gliddon 1998). However, an understanding of the ecology, vegetation structure, and species distribution of halophytic plants in salinised soils and in naturally occurring saline flats, particularly sabkhat and neighboring environments, is needed before any agricultural management is considered. To date, only limited plant ecological and vegetation studies have been carried out in Bahrain (Abbas et al. 1991, Abbas & El-Oqlah 1992, Abbas 1995, Abbas 1998). This work gives an overview of the floral composition and the main plant communities in sabkhat and bordering saline flats of Bahrain.

Study Area

Bahrain is an archipelago of 36 islands located in the mid-western part of the Arabian Gulf about 25 km east of Saudi Arabia at 25° 32' and 26° 20' N, and 50° 20' and 50° 32' latitude E. The largest of these islands is Bahrain Island, with a total area of 706 km^2 (Fig. 1). The short, mild winters and the long, very hot summers characterise its climate. The mean annual temperature is 26.4°C, with an absolute maximum in June of 47.5°C, and an absolute minimum in January of 2.8°C. The mean annual rainfall is 74 mm. The climatic diagram of Bahrain (Fig. 2) shows that the dry period extends from May to October and the wet period extends from November to April.

Landforms and Geomorphology

Bahrain lies on the eastern part of the Arabian plate, which coincide with the Arabian Peninsula. Five physiographic zones characterise the topography of Bahrain (Fig. 1). These are (a) the central plateau and jabals; (b) the interior basin; (c) the escarpments; (d) the backslope; and (e) the coastal lowlands.

This narrow lowland surrounds the backslope and is widest in the north and west and narrowest in the east. It covers an area of approximately 275 km^2, with a general surface elevation below 5 m, composed mainly of loose or partially cemented aeolian or marine sediments. This plain extends from the backslope of the rocky core, where it is at elevations of approximately 10 m, to sea level at the coastline. Sabkhat are found in the coastal plain only.

Bahrain Island is composed mainly of Eocene and Pleistocene sediments. The Eocene rock sediments are in the centre of the island, forming an oval outcropping. The interior basin is in the form of a border of largely unconsolidated Pleistocene and recent sediments surrounding the outcrop (Doornkamp et al., 1980).

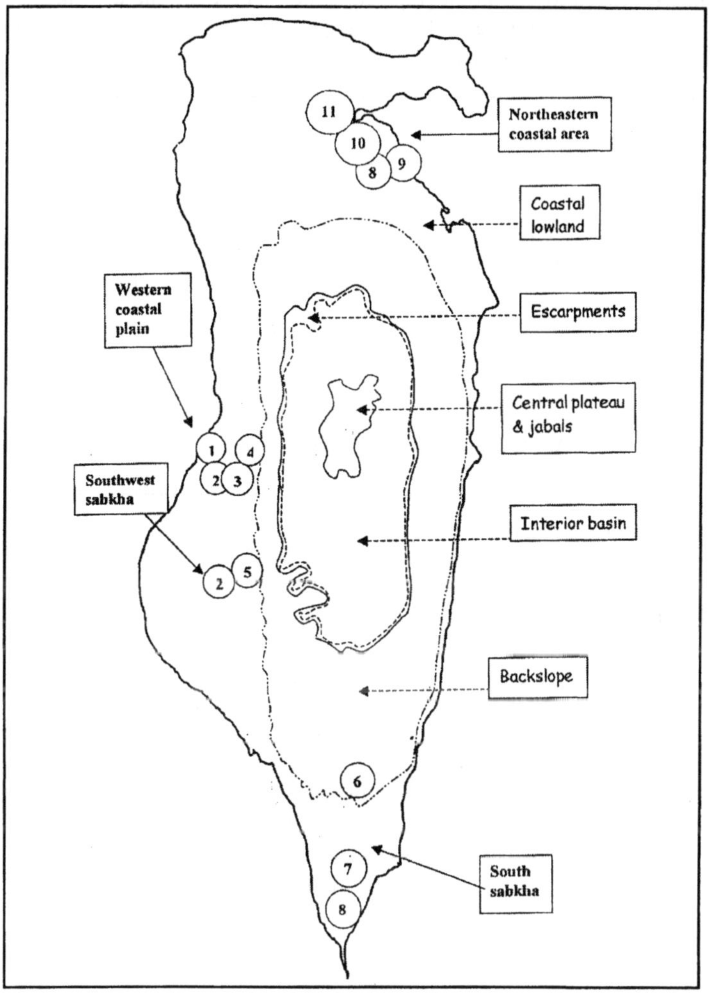

Figure 1. Map of Bahrain showing the physiographic zones, studied areas, and the location of plant communities. The following numbers denote the dominant species in the particular communites: (1) *Aeulropus lagopoides*; (2) *Zygophyllum qatarense*; (3) *Halopeplis perfoliata*; (4) *Sporobolus ioclados*; (5) *Cyperus conglomeratus*; (6) *Stipagrostis plumosa*; (7) *Seidlitzia rosmarinum*; (8) *Arthrocnemum macrostachyum*; (9) *Avicennia marina*; (10) *Suaeda vermiculata*; (11) *Halocnemum strobilaceum*.

Many features of the landscape of Bahrain are attributable to fluvial processes. Accordingly, Bahrain is broadly divisible into four major groups of drainage basins (catchments) related to the physiographical zones. These are the catchments of the interior basin and central plateau, the multiple escarpment zone, the upper Dammam backslope, and the coastal lowland.

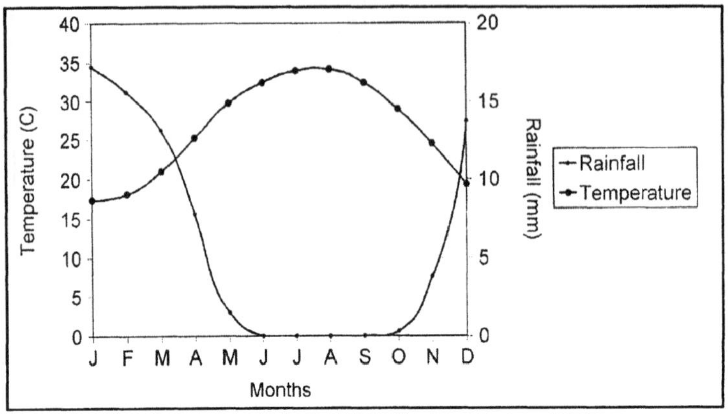

Figure 2. Climatic diagram of Bahrain.

METHODS

The information given in this work is a result of studies of the flora and vegetation of Bahrain Island by the author over a period of fifteen years. The floral list of sabkhat and saline soils was prepared from several collections made over that period. Plant species were identified according to Mandaville (1990). Plant communities were identified as a result of applying different methods in previous studies by the author (Abbas et al. 1991, Abbas & El-Oqlah 1992, Abbas 1995, Abbas & El-Oqlah 1996) The methodology in those studies included transects, stand vegetation cover analyses, and visual observations.

PLANT COMMUNITIES

Dornkamp et al. (1980) classified the coastal lowlands of Bahrain geomorphologically into a number of areas. Plant communities or vegetation in areas with true sabkha or with saline soil supporting halophytic plants are described in this work. The areas included here are: the western coastal plain, the southwest sabkha, the south sabkha, and the northeastern coastal area. A number of plant communities can be recognised in or nearby sabkha areas in the coastal lowlands. Some of these communities are restricted to particular areas, others may be found in more than one area. As in the rest of Arabia, coastal species are either succulent, salt-incorporating halophytes like the representatives of the family Chenopodiaceae, or salt excreting such as *Sporobolus ioclados* (Nees ex Trin) Nees, *Aeluropus lagopoides* (L.) Trin., *Avicennia marina* (Forssk.) Vierh., and *Limonium axillare* (Forssk.) O. Kuntze (Deil, 1998). A list of plant species recorded in these communities is presented in Appendix 1.

The western coastal plain

A strip of coastal plain 1-2 km wide extends over a total distance of 18 km along the coast. It covers an area of 30 km^2. Over much of its length it merges into the gypsiferous sand and gravel mantled slopes. In some places where there is no human activity, inland sabkha with salt-encrusted surface can be found. Four interconnected relatively large halophytic plant communities are found in this plain. These are (1) *Aeluropus lagopoides* community, (2) *Zygophyllum qatarense* Hadidi community, (3) *Halopeplis perfoliata* (Forssk.) Aschers. et Schweinf. community, and (4) *Sporobolus ioclados* community. The saline flats with it's halophytic communities give way landward to small sand dunes supporting communities of glycophytic plants including *Panicum turgidum* Forssk. community, and *Pennisetum divisum* (Gmel.) Henr. community (Fig. 3a).

Aeluropus lagopoides community

This community dominates large areas in the western coastal plain close to the coast but away from tidal inundation effects. The soil is natural solonchak containing deposits of gastropod shells and it is covered with a salt crust. Few individuals of halophytic plants are randomly distributed in this community. These are *S. ioclados*, *Juncus rigidus* Desf., *Limonium axillare*, *Sesuvium sesuviodes* (Fenzl) Verdc., and *Z. qatarense.*

Zygophyllum qatarense community

This species is widely distributed and dominant in communities in a variety of habitats. Abbas (1998) reported significant variations in the *Z. qatarense* total cover, relative cover, relative density, relative frequency, and the importance values among studied communities from various habitats. It has a wide range of distribution in relation to salinity (0.07 – 4.23 mS cm^{-1}). The highest importance value of this species was in the sabkha (38.2%) and the lowest was in the interior basin (11.2%). This species dominates a community in the western coastal plain where the soil is mainly natural solonchak and the crusty surface is covered in many parts by gastropod shells. Associated with *Z. qatarense* are *S. ioclados*, *L. axillare*, *A. lagopoides*, and *Mesembryanthemum forsskalei* Hochst.

Halopeplis perfoliata community

In the western coastal plain to the east of the *Z. qatarense* community, *H. perfoliata* occurs in almost pure stands without tidal inundation. Few individuals of *Z. qatarense* could be seen randomly scattered in this community.

Sporobolus ioclados community

A dense community dominated to a very large extent by *S. ioclados* is found in a natural solonchak soil with salt crusts to the northeast of the *H. perfoliata* community in the western coastal plain. Few individuals of *Z. qatarense*, *Cyperus arenarius* Retz., *A. lagopoides*, *S. sesuvioides*, and *Suaeda vermiculata* Forssk. are randomly scattered in this community.

The southwest sabkha

The western coastal plain widens rapidly in the south and becomes generally lower in elevation forming an area of approximately 60 km^2. Most of it is exposed to tidal inundation during high tides. Evaporation causes the precipitation of salts and the development of a salt crust forming a large salt pan. This salt pan represents the main true coastal sabkha that is almost devoid of vegetation apart from algal mats which become desiccated during dry seasons, and when there is no or little seawater intrusion. More inland, with the decrease in salinity and the change in the soil texture, small sand dunes support *Z. qatarense* and *Cyperus conglomeratus* Rottb. agg. (Fig. 3b).

The south sabkha

The most southerly area of Bahrain island is in the form of a triangle with an elevation range between 1.6 and 3.6 m. The surface sediments are loose, homogenous, poorly-graded fine shell gravels in a matrix of silty fine to medium gypsiferous calcareous quartzose sand (Doornkamp et al. 1980). The soil is saline and only scattered vegetation of *H. perfoliata* occurs. Individuals of *Halocnemum strobilaceum* (Pall.) M. B., *Arthrocnemum macrostachyum* (Moric.) et Delponte, and *S. vermiculata* may be found. This sabkha is delineated from the north by the southern face of the backslope that supports a dense community of *Stipagrostis plumosa* (L.) Munro ex T. Anders that contains scattered individuals of the shrub *Lycium shawii* Roem. et Schult. This sabkha narrows down towards the southern tip of the island. Here, small sand dunes support *Seidlitzia rosmarinus* Ehrenb. ex Bge., and *A. macrostachyum* species (Fig. 3c).

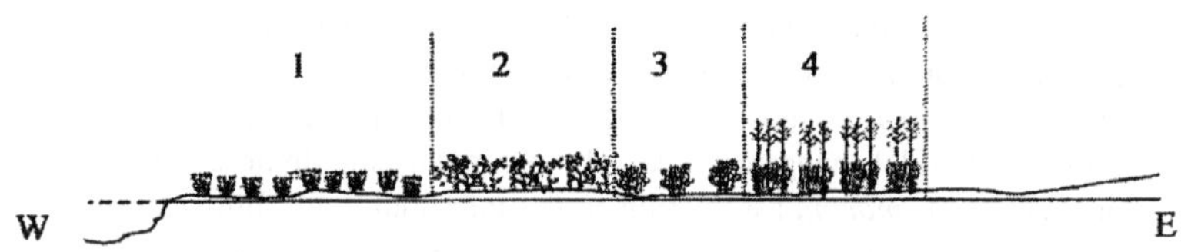

3a: Profile of the western coastal plain covering 2 km.

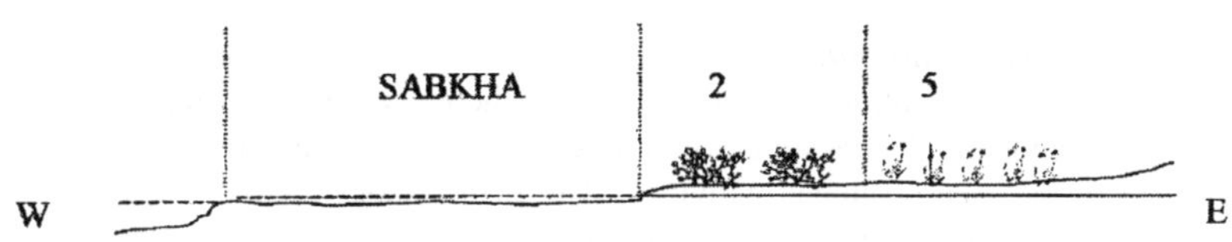

3b: Profile of the southwest sabkha covering 4 km.

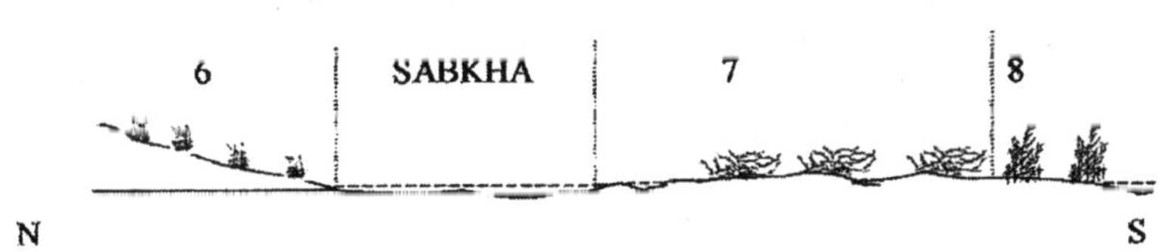

3c: Profile of the south sabkha covering 10 km.

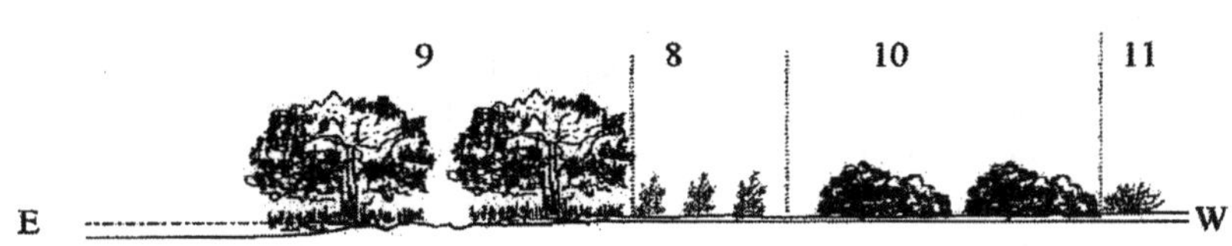

3d: Profile of the northeastern coastal area covering 5 km.

Figure 3. Profiles of studied areas. The numbers denote the dominant species in the particular areas as given in figure 1.

The northeastern coastal area

This low coast is the most sheltered area of the coastal lowland of Bahrain. Fine-grained carbonate mudflats have accumulated in this area. The surface of the neighbouring inland soil is mostly salt-blistered sabkha that gradually changes seaward into an intertidal mangrove area. The perennial halophytic communities in this area include an *A. marina* association in the intertidal zone followed landward by *A. macrostachyum* community, and more inland by *S. vermiculata* community (Fig. 3d). The back boundaries of this area are mostly unvegetated with few scattered individuals of *Cressa cretica* L. More inland away from the mangrove swamp, *H. strobilaceum* dominates a large community in a reclaimed area that used to be part of the mangrove swamp.

Avicennia marina community

In the northeastern littoral muddy shoreline where the soil is loamy cultivated solonchak., stands of *A. marina* are found very close to date palm plantations and runoffs from the nearby human activities. The main associates of *A. marina* are *Suaeda maritima* (L.) Dum. and *A. macrostachyum*. These three species form a zonation with *S. maritima* strips seaward on the banks of tidal channels followed by *A. marina* colonies that are dissected by the tidal channels, and more landward in this zonation is *A. macrostachyum*. Other species that can be found in this community include *Salicornia europaea* L., *Aeluropus littoralis* (Gouan) Parl., and *H. strobilaceum*. Mangrove stands used to extend over large areas but were reduced largely due to human activities and urban developments.

Arthocnemum macrostachyum community

This species dominates communities in the northeastern coasts behind the *A. marina* community where it is occasionally swamped by high tides. The soil is loamy cultivated solonchak. The main species that are found with *A. macrostachyum* include *S. maritima* and *H. perfoliata*.

Suaeda vermiculata community

Further inland and away from tidal effect, thickets of *S. vermiculta* occupy loamy cultivated solonchak soil with a high salinity (76.6 mS cm^{-1}) in the waste agricultural land at the back of the mangrove swamps. The community is found inland. Scattered individuals of *Salsola villosa* Del. Ex Roem., *Prosopis farcta* (Banks et Sol.) Macbride, *Atriplex leucoclada* Boiss., and *Alhagi graecorum* Boiss. are found. Within the boundaries of this community *Phragmites australis* (Cav.) Trin. ex steud. is a common associate.

Halocnemum strobilaceum community

H. strobilaceum is widely distributed along the eastern side of the island in loamy cultivated solonchak. It dominates an inland community in the northeastern region of the island with other halophytic species including *S. vermiculata*, *S. villosa*, and *Z. qatarense*. The land that is now dominated by *H. strobilaceum* used to be part of the northeastern mangrove swamp but the major land reclamation in the northeastern coastal areas changed this part into a wasteland.

DISCUSSION

The total number of plant species in Bahrain Island is around 307 (El-Oqlah & Abbas 1994). Many of these species are found in saline habitats. Some of them such as *A. marina* are restricted to certain habitats (Abbas & El-Oqlah 1992). It seems that due to the small size of the Island and the relatively large area occupied by the coastal lowland, which does not exceed more than 10 m above sea level, many of these plants are found in saline habitats in the coastal lowlands. The flora is similar to that of Qatar and eastern Saudi Arabia (Batanouny 1981, Mandaville 1990). In general, species distribution and community structure in the different types of sabkha and in the saline flats are similar to those reported in other studies in the Arabian Peninsula (Böer & Warnken 1992, Al-Hubaishi & Müller-Hohenstein 1984, Halwagy & Halwagy 1977, Ghazanfar 1995, Babikir 1984).

ACKNOWLEDGEMENTS

The author would like to acknowledge the technical assistance provided by Dr Nedzad Mehic, and the manuscript reading by Dr. Gordon Hoad.

References

Abbas, J.A. 1995. Monthly variations of chloride accumulation by *Zygophyllum qatarense* from saline and nonsaline habitats of Bahrain. In: Ajmal Khan, M. & Ungar, I.A. (eds.), Biology of salt tolerant plants. Book Crafters Publishers, Michigan. pp. 183-189.

Abbas, J.A. 1998. Distribution and structure of *Zygophyllum qatarense* communities in Bahrain Island. *Arab Gulf Journal of Scientific Research* 16: 555-568.

Abbas, J.A. & El-Oqlah, A.A. 1996. Distribution of plant species along a transect across Bahrain Island. *Journal of Arid Environments* 34: 37-46.

Abbas, J.A. & El-Oqlah, A.A. 1992. Distribution and communities of halophytic plants in Bahrain. *Journal of Arid Environments* 22: 205-218.

Abbas, J.A., Mohamed, S.A. & Saleh, M.A. 1991. Edaphic factors and plant species distribution in a protected area in the desert of Bahrain Island. *Vegetatio* 95: 87-93.

Abbas, J.A., El-Oqlah, A.A. & Mahasneh, A. 1992. Herbal plants in the traditional medicine of Bahrain. *Economic Botany* 46: 158-163.

Al-Hubaishi, A. & Müller-Hohenstein, K. 1984. An introduction to the vegetation of Yemen. Ecological basis, floristic composition, human Influence. Deutsche Gesellschaft fur Technische Zusammenarbeit, Eschborn.

Al-Saleh, F.A., Gamal El-Din, A.Q., Abbas, J.A. & Saeed, N.S. 1997. Phytochemical and biological studies of medicinal plants in Bahrain: The family Chenopodiaceae. *International Journal of Pharmacognosy* 35: 38-42.

Babikir, A.A. 1984. Vegetation and environment on the coastal sand dunes, and playas of Khor El-Odaid Area, Qatar. *Geo Journal* 9: 377-385.

Batanouny, K.H. 1981. Ecology and flora of Qatar. Oxford: Alden Press.

Böer, B. & Warnken, J. 1992. Qualitative analysis of the coastal and inland vegetation of the Dawkat ad-Dafi and Dawkat al-Mussalamiya region. In: Establishment of a marine habitat and wildelife sanctuary for the Gulf region. Final report for phase I. CEC/NCWCD, Frankfurt & Jubail. pp.81-101.

Böer, B. & Gliddon, D. 1998. Mapping of coastal ecosystems and halophytes (case study of Abu Dhabi, United Arab Emirates). *Marine & Freshwater Research* 49: 297-301.

Deil, U. 1998. Coastal sabkha vegetation. In: Ghazanfar, S.A. & Fisher, M. (eds.): vegetation of the Arabian Peninsula. *Geobotany* 25: 209-228.

Doornkamp, J.C., Brunsden, D. & Jones, D.K.C. 1980. Geology, geomorphology and pedology of Bahrain. Geo Abstracts, Norwich, U.K.

El-Oqlah, A.A. & Abbas, J.A. 1994. A checklist of vascular plants of Bahrain. *Dirasat* 21:95-118.

GCC 1999. The Cooperation Council for the Arab States of the Gulf. Secretariat General. Statistical Bulletin. Information Center- Statistical Department. Volume 9.

Ghazanfar, S.A. 1995. Vegetation of coastal sabkhat: an analysis of the vegetation of Barr Al Hikman, Sultanate of Oman. In: Ajmal Khan, M. & Ungar, I.A. (eds.): Biology of salt tolerant plants. Book Crafters Publishers, Michigan. pp. 277-283.

Halwagy, R. & Halwagy, M. 1977. Ecological studies in the desert of Kuwait. III. The vegetation of the coastal salt marshes. Journal of the University of Kuwait (Science) 4: 33-73.

Mandaville, J.P. 1990. Flora of Eastern Saudi Arabia. Keagan Paul International, London.

Rozema, J., Decae, R., Niazi, B., Abdulla, Z.U., Zaheer, H., Ahmad, R., Linders, H. & Broekmann, R. 1995. Salt tolerant crops, halophytes and crop cultivation in salt affected arable land. In: Ajmal Khan, M. & Ungar, I.A. (eds.): Biology of salt tolerant plants. Book Crafters Publishers, Michigan. pp. 380-387.

Appendix

List of plant species recorded in or adjacent to the sabkhat studied.

Aizoaceae
Aizoon canariense L.
Mesembryanthemum forsskalei Hochst.
Sesuvium sesuvioides (Fenzl) Verdc.
Asclepiadaceae
Glossonema varians (Stocks) J. D. Hooker
Avicennaceae
Avicennia marina (Forssk.) Vierh.
Boraginaceae
Heliotropium bacciferum Forssk.
Heliotropium curassavicum L.
Heliotropium ramosissimum (Lehm.) DC.
Moltkiopsis ciliata (Forssk.) Johnst.
Capparaceae
Capparis spinosa L.
Caryophyllaceae
Spergularia marina (L.) Griseb.
Chenopodiaceae
Anabasis setifera Moq.
Arthocnemum macrostachyum (Moric.) Moris et Delponte
Arthrocnemum salicornicum (Moric.) Moris & Delponte
Atriplex leucoclada Boiss
Bienertia cycloptera Bunge
Chenopodium murale L.
Cornulaca monacantha Del.
Halocnemum strobilaceum (Pall.) M. Bieb.
Halopeplis perfoliata (Forssk.) Aschers. et Schweinf.
Haloxylon persicum Boiss.
Haloxylon salicornicum (Moq.) Bge.
Salicornia europaea L.
Salsola villosa Del. Ex Roem. & Schultes
Seidlitzia rosmarinus Erenb. ex Bge.
Suaeda aegyptiaca (Hasselq.) Zoh.
Suaeda maritima (L.) Dum.
Suaeda vermiculata Forssk. ex J.F. Gmel.
Compositae
Koelpinia linearis Pall.
Convolvulaceae
Cressa cretica L.
Cyperaceae
Cyperus arenarius Retz.
Cyperus conglomeratus Rottb.
Cyperus laevigatus L.
Gramineae
Aeluropus lagopoides (L.) Trin. Ex Thw.
Aeluropus littoralis (Gouan) Parl.

Cutandia dichotoma (Forssk.) Trabut
Cynodon dactylon (L.) Pers.
Phragmites australis (Cav.) Trin. ex Steud.
Sporobolus ioclados (Nees ex Trin.) Nees
Sporobolus spicatus (Vahl) Kunth

Juncaceae

Juncus rigidus Desf.

Leguminosae

Alhagi graecorum Boiss.
Prosopis farcta (Banks & Sol.) Macbr.

Orobanchaceae

Cistanche tubulosa (Schenk) Hook.f.

Plumbaginaceae

Limonium axillare (Forssk.) Kuntze

Tamaricaceae

Tamarix arabica Bunge

Zygophyllaceae

Zygophyllum qatarense Hadidi

THE COASTAL AND SABKHA FLORA OF QATAR: AN INTRODUCTION

BENNO BÖER[1] & SAIF AL HAJIRI[2]

[1]*The UNESCO Regional Office in the Arab States of the Gulf, P.O. Box 3945, Doha, Qatar*

[2]*The Friends of the Environment Centre, Qatar*

Abstract

Based on the existing scientific literature, dealing with the flora of Qatar, an information survey was conducted, which resulted in a preliminary picture of the flora of the country's coastal and sabkha ecosystems; the flora consists of seawater tolerant halophytes, halophytes, and semi-halophytes, subject to different degrees of salinity. The flora of the country needs to be studied in more detail, since no comprehensive recent documentation on the flora of Qatar exists. This knowledge, however, is urgently required, in order to allow for professional environmental management.

INTRODUCTION

To date the most comprehensive introduction to the flora of Qatar is still the volume which was published by Batanouny twenty years ago (Batanouny 1981). The volume was much based on previous surveys by Boulos (Boulos 1978), as well as by surveys of the author himself. Batanouny provided an update on flora findings and knowledge for Qatar only recently (Batanouny 1999), and Abulfatih *et al.* compiled a recent volume on the vegetation of Qatar (Abdulfatih et al. 2001). Furthermore there is the contribution from Deil (1998), covering the most important aspects of the coastal and sabkha flora of the Arabian peninsula, with some important information regarding the halophyte flora of Qatar. An important contribution is currently being prepared at the University of Qatar, an updated checklist of the flora of Qatar, which will consist of approximately 350 plant species (Abdel Bari in prep.).

Barth & Böer (eds.), SABKHA ECOSYSTEMS, 63-70

This paper on Qatar, provides an overview of the sabkhat occurrence in Qatar as well as an introduction of the existing coastal and sabkha flora of the country. Comprehensive knowledge on the halophyte flora of Qatar has not yet been generated, and the documentation of the exact distribution of plant species along the Qatari Gulf Coast requires urgent attention. The territory of Qatar, 11.000 km^2 in size, and in particular the coastal zone, undergoes massive landuse development processes, and the prevailing anthropogenic impacts on the environment are similar in quality and intensity when compared with other countries in the region. Habitat loss, and a decline in biological diversity (which has not even been fully documented), of the indigenous halophyte resources, are two of the undesired impacts. Professional marine, coastal, and terrestrial land use management plans are required, for the sustainable utilisation of these resources. The exact knowledge of the composition of the halophyte flora is a precondition for the compilation of environmental management plans. Especially, towards the development of coastal ecosystem restoration techniques, and for the development of biosaline agriculture, the sound knowledge on the halophyte flora of Qatar is essential. This paper provides a preliminary list and categorises the native halophyte resources of Qatar.

OCCURRENCE OF SABKHAT IN QATAR

The peninsula of Qatar consists geomorphologically mainly of four different landforms, namely gravel desert, sabkhat, sand desert, and rocky outcrops. The majority of the territory is made up of flat or undulating gravel desert, which is widespread throughout the country. In the south-west rocky outcrops of 70 to almost ca. 100 m elevation above sea-level occur, and large sand dunes are to be found mainly in the south-east of Qatar, between Umm Said, and Khor al Udayd. Coastal sabkhat occur along the eastern coast of the country, and there are some smaller patches at the western coast at the Gulf of Salwa. A large area of inland and coastal sabkhat is to be found between Umm Said and Khor Al Udayd, and they are interspersed with large sand dunes. These sand dunes are mobile, and moving with the major wind direction from NNW, to SSE (Al-Sheeb 1998), where they may eventually cause sabkha-development in the Khor Al Udayd. Furthermore, there is a second large inland sabkha just south-east of Dukhan, which stretches for approximately 20 KM in the north-south direction (Fig. 1).

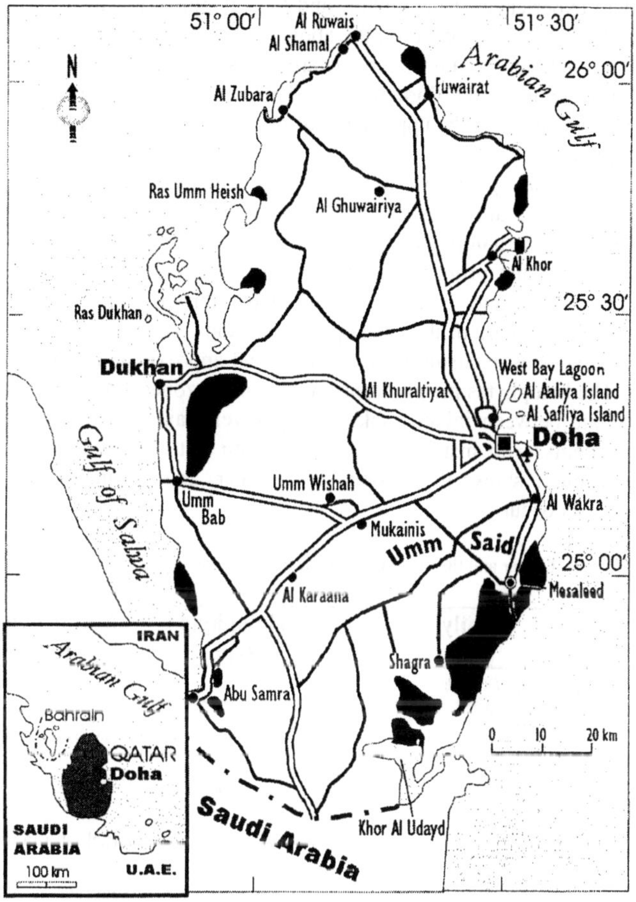

Figure 1. Sabkhat locations (black areas) in Qatar.

THE COASTAL AND SABKHA FLORA

Based on the literature sources as cited in Batanouny (1981, 1999), information within Deil (1998), the recently released book "Vegetation of Qatar" (Abdulfatih et al. 2001), as well as on verbal information from one the nation's leading botanists, Dr. Abdel Bari, a list of all known halophytes in the country was compiled. The species nomenclature was updated according to Jongbloed *et al.* (2000). The species were then so categorized, that the degree of salinity tolerance can be evaluated, based on field experience of the author, and the categories as suggested in Böer & Gliddon (1997). The list is based on the above methods, and can not be regarded as absolute. More detailed botanical surveys are necessary, in order to provide a more comprehensive picture of the flora of Qatar. This paper is merely a first guideline to those species, which are in the upper level of salt tolerance, and promising for saline habitat manipulation.

A total of 49 plant species were identified as halophytes of Qatar, including one mangrove species. 12 are classified as seawater tolerant halophytes (high salinity tolerant = HH), 17 as halophytes (tolerant to salinity = H), 20 as semi-halophytes (tolerant to salinity, but probably dependent on brackish to fresh water for normal development = LH), and 2 are parasitic plants thriving on the roots of true halophytes, such members of the Chenopodiaceae and Zygophyllaceae family. More than 300 plant species occur in the study area, and approximately 15% of these are halophytes. The halophytes mainly occur in the coastal zones, around inland sabkhat, the intermediate between inland sabkhat and gravel plains, as well as in irrigated man-made systems, with insufficient drainage, with salt accumulation in the upper soil layers. Table 1 shows the halophytes of Qatar, and the major ecosystems in which the species occur.

Table 1. Halophytes of Qatar and the major ecosystems in which they occur.

Species	Family	Halophyte type	Major Ecosystem Remarks
	MONOCOTYLEDONAE		
	Arecaceae		
Phoenix dactylifra *L.*		LH	Western coastline
	Cymodoceaceae		
Halodule uninervis (Forssk.) Asch.		HH	Seagrass beds
	Cyperaceae		
Cyperus laevigatus L.		LH	Brackish plantations
	Hydrocharitaceae		
Halophila stipulacea (Forssk.) Aschers		HH	Seagrass beds
	Poaceae		
Aeluropus lagopoides (L.) Trin.		H	Brackish plantations
Arundo donax L.		LH	Rare; brackish plantations
Halopyrum mucronatum		H	Coastal dunes
Phragmites australis (Cav.) Trin. Ex Steud.		LH	Brackish plantations
Sporobolus iocladus (Nees ex Trin.) Nees		H	Salt marshes
Sporobolus spicatus (Vahl) Kunth		H	Brackish plantations

DICOTYLEDONAE		
Aizoaceae		
Mesembryanthemum nodiflorum *L.*	LH	Brackish moist ground
Sesuvium portulacastrum *L.*	HH	Brackish plantations
Asclepiadaceae		
Calotropis procera (Aiton) W.T.Aiton	LH	Introduced; brackish moist ground
Asteraceae		
Pluchea dioscorides (L.) DC.	LH	Introduced; brackish plantations
Avicenniaceae		
Avicennia marina (Forssk.) Vierh.	HH	Mangroves
Boraginaceae		
Heliotropium bacciferum	LH	Various
Chenopodiaceae		
Anabasis setifera Moq.	LH	Sabkha edges; landfill
Arthrocnemum macrostachyum (Moric.) Moris & Delponte	HH	Salt marshes; sabkha edges
Atriplex leucoclada Boiss.	H	Sand sheets
Bassia eriophora (Schrad.) Asch.	LH	Road sides with sand sheets
Bassi muricata (L.) Asch.	LH	Road sides with sand sheets
Halocnemum strobilaceum (Pall.) M. Bieb.	HH	Salt marshes; sabkha edges
Halogeton alopecuroides (Del.) Moq.-Tand	H	Rocky but saline ground
Halopeplis perfoliata (Forssk.) Bunge ex Asch.	HH	Salt marshes; sabkha edges
Haloxylon salicornicum (Moq.) Bunge ex Boiss.	LH	Gravel plains; sand sheets
Salicornia europaea L.	HH	Salt marshes, mangroves
Salsola imbricata Forssk.	H	Sabkha edges; landfill
Salsola soda L. (possibly noted as *Suaeda maritima* (L.) Dum. or vice versa – needs clarification)	HH	Salt marshes; mangroves
Salsola vermiculata L. (needs checking)	H	Saline soils
Seidlitzia rosmarinus Ehrenb. Ex Bunge	HH	Sabkha edges; coastal dunes
Suaeda aegyptiaca (Hassel.) Zohary	H	Sabkha edges; brackish water influence

Suaeda vermiculata Forssk. Ex J.F.Gmel.	HH	Salt marshes, rocky headlands; saline moist ground
Convolvulaceae		
Cressa cretica L.	H	Brackish moist ground
Ipomea pes-caprae (L.) R.Br.	LH	Brackish moist ground
Cynomoriaceae		
Cynomorium coccineum L.	Glycophyte ?	Parasite on halophytes
Fabaceae		
Alhagi maurorum Medik.	LH	Brackish moist ground
Lotus garcinii DC.	LH	Brackish moist ground
Frankeniaceae		
Frankenia pulverulenta L.	H	Brackish moist ground
Mimosaceae		
Acacia ehrenbergiana Hayne	LH	Saline gravel plains
Prosopis cineraria (l.) Druce	LH	Road sites
Prosopis juliflora (Sw.) DC.	H	Introduced; a true halophyte germinates in brackish moist ground
Orobanchaceae		
Cistanche tubulosa (Schenk) Wight	Glycophyte ?	Parasite on halophytes; occurs in two totally different ecosystems: in seawater inundated salt marshes, as well as in dry sand dune areas; possibly sub-species ?
Plumbaginaceae		
Limonium axillare (Forssk.) Kuntze syn. *L. stocksii* (Boiss.) Kuntze	HH	Coastal areas and adjacent to sabkha
Portulacaceae		
Portulaca oleracea L.	LH	Brackish ground in plantations
Solanaceae		
Lycium shawii Roem. & Schult.	LH	Low salinity gravel plains (a flagship species for Qatar, locally known as

		"awsaj")
Tamaricaceae (needs careful taxonomic checking)		
Tamarix aphylla (L.) Karst.	H	Planted; saline moist sub-surface
Tamarix passerinoides Del. Ex. Desv.	H	Brackish moist ground
Tamarix ramosissima Ledeb.	H	Brackish moist ground
Zygophyllaceae		
Fagonia indica Burm.f.	LH	Dry saline gravel plains
Zygophyllum qatarense Hadidi	H	Coastal zones
Zygophyllum simplex L.	H	Saline gravel plains, and sand adjacent to sabkhat

RESEARCH, ECONOMIC VALUE, AND CONSERVATION

Currently a total of ca. 300 species is known to occur in Qatar. This figure was given from Batanouny in 1981 and again in 1999 (Batanouny 1989, 1999). Distribution maps of most of the above species are either non-existent or not available, and they are urgently required for science based environmental management. Furthermore, there are still gaps regarding the knowledge of ephemeral species (Batanouny 1999), as well as aquatic and semi-aquatic species (for example *Halophila ovalis* and members of the genus *Potamogeton* have not yet been recorded, although they are highly likely to occur within Qatar's territory. The same is true for mosses, and the knowledge of marine and freshwater algae (pers. comm. Al Muftah and Kornprobst). One can still expect the discovery of numerous new species, as indicated by Abdel Bari (in prep.), and it would be useful, to have a completely new flora written, professionally compiled, illustrated, and with up-to-date nomenclature. Jongbloed *et al.* (2000) recently showed, that over the last decade, a more co-ordinated approach towards the survey of the UAE's flora, and the compilation of a national checklist resulted in the discovery of almost 300 new species, previously unknown to occur in the country. Some of the species were even new record for the Arabian peninsula. This tendency is also true for Qatar. All the floristic data should be collected, and made available in a national herbarium and environmental database. It is also important to clarify taxonomic uncertainties, especially with the Chenopodiaceae, and Poaceae. One of the prevailing impediments towards biological diversity conservation is the lack of reliable taxonomic products; in Qatar, the correct and up-to-date biosystematic nomenclature (including the correct utilisation of synonyms) is still not been applied. The correct taxonomic knowledge however, is absolutely required, in order to enable environmental managers to professionally conduct biological diversity conservation, habitat restoration, and other disciplines within environmental management. Most of the volumes cited in this paper, dealing with the flora of Qatar, are taxonomically either incorrect, and/or not up-to-date; for example is *Haloxylon salicornicum* still being noted as *Hammada elegans, Limonium axillare* is being mixed up with *L. stocksii, Salsola soda* with *Suaeda maritima, Arthrocnemum macrostachyum* with *A. glaucum, Sporobolus arabicus* with *S. iocladus* etc. etc.

Considering the immense ecological and potential economic importance of the halophyte flora, it is suggested to immediately carry out detailed botanical surveys, and in particular along Qatar's coastline, and the beaches should be monitored for changes in the vegetation. It is also recommended to protect these halophytes within nature reserves, and in a living genetic resources collection. With the results presented in this paper we have now a better understanding of the lack of knowledge towards the national halophyte resources.

ACKNOWLEDGEMENTS

We are thankful to Dr. Homaid Al Midfa from the Scientific & Applied Research Centre (SARC) at the University of Qatar, as well as Dr. E.M. Abdel Bari from the Faculty of Science at the University of Qatar for providing assistance and valuable comments on the manuscript.

References

Abdel Bari, E.M. (in prep.). The checklist of the flora of Qatar.

Abulfatih, H.A., E.M. Abdel Bari, A. Alsubaey & Ibrahim, Y.M. 2001. Vegetation of Qatar. Scientific and Applied Research Center (SARC), University of Qatar. 360 pp.

Al-Sheeb, A.I. 1998. Sand movement in the state of Qatar - The problem and solution. In: Omar, A.S., Misak, R. & Al-Ajmi D. (eds.), Sustainable development in arid zones, Vol. 1. Balkema, Rotterdam, pp. 223-239.

Batanouny, K.H. 1981. Ecology and flora of Qatar. Center for Scientific and Applied Research, University of Qatar. Alden Press, Oxford. 245 pp.

Batanouny, K.H. 1999. On the flora of Qatar: Characteristics and endangered species. In: A. Abuzinada & Joubert, E. (eds.), proceedings of the workshop on the conservation of the flora of the Arabian peninsula. Arabian Plant Specialist Group with NCWCD & IUCN, Riyadh 1999, pp. 30-32.

Böer, B. & Gliddon, D. 1997. Abu Dhabi's coastal zone - Conservation requirements and land use potentials. International conference on water management, salinity and pollution control towards sustainable irrigation in the meditarranean region. Instituto di Bari, Italy, 22-26 September, 1997.

Boulos, L. 1978. Materials for a flora of Qatar. *Webbia* 32 (2): 369-296.

Deil, U. 1998. Coastal and sabkha vegetation. In: Ghazanfar, S. & Fisher, M (eds.), Vegetation of the Arabian Peninsula. *Geobotany* 25. Kluwer Academic Publishers, Dordrecht/Boston/London, pp. 209-228.

Jongbloed, M., Western, R.A. & Böer, B. 2000. Annotated check-list for plants in the U.A.E. Zodiac Publishing, Dubai, 91 pp.

SABKHAT AND HALOPHYTES IN KUWAIT

SAMIRA A. S. OMAR, RAAFAT F. MISAK &
SHABBIR A. SHAHID

Kuwait Institute for Scientific Research, P.O. Box 24885, 13109 Safat, Kuwait

Abstract

The distribution of sabkhat in Kuwait is described. Studies carried out by the Kuwait Institute for Scientific Research regarding soils and vegetation of sabkhat showed that most soils are aquisalids. The major part of the coastal sabkhat were classified under the order of *Aridisols* and the suborder *salids*, which were further recognised mainly as *gypsic aquisalids* and to a minor extent as *typic aquisalids*. To a minor extent sabkhat were also mapped as *torriorthents*. Five major plant communities were identified in the sabkhat of Kuwait. These are: *Nitraria resuta, Tamarix aucheriana, Zygophyllum qatarense, Halocnemum strobilaceum* and *Seidlitzia rosmarinus*. Since the sabkhat of Kuwait are threatened by different anthropogenic activities conservation measures were necessary. These are shortly described together with recommendations for future research concerning the subject.

INTRODUCTION

In Kuwait, sabkhat with their picturesque features constitute one of the most common landforms along the coastal plains and in some inland areas. In the past years the Kuwait Institute for Scientific Research has carried out intensive soil and vegetation surveys in sabkhat environments. In this paper some of these results will be presented.

Barth & Böer (eds.), SABKHA ECOSYSTEMS, 71-81

SABKHA DISTRIBUTION

Sabkhat in Kuwait cover about 10% of the country area (about 1800 km²). Geographically they are differentiated into coastal sabkhat (about 7% of the country) and inland sabkhat (about 3% of the country area). The coastal sabkhat are distributed in the northern coastal plain which includes the western coast of the Khor As Sabiyah and Bubiyan Island (Fig. 1). Moreover, coastal sabkhat occur at the northern coast of Kuwait Bay and along the southern coastal plain (Julayah - Az Zour strip) (Fig. 1). These coastal sabkhat are developed as a result of the intersection between the highly saline groundwater table (resulting from sea water intrusions) and the low ground surface. Wetting and drying of these surfaces are very common throughout the year.

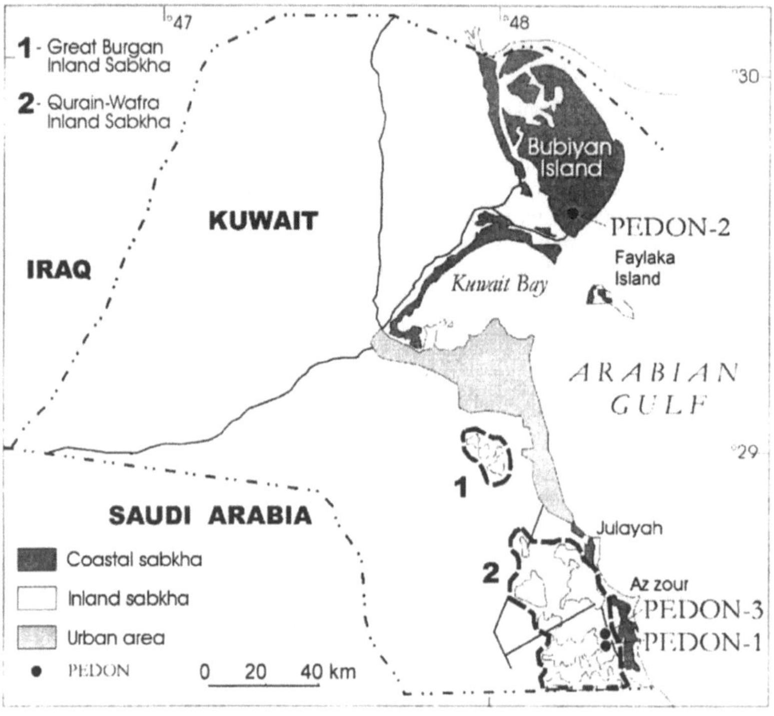

Figure 1. Map of Kuwait showing the geographic distribution of sabkhat and the location of soil pedons 1-3.

Morphologically, lithologically, geochemically and biologically the coastal sabkhat are highly variable. In the northeastern part of the country the sabkhat are dominated by laminated silt and clay with a high gypsum content. The fine sediments are derived mainly from the Shatt Al Arab. The elevation of the sabkhat at the northeastern part is around 4 m above sea level and the groundwater table is between 2 and 3 m below the surface level. In Kuwait Bay the sabkha is dominated by fine to medium sand with high salt content. The average groundwater level is at a depth of around 145 cm at Bahra (central part of Kuwait Bay). The sabkhat along the southern coastal plain are composed of sands, carbonate grains and evaporites. The average groundwater table is located at a depth of about 120 cm at Nuwassib sabkha (extreme southern part).

In the southern coastal plain, and at local parts of Kuwait Bay, the seaward (Gulfward) side of the coastal sabkhat are skirted by off shore bars, oolitic ridges and aeolian accumulations. In the northern coastal plain and the majority of Kuwait Bay, the sabkhat are not separated from intertidal flats.

Along the coastal sabkhat, a wide variety of micro-landforms are prominent. These include aeolian features (nabkhas, sand sheets, deflational depressions and soft and hard yardangs), fluvial forms (rills, gullies and alluvium cones) and evaporitic salt crusts.

The inland sabkhat occupy the floors of enclosed depressions. During heavy rain storms, the runoff water from the surrounding higher terrain accumulates forming water ponds (30-50 cm depth, 0.5 - 2 km^2 area). The inland sabkhat are scattered within the Great Burgan oil field (south of Kuwait City), and around Qurain area (see Fig. 1). The sabkhat constitute depressions of different sizes. Generally the long axis of one of these sabkhat attains about 3 km in NW-SE direction with a width ranging between 0.5-2 km. The depressions are in their central parts up to 3 m lower than the surrounding terrain. Their ground elevation varies between 47-58 m above sea level.

The inland sabkhat of Qurain-Wafra cover a wide area in comparison with those of the Great Burgan oil field. The long axis of these sabkhat attains about 4 km in NW-SE direction with a width varying between 0.5 and 1.5 km. The depressions here are in their central parts up to 6 m lower than the surrounding area. The rims of the sabkhat are cut by shallow and short rills which drain into the sabkha floors. One of the sabkhat at the northwestern part of Qurain-Wafra receives surface runoff by a wadi (Khar Al-Qurain) which extends around 5 km into surrounding terrain to the west. This fact would classify the landform as a playa. But since it's origin seems to be purely deflational and the drainage into the wadi is only that of a very small area, we still define it as a sabkha.

Gunatilaka & Mwango (1987) studied the sabkhat at the southeastern part of Qurain-Wafra. They stated that the sabkhat are made up of quartz, calcite and gypsum. Regarding the origin of the inland sabkhat, it seems that they were developed as a result of the intersection with shallow groundwater bodies in Quaternary sand and gravel beds. Strong winds in the past excavated these depressions.

SOIL CHARACTERISTICS

The soils of the state of Kuwait have been mapped at a scale of 1:100.000 (KISR 1999) using the soil classification system of the United States Department of Agriculture (Soil Survey Staff, Soil Conservation Service 1994). The soil profiles were described as per Soil Survey Manual (Soil Survey Division Staff 1993). Most sabkhat are mapped as *aquisalids* (Soil Survey Staff, Soil Conservation Service, 1994). The *aquisalids* have a salic horizon and water table fluctuations within one meter from the soil surface. The sabkha sites at the islands are close to the sea level and therefore, seawater intrudes to these areas, evaporates and salts are concentrated to form marine evaporites. The major part of the coastal sabkhat were classified under the order of *Aridisols* and the suborder *salids*, which were further recognised mainly as *gypsic aquisalids* and to a minor extent as *typic aquisalids* (Soil Sruvey Staff, Soil Conservation Service 1994). The salinity levels are very high and even halophytes rarely establish. To a minor extent sabkhat were also mapped as *torriorthents*.

The chemical and physical characteristics of two representative pedons are shown in tables 1 and 2. The morphologies of the soil families are given in table 3. The location of pedons as described in table 3 is shown in figure 1.

The pH was measured of the soil saturated paste, and the electrical conductivity of the soil saturation extract obtained under vacuum from the saturated soil paste (Richards 1954). The soil saturation extract was analysed for solution chemistry (soluble anions and soluble cations). The cation exchange capacity was determined by saturating the soil exchange complex with 1M sodium acetate (pH 8.2) solution, followed by ethanol washing, and displacing the Na with 1N (pH 7) ammonium acetate solution. The displaced Na was measured by atomic absorption spectrophotometer. Gypsum was measured by acetone precipitation method, and calcium carbonate by calcimeter method. In the field $CaCO_3$ was measured through effervescence test (fizzing reaction caused when carbonate mineral is treated with acid). For all the above measurements, methods as described by the United States Department of Agriculture (USDA 1996) were used.

In table 1 the slightly to moderate alkaline pH values of all the layers can be attributed to the presence of "halite" (NaCl) in dominant form throughout the profiles. The relatively higher values of SAR in both the pedons are attributed to the higher concentrations of Na compared to Ca and Mg in the saturation extract.

Analytical data for the particle size distribution is presented in table 2. It is apparent that pedon 1 is sandy in texture throughout the profile, pedon 2 is loam (0-40cm) and sandy loam (40-100 cm). The total percent loss (soluble salts, gypsum and carbonates) is maximum in pedon 2 (decomposition of calcareous shells and calcium carbonates). Total clay is maximum in pedon 2. This is due to the deposition of fresh material through tidal coverage. More detailed information on occurrence of different soil units is given in the soil survey of the State of Kuwait (KISR 1999).

Table 1. Analytical properties of the representative pedons.

Horizon	Lower Depth (cm)	pHs	ECe dS/m	Na	K	Ca	Mg	HCO_3	Cl	SO_4	SAR	CEC	SP	Gypsum	$CaCO_3$
				(………… me/l …………)							(mmoles/l)$^{0.5}$	(me/100g)	(……%……)		(%eq.)
PEDON 1: Gypsic Aquisalids, sandy, mixed, hyperthermic (236057 mE 3173153 mN Zone 39)															
Akz	2	7.4	358	1247	253	78	101	0.65	1600	74	131	0.88	17.2	10	4
Bkz	25	8.0	64	571	37	89	77	0.09	674	87	63	1.45	17.6	3	6
Bkyz1	60	7.9	46	345	13	70	41	0.14	381	74	46	1.79	21.6	14	5
Bkyz2	90	7.9	31	240	9	59	31	0.10	258	75	36	2.07	26.0	14	7
Bkyz3	120	7.8	44	437	18	93	54	2.51	504	90	51	1.60	22.0	5	9
PEDON 2: Typic Torriorthent, sandy, skeletal, carbonatic, hyperthermic (229140mE 3280949 mN zone 39)															
A	25	8.2	8	51	1.5	38	9	1.64	48	49	11	7.19	33.9	37	29
C1	40	7.4	29	188	1.7	51	16	0.88	242	45	32	6.26	36.0	8	53
C2	70	7.6	28	163	1.8	47	18	0.59	200	60	28	5.00	41.0	3	55
C3	100	7.6	27	144	1.6	48	17	0.78	180	58	25	4.36	20.0	4	56

pHs = pH of saturated soil paste; ECe = Electrical conductivity of soil saturation extract; SAR = sodium adsorption ratio; CEC = cation exchange capacity; SP = saturation percentage

Table 2. Physical characteristics of the representative pedons.

Horizon	Lower Depth (cm)	Total Clay <.002	Total Silt .002-.05	Total Sand .05-2	Silt Fine .002-.02	Silt Coarse .02-.05	Sand VF .05-.10	Sand F .1-.25	Sand M .25-.5	Sand C .5-1	Sand VC 1-2	TPL	Coarse Fraction F 2-5	Coarse Fraction M 5-20	Coarse Fraction C 15-75
		(.......				% of < 2mm (carbonate free)					.......)	% of < 2mm	(.....% of < 75		...)
PEDON - 1: Gypsic Aquisalids, sandy, mixed, hyperthermic (236057 mE 3173153 mN Zone 39)															
Akz	2	1.3	2.5	96.2	1.9	0.6	7.8	20.0	26.8	28.4	13.2	22.5	0	0	0
Bkz	25	0.5	3.8	95.7	2.7	1.1	8.5	22.1	30.4	26.9	7.8	9.3	0	0	0
Bkyz1	60	2.6	3.1	94.3	1.3	1.8	11.5	25.8	32.6	18.5	5.0	23.2	0	0	0
Bkyz2	90	2.0	5.9	92.1	2.6	3.3	11.7	30.5	34.6	12.9	2.4	23.9	0	0	0
Bkyz3	120	1.2	3.8	95.0	2.9	0.9	8.0	30.7	47.8	7.5	1.0	14.5	0	0	0
PEDON - 2: Typic Torriorthent, sandy, skeletal, carbonatic, hyperthermic (229140mE 3280949 mN zone 39)															
A	25	24.9	33.0	42.1	20.6	12.4	26.2	8.7	6.1	1.1	0.0	53.7	8	3	0
C1	40	20.7	42.1	37.2	23.4	18.7	24.5	5.5	5.5	1.1	0.6	64.0	25	16	0
C2	70	14.8	11.8	73.4	7.4	4.4	6.8	8.3	15.4	27.5	15.4	66.8	33	20	0
C3	100	11.8	15.4	72.8	10.4	5.0	6.5	9.5	14.8	23.5	18.3	66.7	33	31	0

TPL = total pretreatment loss; VF = very fine; F = fine; M = medium; C = coarse; VC = very coarse

Table 3. Morphological properties of representative pedons.

Horizon	Lower depth (cm)	Munsell color dry	Munsell color moist	Texture[#]	Structure[†]	Consistence[‡]	Effervescence[§]	Lower[¶] boundary
PEDON 1: Gypsic aquisalids, sandy, mixed, hyperthermic (236057 mE 3173153 mN Zone 39)								
Akz	2	2.5Y 6/4	2.5Y 5/3	cs	m	vfr	ste	as
Bkz	25	2.5Y 6/4	2.5Y 5/3	cs	m	vfr	ste	as
Bkyz1	60	2.5Y 6/4	2.5Y 5/3	s	m	vfr	ste	cw
Bkyz2	90	2.5Y 7/4	2.5Y 5/3	s	m	vfr	ste	cw
Bkyz3	120	2.5Y 7/4	10YR 5/4	s	m	fi	ste	-
PEDON 2: Typic torriorthents, sandy, mixed, hyperthermic (229140mE 3280949 mN zone 39)								
A	25	2.5Y 7/3	2.5Y 5/3	l	wtp	fr	ste	gw
C1	40	10YR 5/2	10YR 6/2	gcs	m	fi	vie	gw
C2	70	10YR 5/2	10YR 6/2	gcs	m	fi	vie	dw
C3	100	10YR 5/2	10YR 6/2	gcs	m	fi	vie	-
PEDON 3: Typic aquisalids, sandy, mixed, hyperthermic (235999 mE 3173123 mN zone 39)								
Az	5	2.5Y 6/4	10YR 5/3	s	m	fr	se	cs
Bkz	25	2.5Y 6/4	2.5Y 5/3	s	m	vfr	ste	cs
Bk1	60	2.5Y 6/3	2.5Y 5/3	s	m	vfr	ste	cw
Bk2	90	2.5Y 7/3	2.5Y 5/3	cs	m	vfr	ste	cs
Bk3	120	2.5Y 7/3	2.5Y 5/3	cs	m	l	vse	-

Texture abbrev. l = loam; s = sand; cs = coarse sand; gcs = gravely coarse sand
† Structure abbrev. m = massive; wtp = weak thick platy
‡ Consistence moist abbrev. l = loose; fr = friable; vfr = very friable; fi = firm
§ Effervescence abbrev. vse = very slightly effervescent; se = slightly effervescent; ste = strongly effervescent; vie = violently effervescent
¶ Lower boundary abbrev. as = abrupt smooth; cw = clear wavy; cs = clear smooth; gw = gradual wavy; dw = diffuse wavy

HALOPHYTES IN SABKHA ENVIRONMENTS

Halophytes are widespread in low lying lands that have been waterlogged and salt affected by surface and sub-surface flooding. Their cells maintain a higher osmotic potential to prevent water loss. The roots can withstand low oxygen levels without losing control over ion uptake. Some halophytes develop root air channels and adventitious roots, which can assist them in surviving under sabkha conditions. Some absorb salts while others lose salts via leaf tissues.

Five major plant communities are identified in the sabkhat of Kuwait. These are: *Nitraria resuta, Tamarix aucheriana, Zygophyllum qatarense, Halocnemum strobilaceum* and *Seidlitzia rosmarinus*. Associated species are: *Aeluropus lagopoides, Aeluropus littoralis, Frankenia pulverulenta, Bienertia cycloptera, Cressa cretica, Aizoon canariense, Aizoon hispanicum* and *Mesembryanthemum nodiflorum*. The site locations of these plant communities are shown in Fig. 2.

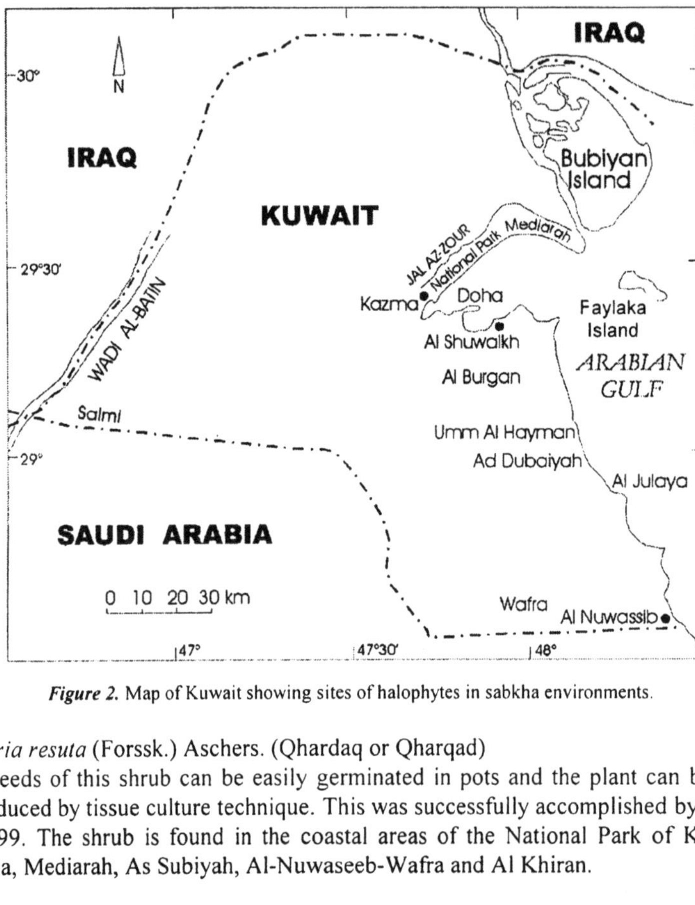

Figure 2. Map of Kuwait showing sites of halophytes in sabkha environments.

Nitraria resuta (Forssk.) Aschers. (Qhardaq or Qharqad)
The seeds of this shrub can be easily germinated in pots and the plant can be also reproduced by tissue culture technique. This was successfully accomplished by KISR in 1999. The shrub is found in the coastal areas of the National Park of Kuwait, Kazma, Mediarah, As Subiyah, Al-Nuwaseeb-Wafra and Al Khiran.

Tamarix aucheriana (Decne.) baum (Athl)
This shrub grows in saline grounds at sabkha edges. It is found in Doha, Ad Dubaiyah and Al Burgan.

Zygophyllum qatarense Hadidi (harim)
This shrub is found along the coastal zones in the north and south of Kuwait. It occurs also inland in saline grounds at sabkha edges.

Halocnemum strobilaceum (Pall.) M.B. (Thullayth)
This shrub grows on mudflats which are periodically submerged by sea-water along the coastal line in the north and south of Kuwait.

Seidlitzia rosmarinus Ehrenb. Ex. Bge. (Shinan, Ushnan)
This shrub grows in few locations on islands and off shores along the coast in the north. It also occurs in the Al Shuwaikh area.

Associated species:
Aeluropus lagopoides (L.) Trin. Ex. Thwaites (Ikrish)
This perennial grass is commonly found in Kazma, Umm Al Hayman and Al-Khiran.

Aeluropus littoralis (Goauan) Parl. (Sheraib, Ikrish)
This perennial grass occurs occasionally in the east of Kuwait and was reported in Kazma.

Frankenia pulverulenta L. (Mulaih, Abuthurayb)
This annual herb is common in sabkha areas with Tamarix shrubs, at Ad Dubaiyah, Al Khiran, Al Shuwaikh and Doha.

Bienertia cycloptera Bunge (Golloman)
This annual herb is found in low saline lands in Ad Dubaiyah, Al Shuwaikh, Doha and the National park coastal zone.

Cressa cretica L. (Shuwayla Nadwa)
This short herb is common in sabkhat in Al Nuwassib, Doha, As Subiyah and the National park of Kuwait.

Aizoon canariense L. (Hudq)
This herb is found in scattered localities near sabkha areas, i.e. Ad Dubaiyah, Al Shuwaikh and Doha

Aizoon hispanicum L. (Mullayh)
This annual herb appears in coastal areas on hard desert soils and sometimes in inland areas.

Mesembryanthemum nodiflorum L. (Qasool)
This succulent herb is found in sandy shores in the south and north of Kuwait.

MEASURES FOR CONSERVING SABKHA ECOSYSTEMS IN KUWAIT

The sabkha ecosystems in Kuwait are threatened by activities and processes like excessive grazing, off road driving (Fig. 3), oil pollution (with reference to the Gulf war 1991), dumping and sand encroachment. KISR carried out several studies to assess the impacts. Therefore measures for conservation of sabkha ecosystems were necessary. As a result of that many areas in Kuwait were set aside to protect the fragile habitats. The following areas were allocated for their importance in respect to vegetation and wildlife:
Doha-Sulaybikhat Bay, Jahra Pond, the National Park of Kuwait and the Botanical Garden and Eco Park at the coastal zone of Sulaybikhat Bay near Doha.

Figure 3. Sabkha soil affected by off road vehicle use in Ad Dubaiyah.

RECOMMENDATIONS

Because knowledge of sabkha ecology is very limited the following research studies should be carried out as the next step towards a better understanding the processes:

- The seasonal variations of the water table which has an impact on the vigor and growth of sabkha plants.

- Monitoring of soil salinity in order to understand the salinity build up problem in sabkhat.
- Study of the mechanism by which the watertable and logging influence root growth and performance of *Nitraria resuta* and *Tamarix aucheriana.*
- Study the sand dynamics of mobile sand.
- Study of microbial activities in sabkhat. The type of microbes surviving under severe salinity may have valuable benefit that is worth investigating.
- Wildlife such as birds, reptiles and insects that survive in sabkha areas as their niche need to be identified.
- The impact of salinity on different seeds of native plants needs to be investigated in order to understand the mechanism by which plants propagate under saline conditions. Tissue culture techniques are recommended for multiplication of sabkha plants, particularly species that are under pressure.
- Study of plant adaptation and mutation. The mechanisms by which plants resist salinity needs further research and assessment.
- Grazing impact and sand encroachment need to be surveyed and assessed.

References

Gunatilaka, A. & Mwango, S. 1987. Continental sabkha pans and associated nebkhas in Southern Kuwait, Arabian Gulf. Desert sediments. Ancient and Modern. *Geological Survey Special Publication* 35: 187-203.

KISR 1999. Soil Survey for the State of Kuwait. 5 Volumes. AACM International: Adelaide, Australia.

Richards, L.A. 1954. Diagnosis and Improvement of Saline and Alkali Soils. United States Salinity Laboratory Staff, Handbook No. 60.

Soil Survey Division Staff. 1993. Soil Survey Manual. United States Department of Agriculture. Handbook No. 18. U. S. Govt. Print. Office, Washington, DC.

Soil Survey Staff, Soil Conservation Service. 1994. Keys to Soil Taxonomy-6th Edition. United States Department of Agriculture. U. S. Govt. Print. Office, Washington, DC.

USDA. 1996. Soil Survey Laboratory Methods Manual. Soil Survey Investigation Report No. 42. Version 3.0. United States Department of Agriculture. U. S. Govt. Print. Office, Washington, DC.

AN OVERVIEW OF AN INLAND SABKHA IN JORDAN: THE TABA SABKHA, SOUTHERN WADI ARABA

ABDULKADER M. ABED

Department of Geology, University of Jordan, Amman, Jordan

Abstract

The Taba Sabkha is situated in the southern part of the Dead Sea-Wadi Araba, 32 km north of the Gulf of Aqaba. It occupies an area of about 55 square km west of the Precambrian granitic basement. It forms the distal facies of several alluvial fans emerging from the basement. Twenty shallow pits (1-1.5 m deep) and 9 boreholes up to 17 m depth were drilled into the Sabkha, samples and cores were obtained and some groundwater samples were collected. Zonation in the sabkha is rather clear both in plant habitat and evaporite minerals as a function of increasing salinity of groundwater towards in the centre. Brown clays dominate its centre, which are grading eastwards into more sandy facies. Detrital minerals present include quartz, feldspars, chlorite, mica, kaolinite, illite, smectite and illite/smectite mixed layers. Calcite, although partly authigenic, is dominantly wind blown into the sabkha. Gypsum (transferred partly into anhydrite in summer) is present at the outer parts while halite dominates the centre. Dolomite and sepeolite seem to have formed from schizohaline water not attaining the salinity of coastal sabkha.

INTRODUCTION

The aim of this work is to give an overview of the sabkhat in Jordan typified by the Taba Sabkha located towards the southern end of the Wadi Araba in the Dead Sea Transform. Due to the fact that the vast majority of the salt flat is never flooded by runoff water coming from the wadis to the east, and the flooding itself persists

Barth & Böer (eds.), SABKHA ECOSYSTEMS, 83-97

only for a few days it is referred to as sabkha and not playa as one might expect regarding the geological situation (see introduction of this book).

DISTRIBUTION OF SBKHAT IN JORDAN

No coastal or supratidal sabkhat exist in Jordan due to the coast of Jordan on the Gulf of Aqaba is small and highly urbanised. Consequently, only inland sabkhat are present in Jordan (Kendall 1979). Inland sabkhat in Jordan are found at the following three localities (Fig. 1):

1. There are three sabkhat in the southern Wadi Araba. They are from south: Ed-Dafiyya (just north of Aqaba airport), Megreh further north, and Taba some 32 km north of Aqaba (Figs. 1 & 3).
2. One ancient sabkha, Mudawwara sabkha, is situated in the extreme south Jordan near Saudi Arabian borders (Fig. 1). It is around 100ka old and is associated with the much wetter climate than now (isotope stage 5 of the Pleistocene) (Abed et al. 2000). It is no more active now because the water table is too deep these days.
3. The Azraq Depression in the east has complex geomorphic features (Fig. 1). Permanent fresh water springs, in its western part, indicates an oasis. This shallow, fresh aquifer is underlain by a saline aquifer, which is 2-3 m deep further east. Salinity is 260mg/l. This eastern part is truly an inland sabkha where the locals extract table salt from the saline, near surface groundwater (Fig. 2). The sabkha occupies the lowest area of this large basin. *Nitraria* and *Tamarix* nabkhas up to 6 m high, around the barren surface of the sebkha are good indicators of the deflation processes. However, in good rainy years, the sabkha and its periphery are flooded with rain water for a few months, thus forming a short-lived saline lake. There is prehistoric evidence (Garrard et al. 1988, Ibrahim 1996) that the Azraq Basin had permanent lakes, both saline and fresh. Under these conditions, this area is possibly more of a playa rather than an inland sabkha.

GAA` VERSUS INLAND SABKHA

Amiel & Friedman (1971) while describing the western parts of the Taba Sabkha mentioned the presence of an other larger sabkha near Gharandal some 35 km further north in central Wadi Araba (Fig. 3). This geomorphic feature is in fact a Gaa` (Arabic for mud flat). Careful inspection of this Gaa` Es-Saideyeen showed no evaporites and the local bedouin tribe, Es-Saidyeen, are cultivating it in the rainy years. Gaa` is a very flat area inclined towards its centre. In most cases it has no outlet, but some has. In the rainy years it is covered with shallow water which disappears soon after. The sediments are laminated silt to fine sand deposited by runoff water and partly by wind. No evaporites are present in the Gaa`. That is why they are cultivated in several parts of Jordan especially in the south where groundwater is available (e.g. Gaa` Disi, Gaa` Khreim).

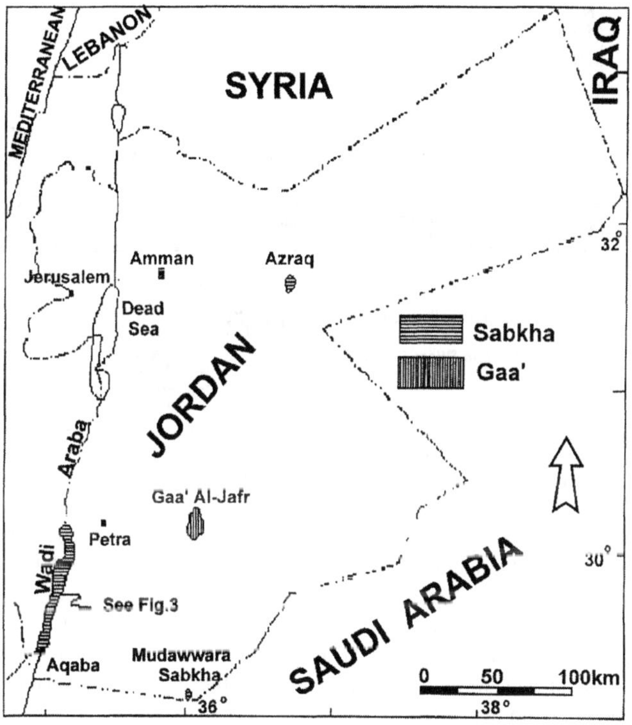

Figure 1. Location map of the sabkhat in Jordan.

TABA SABKHA

General Description

Taba Sabkha is the best example of an inland sabkha in Jordan. Amiel & Friedman (1971) studied the western part of it west of the truce line. It is 32 km north of the Aqaba (Figs. 1 & 3) and occupies about 55 km^2 in the southern Wadi Araba. It is bordered in the east by a mountain range (>1000 m asl) comprised essentially of granitoid rocks. It is drained by E-W wadis that have their alluvial fans on the eastern side of Wadi Araba (Fig. 3). The western mountain range consists of Paleozoic clastics and Mesozoic carbonates and clastics (Weissbrod 1969, Bender

1974). The eastern mountain range is much higher than the western range (Quennell 1958, Freund et al. 1970).

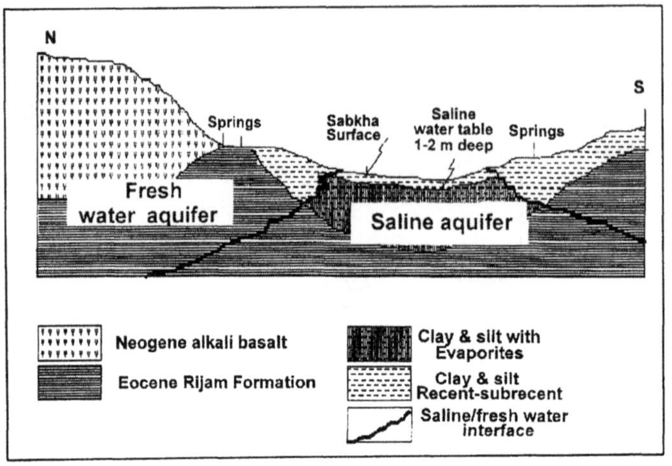

Figure 2. Generalised cross section in the Azraq basin showing the position of the sabkha, the two groundwater regimes, and the bed rock. For location see Fig. 1.

North of the sabkha, sand sheets prevail for about 35 km (Fig. 3). Consequently, and because of the dominant northerly winds, the northern part of the sabkha is invaded by sand.

Seventeen pits, 70-225 cm deep, and 8 boreholes to a maximum of 17 m depth were dug into the sabkha (Fig. 4). Water samples were collected from the pits, boreholes and ancient wells. Sediments were subjected to the various types of chemical and mineralogical analysis.

The surface inclination of the sabkha is only a few seconds to the west and southwest. In the east, it laterally interfingers with several alluvial fans where wadis cut across the Precambrian basement. The sabkha sediments are mainly the result of the geological work of these wadis. At the mountain foot, the coarse, proximal alluvial fan deposites are present. Sand then prevails at the eastern (pits 1 and 2), northern and northeastern (borehole, BH 1) periphery of the sabkha (Fig. 4); massive clays occur at the distal part of the fan (Yagmurlu & Helvaci 1994, Inci 1991), composing the centre of the sabkha (Figs. 3, 4). Sometimes sand tongues where found in certain pits and boreholes in the interior of the sabkha. Part of the sandy deposits is certainly wind blown from the northerly sand dune field. Nabkha dunes characterise the northern part of the sabkha (Guantilaka 1986).

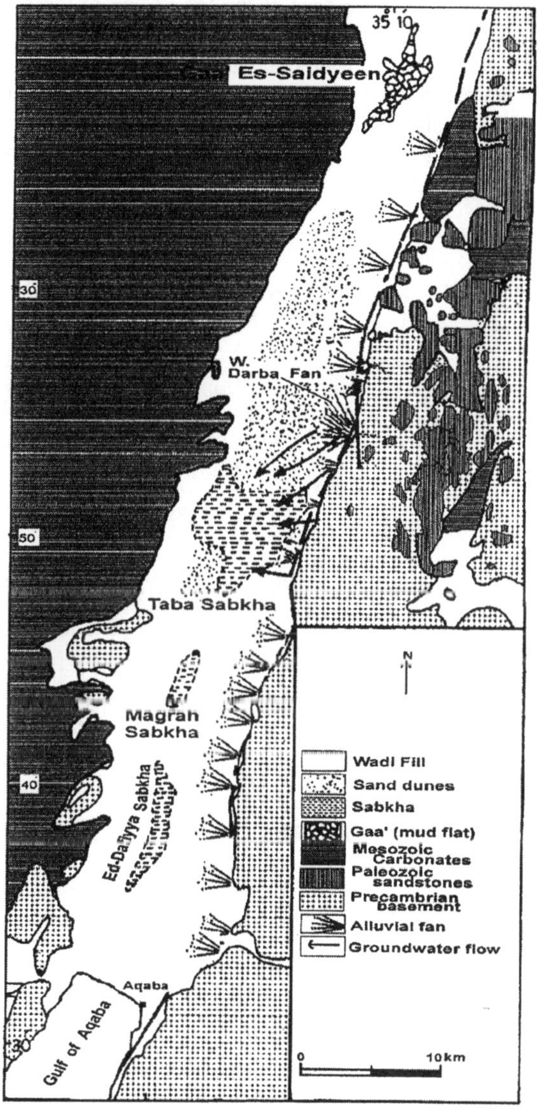

Figure 3. Geological setting of the area around Taba, Magrah and Ed-Defiyya sabkhat in southern Wadi Araba. For location see Fig. 1.

Groundwater is only one metre deep at the toe of one of the alluvial fans. Up until now, bedouins used these shallow wells for their camels and goats. These are the historic Taba wells (Fig. 4). In winter, the groundwater of this fan emerges on the surface some 200 m westwards of the wells near the sand-clay boundary supporting a palm tree oasis (Taba Oasis). Also, water coming from the eastern wadis as well as rain water was seen flowing on the relatively low surfaces of the sabkha precipitating halite.

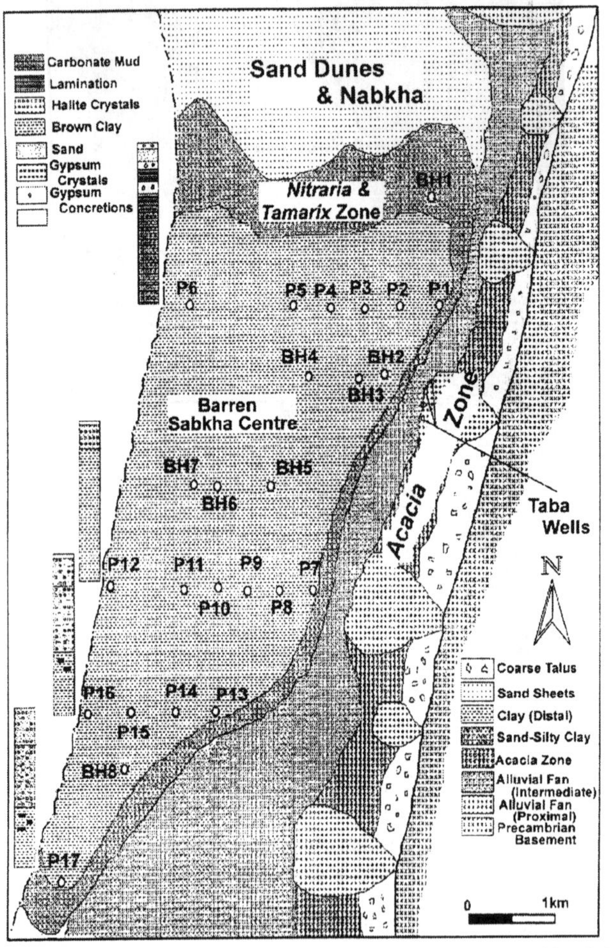

Figure 4. The surface Taba Sabkha sediments, alluvial fans, plant zonation, pits and boreholes and their sediments. See Fig. 3 for location.

The barren interior of the sabkha consists of two colours (Fig. 5): a) whitish involving the relatively low areas (1-3 cm) with a salt crust and desiccation cracks, b) brownish, a little higher, with older desiccation cracks that have holes (2-7 cm wide) and salt cake beneath the holes. The former areas are the currently active parts of the sabkha.

The southern Wadi Araba is a hot, arid region. Average air temperature range is 15°C - 31°C. The daily maximum can reach 47.6°C during the summer. Annual rainfall averages 31 mm. The eastern mountain ranges receives up to 200 mm/year; flood water flows westwards via numerous wadis to Taba Sabkha. The western mountains receive only about 50 mm/year. Annual mean wind speed at the Aqaba airport, south of the study area, is 11 knots and 70% of the prevailing winds are norhterly and 8% are southerly (Aqaba Meteorological Station 1955-1985).

Figure 5. The clayey, vast, barren centre of the sabkha. a) the transition between the relatively moist whitish part (A) and the older brownish part (B) of the Taba Sabkha interfingering.

Plant Zonation

Four plant zones exist in the outer parts of the sabkha, especially the E and N (Fig. 4). Plants were identified by Prof. D. Al-Eisawi. The zones are from the sabkha edge to the central parts.

1. The *Acacia* zone (*Acacia radiana, Acacia tortilis*) on the alluvial fans.
2. The palms zone (*Phoenix dactylifera, Junctus acutus*).
3. The *Nitraria retusa* and *Tamarix* sp. Zone
4. The *Cressa cretica* cf. Zone is only patchy.

The four plant zones are present on the periphery of the sabkha. The vast area occupied by the sabkha centre is completely barren (Fig. 5).

Groundwater

The major ions were analysed by flame photometry, spectrophotometry, and atomic absorption spectrophotometry. Table 1 shows the distribution of the major ions and salinity of the groundwater of the sabkha which is different from sea water (Butler 1969, Ali & West 1983, Lewy 1980, Gavish 1974, Abu Zir 1989). The water table is deepening in the W and SW to about 10 m (Fig. 6). The salinity increases from 3422 to 15.428 mg/l from the toe of the alluvial fan (Taba Wells in Fig. 4) to the *C. cretica* zone some 400 m west (the first 4 samples in Tab. 1).

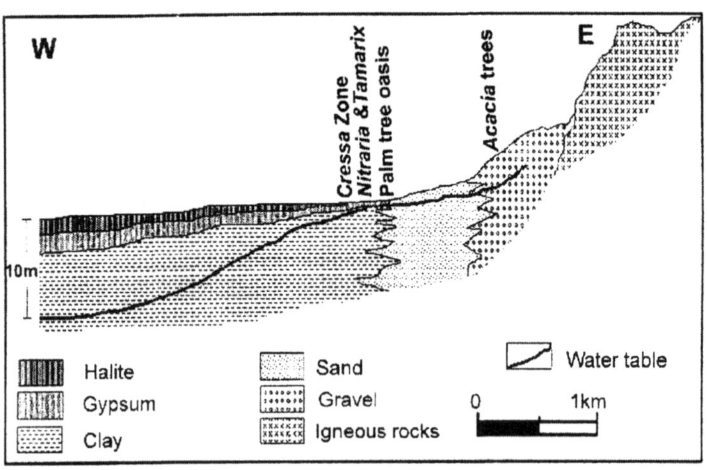

Figure 6. A generalised E-W cross section in the northern part of Taba Sabkha showing the approximate position of the groundwater table, the gypsum and halite below the sabkha surface.

The HCO_3 content increased first (sample 1) to Taba Oasis (sample 2) due to evaporation, then dropped westwards due to the precipitated lime mud in pit P1 (Fig. 4). Ca and SO_4 first increased and then slowly decreased in pit P2 due to the starting gypsum precipitation. The remaining four ions Na, K, Mg, and Cl continue to increase towards the centre of the sabkha (Dean 1978). Sample SW is a surface rain water collected from the low, whitish part of the sabkha. Its salinity is 252.297 mg/l or about 25.2% and precipitating halite. Consequently, the barren surface of the sabkha is most probably due to the salinity of this type of surface water.

Table 1. Major ions and salinity of some groundwater samples in Taba Sabkha (mg/l).

Sample No.	Na^{+}	K^{++}	Ca^{++}	Mg^{++}	Cl^{-}	HCO^{3-}	$SO4^{--}$	pH	Salinity
1	644	32	55	41	2198	143	309	7.20	3422
2	1275	47	1075	90	4523	716	753	7.35	8489
P1	2079	69	2050	179	8179	256	1602	7.55	14390
P1	2464	114	1870	196	8614	267	1560	7.75	15428
SW	18360	2040	15880	2229	212700	128	957	7.15	252293
AW	1387	129	810	845	3935	540	1448	7.75	9194
DF	192	5	68	37	218	155	176	-	870

1 The historic wells of Taba Oasis (1 m deep) at the western edge of an alluvial fan bordering the sabkha from the east (Fig. 2).

2 Some 100 m to the west of 1 below the palm trees in the oasis where groundwater emerges at the surface (Fig. 2).

P1 Some 200 m west of 2 in pit no. 1 and *Nitraria/Tamarix* plant zone (Fig. 6).

P1 Some 100 m west from the second pit and *Cressa cretica* plant zone (Fig. 6).

SW Surface water taken from the centre of the sabkha in winter.

AW An ancient well 3 m deep at the SE boundary of the sabkha about 7 km south of Taba well no. 1 (Fig. 2).

DF Darba fan wells groundwater (mean of 8 samples). From Abu-Zir (1989).

Salinity = the sum of the total dissolved salts.

MINERALOGY AND GEOCHEMISTRY

Based on thin section, x-ray diffraction, and chemical investigations on the samples obtained from pits and boreholes, the following general trends are drawn. Detrital minerals like quartz, feldspars and micas decrease westwards towards the centre of the sabkha away from their source in the east. However, wind blown sand found deeper in the centre of the sabkha can not be excluded.

The clay minerals, kaolinite, smectite, illite, chlorite and sepeolite form up to 60% of the sediments in the sabkha centre. Except sepeolite which is concentrated towards the centre of the sabkha, the other clay minerals are believed to be derived from the basement in the east either directly (chlorite) or with slight alteration (illite and smectite from mica). Sepeolite is reported only in few samples with dolomite in the inner parts of the sabkha and seems to be authigenic (Papke 1972, Khoury et al. 1982). Most of the calcite seems wind-blown from Mesozoic limestone strata in the western mountains. However, some of the calcite must be precipitating from groundwater at least in the peripheral part of the sabkha as explained previously. Dolomite is far less abundant. Part of it is derived from the western mountains, but an authigenic origin for part of the dolomite is indicated by a) the presence of clear and perfect dolomite crystals with sharp edges, b) the presence of a crude but definite trend in the distribution of dolomite. It is not present (or very minor) in the gypsum zone, while it is more abundant further west within the sabkha as indicated by borehole data.

The evaporite minerals are gypsum and halite. Gypsum (sometimes anhydrite and/or basanite in summer) is found in two settings. a) As a surface or near surface, concretionary to powdery deposit in the outer 600 m of the sabkha especially around the plants in zone 3 and 4. This type is due to direct precipitation from the groundwater. b) As a deeper horizon overlain by halite towards the centre of the sabkha. It is present as small almond-shape concretions with fibrous gypsum crystals occupying the space of the concretions perpendicular to the cavity wall (Fig. 7). Gypsum is not encountered much deeper in the same wells (e.g., 14 to 15 m). This type is precipitated by the rising groundwater due to capillarity. The overlying halite does testify to this interpretation (Schreiber & El-Tabakh, 2000).

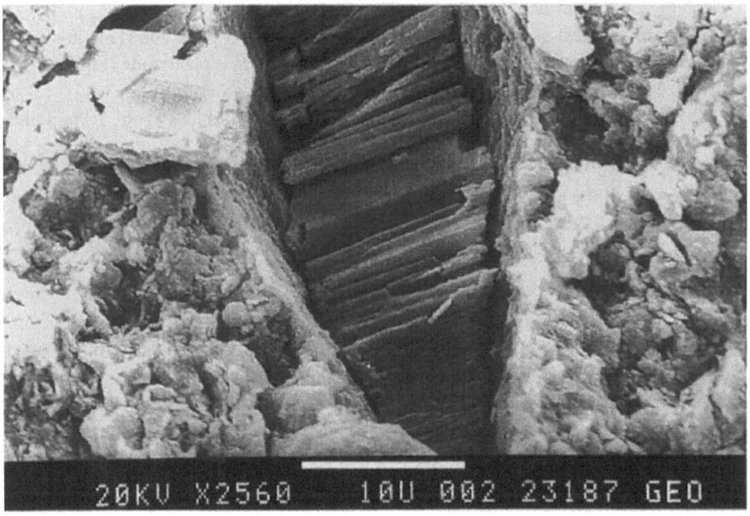

Figure 7. Gypsum filling almond-like vugs within the clayey matrix. Fibrous gypsum is perpendicular to the cavity wall (SEM).

Halite predominates towards the inner parts of the sabkha and is also present in two settings. a) As a desiccated whitish crust, about 1 cm thick, precipitated from the highly saline surface water indicated previously. In the older, reddish parts of the sabkha "salt-rich cakes" is found underlying "dissolution holes" at the centre of the shrinkage polygons indicating downward movement of the surface salt (Fig. 8). b) As a subsurface horizon, several cm below surface and 15-40 cm thick underlain by gypsum in the centre of the sabkha. It is composed of discrete crystals, few mm in diameter, scattered within the clay matrix (Fig. 4). This type has a similar origin like the underlying gypsum.

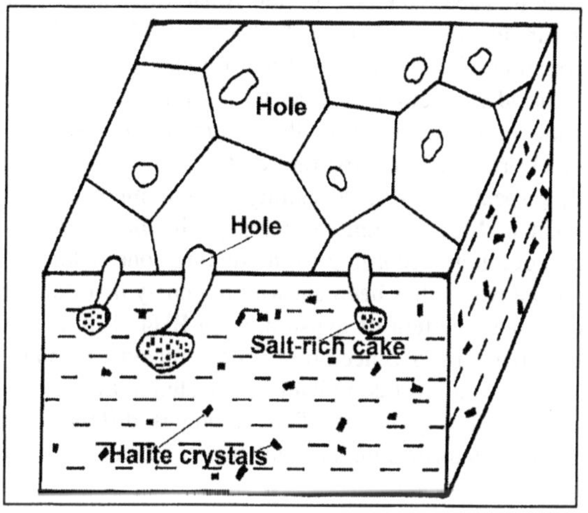

Figure 8. The rather abundant dissolution holes in the brownish, older part of the sabkha (B of Fig. 5). Schematic representation of the "dissolution holes" and the halite-rich "cake" below them and their relationship to the polygons of the brownish part of the sabkha.

Cluster analysis on the chemical data (see Abed 1998) has shown that two phases of sediments are governing the distribution of the analysed elements: the detrital and evaporite phases. The detrital phase consists of the alumino-silicates (clays, quartz, feldspars, etc.) and explains the distribution of the elements Si, Ti, Al, Fe, Mn, K, Rb and partially Mg and Ca. The latter two elements are also involved in the evaporite sediments (Handford 1982, 1988). The minerals halite, gypsum, dolomite, calcite, and sepeolite are the constituents of the evaporite phase which explains the distribution of Na, Ca, Mg, Cl, HCO_3, CO_2, and SO_4. However, the inter-elemental relationships are not as obvious as in the previous phase. The precipitation of halite and gypsum and their relationship with Na, Ca, Cl and SO_4 is rather conspicuous and needs to be discussed further.

Both CaO and CO_2 correlate highly negatively with the elements constituting the detrital phase. This might indicate an evaporite chemical origin for the calcite. Although part of the calcite is certainly of this origin, there are indicators of its detrital, wind blown origin from the western mountains as discussed previously.

MgO correlates positively with elements in the detrital phase, which masks its role in the formation of dolomite and sepeolite. As mentioned previously, dolomite forms authigenically in the centre of the sabkha. This view is based on the perfect euhedral crystal, their small size (< 10 μm) and the association with the Mg-rich clay mineral, sepeolite. Also, authigenic dolomite is known from inland basins and sabkhat comparable to Taba Sabkha, e.g.: the Amargoza flat (Khoury et al., 1982) and Amboseli basin in Kenya (Guven & Karney 1979). Folk & Land (1975) explained the formation of dolomite in sabkhat flooded by meteoric water. Salinity is rather low and the Mg/Ca ratio as low as 1 : 1. Although our data are incomplete on the Mg/Ca ratio of groundwater in the centre of the sabkha, the recharging groundwater of the Darba wells and the fans east of the sabkha have a ratio of about 1 : 1 and 2 : 1, respectively with a salinity of 870 mg/l and 3400 mg/l. Such meteoric water rich in Mg, flooding or mixing with higher salinities within the sabkha might form dolomite according to the mixing zone dolomite model of Folk & Land (1975). Moreover, it has been shown previously that dolomite increases in abundance after the precipitation of gypsum, as in BH4, 7, 8, and 9. This leads to the increase in the Mg/Ca ratio. Since no or little competing ions are present in this water, and because SO_4 is almost completely exhausted in the formation of gypsum, dolomite formation is most probable. This setting is different from the coastal sabkha dolomite, which is formed under much higher salinities and Mg/Ca ratios.

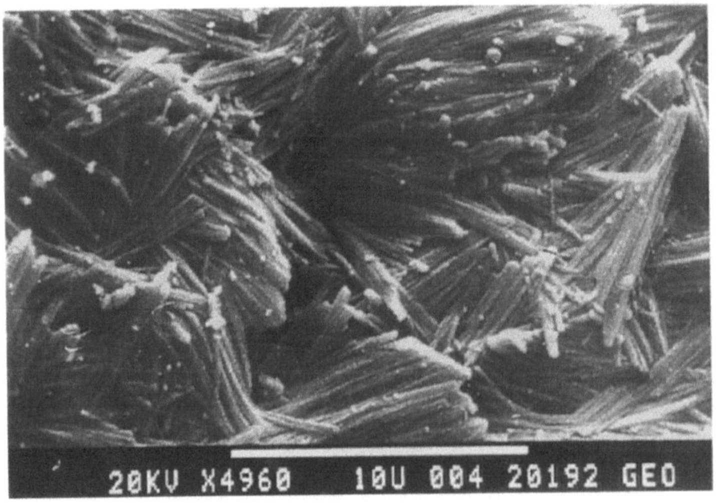

Figure 9. Fibrous sepeolite (SEM) from the centre of the sabkha.

The data are incomplete for discussing the origin of sepeolite, a needle-like, Mg-rich clay mineral (Fig. 9). However, the Amargosa desert flat in Nevada is a close analogy to the Taba Sabkha where sepeolite is forming from non-saline solutions without extensive evaporation concentrations (Khoury et al., 1982). The salinity at the outer zone is only 15.000 mg/l with a pH of 7.75. The presence of halite in the inner parts of the sabkha would suggest more evaporative concentration, which might raise the pH to approach the alkalinity needed for the formation of sepeolite (Parry & Reeves 1968, Post 1978, Starkey & Blackman 1984, Gehring et al. 1995). This high pH would also facilitate the dissolution of silica to form these minor amounts of this mineral. The presence of authigenic quartz in the centre of the sabkha would also support this interpretation.

CONCLUSIONS

1. Examples of sabkhat in Jordan are restricted to inland sabkhat. No coastal sabkhat are present.
2. Five inland sabkhat are present in Jordan, three in southern Wadi Arab, one near Mudawwara, and one in the Azraq Basin.
3. The Taba Sabkha is a deflation salt flat situated in the southern Wadi Araba and it is interfingering with sand dunes especially from the north and alluvial fans from the east.
4. Low salinity groundwater from the fans seeping westwards through the sabkha sediments is precipitating gypsum first and then halite and producing four vegetation zones. High evaporation and extreme aridity superimposed on a shallow water table (about 1 m) are responsible for these features. Relatively deeper gypsum and halite are due to precipitation from groundwater by capillary action.
5. Dolomite is forming authigenically under much lower salinities and Mg/Ca ratios than encountered in coastal sabkha; mixing zone dolomite.
6. Sepeolite is forming under similar conditions to dolomite possibly with higher pH and silica availability.

References

Abed, A.M. 1998. The geology and mineral genesis of the Taba inland sabkha, southern Dead Sea Transform, SW Jordan. In: Alsharhan, A., Glennie, K. & Kendall, C. (eds.), Quaternary Deserts and Climatic Changes, Balkema, pp. 71-87.

Abed, A.M., Carbonel, P., Collin-Girard, J., Fontugne, M. Petit-Maire, N., Reyss, J. & Yasin, S. 2000. Un paleolac du dernier interglaciaire Pleistocene dans l'extreme sud hyperaridede la Jordanie. *C. R. Acad. Sci.* Paris 330: 259-264.

Abu-Zir, I.M. 1989. Hydrochemistry and hydrogeology of the central part of Wadi Araba. M. Sc. thesis, University of Jordan.

Ali, Y.A. & West, I. 1983. Relationships of modern gypsum nodules in sabkha of loess to composition of brines and sediments of northern Egypt. *Jour. Sed. Petrol.* 53: 1151-1168.

Amiel, A.I. & Friedman, G.M. 1971. Continental sabkha in Arava valley between Red Sea and Dead Sea: Significance for the origin of evaporites. *AAPG Bull.* 55: 582-592.

Aqaba Meteorological Station 1955-1985. Unpubl. reports of the Meteorological Department, Amman, Jordan

Bender, F. 1974. Geology of Jordan, Gebrüder Bornträger, Berlin

Butler, G.P. 1969. Modern evaporite deposits and geochemistry of coexisting brines, the sabkha, Trucial Coast, Arabian Gulf. *Jour. Sed. Petrol.* 39: 70-89.

Dean, W.E. 1978. Theoretical versus observed successions from evaporating of sea water. In: Dean, W.E. & Schreiber, B.C. (eds.), *Marine Evaporites*, SEPM Short Course No. 4, Tulsa, Oklahoma, 43-73.

Folk, R.L. & Land, L.S. 1975. Mg/Ca ratio and salinity: two controls over crystallization of dolomite. *Amer. Assoc. Petro. Geol. Bull.* 53: 60-68.

Freund, R., Garfunkel, Z., Zak, I., Goldberg, M., Weissbrod, T. & Derin, B. 1970. The shear along the Dead Sea Rift. *Phil. Trans. Roy. Soc., London* A267: 105-127.

Garrard, A.N., Betts, A., Byrd, B., Colledge, S. & Hunt, C. 1988. Summary of the paleoenvironmental and prehistoric investigations in the Azraq basin. BAR International Series 396 ii: 311-337.

Gavish, E. 1974. Geomorphology and mineralogy of a recent sabkha along the coast of Sinai, Gulf of Suez. *Sedimentology* 21: 397-414.

Gehring, A.U., Keller, P., Prey, B. & Luster, J. 1995. The occurrence of spherical morphology as evidence for changing conditions during the genesis of a sepeolite deposit. *Clay Minerals* 30: 83-86.

Gunatilaka, A. 1986. Kuwait and the northern Arabian Gulf: A study in Quaternary sedimentation. *Episodes* 9: 223-231.

Guven, N. & Carney, L.L. 1979. The hydrothermal transformation of sepeolite to stevensite and the effect of added chlorite and hydroxides. *Clay and Clay Minerals* 27: 253-260.

Handford, C.R. 1988. Depositional interaction of siliclastics and marginal marine evaporites. In: Schreiber, B. (ed.), Evaporites and Hydrocarbons. Columbia University Press, New York, 139-182.

Handford, C.R. 1982. Terrigenous clastics and evaporite sedimentation in a recent continental sabkha playa basin, Bristol Dry Lake, California. In: Handford, C.R., Loucks, R.C. & Davis, G.R. (eds.), Deposition and Diagenetic Spectra of Evaporites, SEPM Short Course No. 4, Tulsa Oklahoma, 65-74.

Ibrahim, K. 1996. The regional geology of Al-Azraq area. Map Sheet No. 3353 I. *Geological Mapping Bull.* 36, 67 pp.., Natural Resources Authority, Amman.

Inci, U. 1991. Miocene alluvial fan-alkali playa lignite-trona bearing deposits from an inverted basin in Anatolia: sedimentology and tectonic controls on deposition. *Sediment. Geol.* 71: 73-97.

Kendall, A.C. 1979. Continental and supratidal (sabkha) evaporites. In: Walker, R.G. (ed.), Facies Models, Geoscience Canada Reprint Series 1, Calgary, 147-157.

Khoury, H.N., Eberl, D.D. & Jones, B.F. 1982. Origin of magnesium clays from the Amargosa desert, Nevada. *Clay and Clay Minerals* 30: 327-336.

Lewy, Y. 1980. Evaporitic environments in northern Sinai. In: Nissenbaum, A. (ed.), Hypersaline Brines and Evaporitic Environments, *Developments in Sedimentology* 28, Elsevier, Amsterdam, pp. 131-143.

Papke, K.J. 1972. A sepeolite rich playa deposits in southern Nevada. *Clay and Clay Minerals* 20: 211-215.

Parry, W.T. & Reeves, C.C. 1968. Sepeolite from Mount Lake, Lynn and Terry Counties, Texas. *Am. Mineralogist* 53: 984-993.

Post, J.L. 1978. Sepeolite deposits in Las Vegas, Nevada area. *Clay and Clay Minerals* 26: 58-64.

Quennell, A.M. 1958. The structures and evolution of the Dead Sea Rift. *Quat. Jour. Geol. Soc. London* 64: 1-24.

Schreiber, B.C. & El-Tabakh, M. 2000. Deposition and early alteration of evaporites. *Sedimentology* 47: 215-238.

Starkey, H.C. & Blackman, P.D. 1984. Sepeolite in Pleistocene Lake Tocopa, Inyo County, California. In: Singer, A. & Gallan, E. (eds.), Palygorskite – Sepeolite, Occurrences and Uses, *Developments in Sedimentology* 37, Elsevier, Amsterdam, pp. 137-149.

Weissbrod, T. 1969. The Paleozoic of Israel and adjacent countries. *Bull. Geol. Surv. Israel* 47, 23 pp.

Yagmurlu, F. & Helvaci, C. 1994. Sedimentological characteristics and facies of the evaporite-bearing Kirmir Formation (Neogene), Beypazari basin, central Anatolia, Turkey. *Sedimentology* 41: 847-860.

THE SABKHA VEGETATION OF OMAN

SHAHINA A. GHAZANFAR

Department of Plant Sciences, Downing Site, Cambridge CB2 3EA, U.K.

Abstract

In the Arabian Peninsula, the coastal and inland sabkhat of Oman are unique in their geological and climatic history, being formed as a result of the lowering of sea levels and the alternation of arid and pluvial climatic periods on the Peninsula. In Oman, the sabkhat are either devoid of vegetation or are covered with a sparse cover of dwarf shrubs and tussock grasses. The sabkhat vegetation shows a pattern of zones and mosaics, composed of species such as, *Atriplex* spp., *Cyperus conglomeratus*, *Halopeplis perfoliata*, *Limonium stocksii*, *Suaeda* spp., *Urochondra setulosa* and *Zygophyllum qatarense*. Biogeographically the sabkhat of Oman lie in the Arabian Regional Subzone and have several species that are endemic to that phytochorion. A few species are also endemic to the Arabian Peninsula. There is very little habitation and no cultivation in the sabkhat, however small stocks of goats and camels graze and browse on the sparse vegetation, and on any annuals that may come up after rain.

INTRODUCTION

Coastal and inland sabkhat are highly stressed habitats that are characterised by poor vegetation cover and a paucity of species. The commonly occurring species possess morphological modifications and are highly specialised in their physiology, metabolism, photosynthetic pathways and seed germination strategies (Gutter man 1993, Ismail 1983, Ismail & El-Ghazaly 1990, Ungar 1987, 1991). Many halophytic plants develop salt excreting glands, accumulate salt or use C4 pathways (Babikir & Kürschner 1992, Frey et. al 1984, 1985). Species living in sabkhat often occur in monospecific stands, and, as a result of the diminishing effects of soil salinity from the centre towards the edge of sabkha, the vegetation typically exhibits a zonation pattern of distinct plant communities. These communities consist of species that are

Barth & Böer (eds.), SABKHA ECOSYSTEMS, 99-107

the most salt tolerant in the central areas to the least salt tolerant at the edges. In Oman, the dominant plant species that occur on sabkhat belong to the families Chenopodiaceae, Plumbaginaceae, Zygophyllaceae, Cyperaceae and Poaceae. A lack of fresh water and the high salinity of the soils render sabkhat uninhabitable, however some habitation may be present at the edges of sabkhat. In Oman, small fishing communities inhabit the fringes of coastal sabkhat, and their livestock, mainly camels, graze on the sparse vegetation. In this paper the typical vegetation, plant communities, vegetation zonation, biogeography and endemism of the coastal and inland sabkhat of Oman are described.

COASTAL AND INLAND SABKHAT OF OMAN

The coastal and inland sabkhat of Oman are amongst the largest occurring on the Arabian Peninsula. The inland sabkha of Umm as Samîm and the coastal sabkha of Barr al Hikmân are major landscapes features of Oman. Sabkhat are also present east of Sayh al Hayma in the At Tabaqah and Subhât al Wad areas and to the east and west of Wadi Mukhayznah on the Jiddat al Harâsîs (Fig. 1).

Coastal sabkhat

The coastal sabkha of Barr al Hikmân covers an area of ca. 1274 km^2, and occupies more or less the entire area of the Barr al Hikmân peninsula. Areas above the high tide level are subjected to flooding by extreme high tides and storms and subsurface seepages from the sea. The Barr al Hikmân peninsula lies 350 km south of Muscat, between latitude 20-21°N and longitude 58-59°E. It is bordered in the north by Wadi Halfayn and Ramlat Wahibah and in the south, east and west by the Arabian Sea (Fig. 1). Most of the area of the peninsula is a flat featureless plain from sea level to about 20 m a.s.l. There are extensive mud flats on the eastern coast which flood at high tide; in the north-east and south-west tidal lagoons and sea-inlets are present that are separated from the sea for most of their length by sand bars. Low sandy dunes are present along the beaches. Inland, there are vast expanses of highly saline areas, encrusted with salt and devoid of vegetation. Historically, the Barr al Hikmân peninsula has been formed as a result of a fall in sea level. Geologically it is unique in its composition of quaternary sediments (the largest in the Arabian Peninsula) with alluvial sandy gravel overlain with a thin layer of shifting aeolian sand. In the south-west this gravel is of marine origin (coralline and shell debris of quaternary origin). The mud flats in the east have high silt content. The Barr al Hikmân is the only sabkha in the Arabian Peninsula that faces the south-west monsoon winds (Glennie1987, Gubba & Glennie 1998).

There are no meteorological data for the Barr al Hikmân peninsula. The nearest station is on the offshore island of Masirah, which is probably representative for the coastal areas of the peninsula. Based on data from 1985-1995, the mean air temperature is 26.4°C (maximum 45.2°C, minimum 12.3°C) with March, April and May being the hottest months and December to January the coolest months. The mean total annual rainfall, which usually falls during March and April, is 67mm (Fisher & Membery 1998).

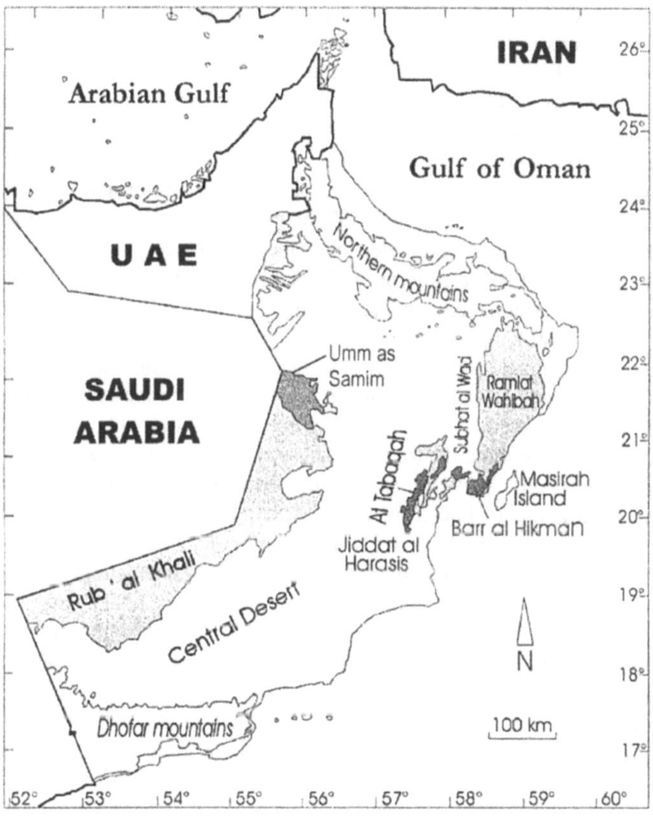

Figure 1. Map of Oman showing the major sabkhat.

Inland sabkhat

The largest inland sabkha in the Arabian Peninsula is Umm as Samîm. It lies at the western border of central Oman, bounded by scarps of Miocene Fars siltstones and limestones to the north and east, and by the sand dunes of the Rub' al Khali to the south and west (Fig. 1). It is ca. 100 km long (north to south) and ca. 50 km wide. Evidence from the sequence of deposits in the sabkha show its climatic and geological history which correlates with the alternate wet and dry periods in the Arabian Peninsula during the past 30.000 yrs (Heathcoat & King 1998). The course of events was as follows: About 30.000 BP a shallow closed depression was eroded by deflation in the gypsiferous Fars sediments; between 30.000 and 20.000 BP a wet period existed, and the closed depression was occupied by a saline lake, re-deposited Fars gypsum and also halite. Much of the water entering the lake was surface flow from the mountains in the north; from 20.000 to 15.000 BP a drier climate existed, and a sabkha condition was established, and sand deposition occurred in the sabkha; between 15.000 to 12.000 BP wetter conditions returned and the lake was re-established; between 12.000 to 9000 BP, a dry period existed, sabkha conditions were re-established with a predominant deposition of sand; during 9000 to 6000 BP cooler, wet conditions returned, and the lake was re-established. The previously deposited halite dissolved but the less soluble gypsum was unaffected. Water inflow was from the north and predominated over the upward seepages from the tertiary limestone. From 6000 BP to present, with the prevalence of a drier climate, sabkha conditions were re-established, with sand as the major deposit and halite as the minor deposit (Heathcoat & King 1998).

Three main drainage systems can be recognised within the Umm as Samîm catchment area: Wadi Aswad in the north, Wadi Umayri in the east and the Umm al Hayat system in the south. The latter drains the southern mountains and the central plains into Umm as Samîm. There is evidence of surface and subsurface water flow in all wadis, however, for Wadi Aswad this flow ceases at about 40 km, and for Wadi Umayri at about 10 km, from Umm as Samîm. The presence of a large delta and soft ground at Wadi Umayri before draining into Umm as Samîm, implies surface flow all the way in the past (Beydoun 1980). Interestingly, an additional supply of water to Umm as Samîm is derived from the underlying aquifer, the Palaeocene Umm ar Radhuma Formation, which is overlain with evaporites of the Eocene Rus Formation, and from which some of the salt is derived (Glennie 1976, Heathcoat & King 1998).

The sabkha deposits consist of a main zone of salt crust, including heaved crust. Brine is present close to the surface, being 0.13 - 0.78 m below ground level. Fresh salt is continually precipitated as a result of evaporation and the expansion breaks the surface into polygonal plates bounded by rims of fresh salt. The crust accumulates wind blown dust and gradually darkens. Continuing evaporation results in the plates to buckle and heave often rising to a height of 0.5 m. Towards the Rub' al Khali, soft, fine-grained sand is saturated with brine 1-1.5 m below ground level. In the east, where the brine level is generally up to 1.6 m deep, an area with wind-blown sand occurs. The outflow of Wadi Umayri east of the Umm as Samîm, consists of an accumulation of soft sediment. There, fresh water has dissolved channels in the salt (Heathcoat & King 1998).

The closest meteorological data available for this region is that for Fahud, which would be generally representative for the Umm as Samîm area. Based on data from

1985-1995, the mean total annual rainfall in Fahud is 24 mm. Mean annual temperature is 28.8°C with maximum and minimum absolute temperatures of 50.7°C and 5.6°C respectively. The mean monthly maximum temperature range is 19.4°C. The coefficient of variations of total annual rainfall and mean annual temperature indicate that there is very little inter-annual variation of temperature and that total annual rainfall is exceptionally variable between years exhibiting little seasonality (Fisher & Membery 1998).

VEGETATION

The vegetation of sabkhat is generally impoverished and is characterised by halophytic shrub communities forming a pattern of vegetation zones and mosaics. Highly saline depressions are devoid of vegetation, but plants are found at the fringes of such depressions. Species and plant communities vary in their composition and richness on different sabkhat, dependent largely on the different substrates, and the occurrence of rain or drainage at these areas.

Vegetation cover is generally poor in sabkhat. An analysis of vegetation cover in the Barr al Hikmân peninsula showed that the vegetation cover and plant species richness is greatest around the coastline, while in much of the central area, vegetation cover is sparse and species low in number. Only 1.07% of the total area of the peninsula is vegetated; however, in a belt 1 km wide around the coast the vegetation cover is 7.2% (Ghazanfar 1995). It is evident that the humid sea air and the low soil salinity (ca. 3%) that is present near the coast contribute to the richness and density of species there. Inland, where, as a result of evaporation, the air is drier and soil salinity higher (4-5%), conditions for plant growth are unfavourable.

In Oman, about 15 halophytic species are common in coastal and inland sabkhat. These are species of *Arthrocnemum*, *Atriplex*, *Halopeplis*, *Halopyrum*, *Limonium*, *Sporobolus*, *Suaeda*, *Urochondra* and *Zygophyllum* (Deil 1998, Ghazanfar 1992 a & b, 1999b, Ghazanfar & Rappenhöner 1994). Species of *Atriplex* and *Suaeda* are the most abundant and common, followed by *Limonium*, *Zygophyllum*, *Sporobolus* and *Urochondra*. Both obligate halophytes, such as species of *Atriplex*, *Halocnemum* and *Arthrocnemum*, and facultative halophytes such as *Cornulaca monacantha*, *Limonium stocksii*, *L. axillare*, *Heliotropium kotschyi* and *Zygophyllum qatarense* occur. Annuals are rare, and when present come up after heavy rain or are found near irrigated areas. Psammophytic shrubs such as *Limonium stocksii*, *Cyperus conglomeratus* and *Urochondra setulosa* are effective sand-binders. Wind drifted sand is collected at the base of these shrubs forming small mounds called *nebkhas*. These *nebkhas* provide shade and moisture from dew, and support a variety of animal life. Ephemerals also tend to germinate and grow on these *nebkhas*.

Plant communities of coastal sabkhat

The vegetation of Barr al Hikmân is broadly classified as the *Halopeplis perfoliata* and *Cyperus conglomeratus* vegetation type (Deil 1998, Ghazanfar 1991, 1994, 1999a & b, Ghazanfar & Rappenhöner 1994). This type of vegetation is typical of the saline plains and saline coastal dunes in the eastern regions of central Oman. It is found throughout the Barr al Hikmân sabkha, extends to Ramlat Wahibah in the north and the sabkhat below the Huqf escarpment in the south. This vegetation type shows a pattern of zonation of plants, which can be classified into three communities:

(1) *Atriplex-Suaeda* community

This community is found on silty or coralline sand with gravel, and is found along coastal strips and inland lagoons (*khawrs*). The dominant species are *Atriplex farinosa*, *Suaeda moschata* and *S. vermiculata*; associates are *A. coriacea*, *Halopeplis perfoliata* and *S. aegyptiaca*. Where the substrate is muddy, *Arthrocnemum macrostachyum* is present. Low coastal dunes are vegetated with *Halopyrum mucronatum*. Rappenhöner (1991) estimated that the cover for this community was 30-50% in the southwest littoral belt of Barr al Hikmân.

(2) *Urochondra setulosa* community

This community is mainly composed of the perennial grass *Urochondra setulosa*. Dense tussocks of this grass form a narrow transition belt between the coastal dunes and the flat inland plains of the sabkha. Often associated with this species, and forming a vegetation mosaic are two species, *Limonium stocksii* and *Cyperus conglomeratus*. The vegetation cover in this community was about 30% (Rappenhöner 1991).

(3) *Limonium-Cyperus-Heliotropium* community

This plant community occupies the central, sandy-gravel plains of Barr al Hikmân. The two dominant species are *Limonium stocksii* and *Cyperus conglomeratus* with *Heliotropium kotschyi* an associate in the drier, stony areas. Other species, such as *Zygophyllum qatarense*, and *Echiocilon jugatum* may also be present on stony substrates. *Halopeplis perfoliata* is often found on the fringes of hyper-saline depressions. Vegetation cover for this community was estimated to be 5-10% (Rappenhöner 1991).

Plant communities of inland sabkhat

The vegetation of inland sabkhat can be broadly classified as the *Acacia-Zygophyllum-Heliotropium* vegetation type (Ghazanfar 1991, 1992, 1999a). The hyper-saline sabkha depressions, such as that of Umm as Samîm, are completely devoid of vegetation, but plants are present on the adjacent plains and wadis that drain into these sabkhat. Grazing and browsing by goats and camels has altered the natural vegetation of such wadis, which are now dominated by spiny and unpalatable species. The common shrubs include *Cornulaca monacantha*, *C. aucheri*, *Fagonia* spp., *Rhazya stricta*, *Tephrosia apollinea* and *Zygophyllum qaterense*, and the common trees *Acacia tortilis* and *Prosopis cineraria*. The trees are heavily browsed and show severe browse lines.

Two plant communities can be recognised, that are present on the gravel plains and sandy wadis of inland sabkhat. In Oman, these communities occur on the edges of the sabkhat of Sayh al Hayma in the At Tabaqah and Subhât al Wad region, the sabkha west of Wadi Mukhayznah on the Jiddat al Harâsîs, and the region north and east of Umm as Samîm. These plant communities are:

(1) The *Zygophyllum-Cornulaca* subshrub community
This plant community is present on the gravel plains of the inland sabkhat. Vegetation cover is sparse with most plants growing in shallow depressions, runnels and beside track roads. The two main species are *Zygophyllum qatarense* and *Cornulaca monacantha*, which may be associated with a few others, such as *Fagonia indica*, *Haloxylon salicornicum* and *Salsola rubescens*.

(2) *Acacia* communities
Acacia communities are present in the wadis of inland sabkhat. These consist of *Acacia tortilis*, *Prosopis cineraria* and *Acacia ehrenbergiana*, and form the main woody component of this community. The composition of this community depends on the soil substrate and subsurface water available. *Prosopis cineraria* is the dominant component on sandy substrates, while *Acacia tortilis* is more prevalent on gravelly and silty substrates. Associates in this community are subshrubs *Rhazya stricta*, *Tephrosia apollinea*, *Haloxylon salicornicum* and *Salsola rubescens*. *Cyperus conglomeratus* is the dominant sedge, *Panicum turgidum* the dominant grass and *Citrullus colocynthis* and *Zygophyllum simplex*, the dominant annuals.

BIOGEOGRAPHY AND ENDEMISM

The sabkhat of Oman lie in the Arabian regional subzone of the Saharo-Sindian regional zone (White & Léonard 1991). The Saharo-Sindian regional zone is the largest phytochorion of the Arabian Peninsula, and is divided into an Arabian regional subzone and the Nubo-Sindian local centre of endemism (Ghazanfar 1992b, Kürschner 1998, Léonard 1989, White & Leonard 1991). In Oman, the Arabian regional subzone extends into central Oman, and in the foothills of northern Oman, where it overlaps with the Nubo-Sindian local centre of endemism. The flora of this phytochorion is poor in species and consists of derivatives of an ancient stock of species formerly distributed along the northern and southern coasts of the Tethys sea (Kürschner 1998). Five percent of the flora of this phytochorion is endemic to the Arabian Peninsula, including species such as *Echiochilon jugatum, Limonium stocksii, Sphaerocoma aucheri* and *Zygophyllum qatarense* (Cope 1988, Deil 1987, 2000). In the sabkhat of Oman, species that are endemic to the larger Saharo-Sindian phytochorion include *Cornulaca aucheri, Haloxylon salicornicum* and *Suaeda aegyptiaca*.

CONCLUSION

The coastal and inland sabkhat of Oman are some of the largest on the Arabian Peninsula. Although their unique geology and climatic history is well studied, there is little detailed work done on the vegetation and flora of these sabkhat. Apart from some basic information on the composition of plant communities, data on soil, ecology, and physiology of the flora are lacking. So far there are no studies conducted on the phenology, seed dispersal strategies, seed germination mechanisms or regeneration of plants of Oman sabkhat.

The extensive coastal and inland sabkhat of Oman provide us with a venue that can be used as an open laboratory in order to understand the dynamics of plants growing in stressed environments, in particular the hyper-arid, saline environments that characterize sabkhat. A study of the ecology and physiology of the commonly occurring halophytic species, such as those of *Atriplex*, *Halopeplis*, *Limonium*, *Suaeda*, *Urochondra* and *Zygophyllum*, can provide us with information and an understanding of the plant-soil relationships in halophytes, which may be used in search for suitable fodder plants in desert ecosystems, and for the restoration of desert habitats.

References

Babikir, A.A. & H. Kürschner 1992. Vegetational patterns within a coastal saline of NE-Qatar. *Arab Gulf Journal of Scientific Research* 10: 61-75.

Beydoun, Z.R. 1980. Some Holocene geomorphological and sedimentological observations from Oman and their palaeogeological implications. *Journal of Petroleum Geology* 2: 427-437.

Cope, T. A. 1988. The flora of the Sands. In: Dutton, R.W. (ed.), The Scientific Results of The Royal Geographical Society's Oman Wahiba Sands Project 1985-1987. *The Journal of Oman Studies Special Report No. 3.* Office of the Adviser for Conservation of the Environment, Muscat, Oman, pp. 305-312.

Deil, U. 1987. Plant life in Dubai. Report to the Dubai Wildlife Research Centre. Unpublished.

Deil, U. 1998. Coastal and Sabkha Vegetation. In: Ghazanfar, S.A. & Fisher, M. (eds.), Vegetation of the Arabian Peninsula. Kluwer Academic, Netherlands, pp. 209-228.

Deil, U. 2000. Halophytic vegetation along the Arabian coast - azonal or linked to climatic zones? *Phytoceonologia* 30: 591-611.

Fisher, M. & Membery, D.A. 1998. Climate. In: Ghazanfar, S.A. & Fisher, M. (eds.), Vegetation of the Arabian Peninsula. Kluwer Academic, Netherlands, pp. 5-38.

Frey, W., Kürschner, H., El-Sheikh, A.M., & Migahid, A.M. 1984. Zonation and photosynthetic pathways of halophytes on the Red Sea coast near Tawwal (Saudi Arabia). *Notes from the Royal Botanic Garden Edinburgh* 42: 45-56.

Frey, W., Kürschner, H. & Stichler, W. 1985. Photosynthetic pathways and ecological distribution of halophytes from four littoral salt marshes (Egypt/Sinai, Saudi Arabia, Oman and Iran). *Flora* 177: 127-134.

Ghazanfar, S.A. 1991. Floristic composition and the analysis of vegetation of the Sultanate of Oman. *Flora et Vegetatio Mundi* 9: 215-227.

Ghazanfar, S.A. 1992a. An Annotated Catalogue of the Flora of Oman. Meise, Belgium, National Botanic Garden of Belgium.

Ghazanfar, S.A. 1992b. Quantitative and biogeographic analysis of the flora of the Sultanate of Oman. *Global Ecology and Biogeography Letters* 2: 189-195.

Ghazanfar, S.A. 1995. Coastal sabkhat: an analysis of the vegetation of Barr al Hikman. In: Khan, M.A. & Ungar, I.A. (eds.), The Biology of Salt Tolerant Plants. Department of Botany, University of Karachi, Pakistan, pp. 277-283.

Ghazanfar, S.A. 1999a. A Review of the Flora of Oman. In: Fisher, M. Ghazanfar, S.A. & Spalton, J.A. (eds.), The Natural History of Oman: a Festschrift for Michael Gallagher. Backhuys Publishers, Leiden, The Netherlands.

Ghazanfar, S.A. 1999b. Coastal vegetation of Oman. *Estuarine, Coastal and Shelf Science* 49: 21-27.

Ghazanfar, S.A. & D. Rappenhöner 1994. Vegetation and flora of the Islands of Masirah and Shagaf, Sultanate of Oman. *Arab Gulf Journal of Scientific Research* 12: 509-524.

Glennie, K.W. 1987. Desert sedimentary environments, present and past: a summary. *Sedimentary Geology* 50: 135-165.

Gubba, I. & Glennie, K. 1998. Geology and Geomorphology. In: Ghazanfar, S.A. & Fisher, M. (eds.), Vegetation of the Arabian Peninsula. Kluwer Academic, Netherlands, pp. 63-98.

Gutterman, Y. 1993. Seed Germination in Desert Plants. Springer-Verlag.

Heathcoat, J.A. & King, S. 1998. Umm as Samim: A sabkha with evidence for climate change, In: AlSharhan, A.S., Glennie, K.W., Whittle, G.L. & Kendall, C.G.St.C. (eds.), Quaternary Deserts and Climatic Change. A.A. Balkema, Rotterdam, Brookfield, pp. 141-154.

Ismail, A.M. & G.A. El-Ghazaly 1990. Phenological studies on *Zygophyllum qatarense* Hadidi from contrasting habitats. *Journal of Arid Environments* 18: 195-205.

Ismail, A.M.A. 1983. Some factors controlling the water economy of *Zygophyllum qatarense* (Hadidi) growing in Qatar. *Journal of Arid Environments* 6: 239-246.

Kürschner, H. 1998. Biogeography and Introduction to Vegetation. In: Ghazanfar, S.A. & Fisher, M. (eds.), *Vegetation of the Arabian Peninsula*. S.A. Kluwer Academic, Netherlands, pp. 63-98.

Léonard, J. 1989. Considérations phytogéographiques sur les phytochorics irano-touranienne, saharo-sindienne et de la Somalie-pays Masai. Jardin Botanik Nat. Belgique.

Rappenhöner D. 1991. Study for Wildlife and Conservation Areas Master Plan for the Coastal Areas of the Barr al Hikman and Masirah Island, Draft Final Manuscript. Ministry of Environment, Sultanate of Oman.

Ungar, I.A. 1987. Population ecology of halophyte seeds. *Botanical Review* 53: 301-334.

Ungar, I.A. 1991. Ecophysiology of vascular halophytes. CRC Press, Boca Raton, USA.

White, F. & Léonard, J. 1991. Phytogeographical links between Africa and Southwest Asia. *Flora et Vegetatio Mundi* 9: 229-246.

SALT DESERTS IN IRAN AND AFGHANISTAN

SIEGMAR -W. BRECKLE

Department of Ecology, Faculty of Biology, University of Bielefeld, P.O. Box 100131, D-33619 Bielefeld

Abstract

Arid areas predominate in the Irano-Turanian region. Climatic aridity causes interrupted water courses, only episodically in function. Nevertheless, arid areas are subject to intensive weathering of rocks and transport of material by water and aeolian processes. Vast areas in Iran and Afghanistan are flat endorrheic basins, large playas, where water evaporates and all soluble ions are accumulating since millennia (evaporites). The smaller basins are mostly formed by deflation processes, the larger ones additionally by tectonic activity. These inland basins are characterised by fine-grained substrate often with high salinity or even thick salt crusts. Vegetation is limited to the margins, often exhibiting a typical zonation of various halophyte types, and only thin cyanobacterial crusts may develop under favourable conditions in the centre of the basin. Geographical distribution, geology, hydrology, climate, soils, flora and vegetation of the various salt deserts and ecological processes are discussed.

GEOGRAPHICAL DISTRIBUTION OF SALT DESERTS

Climatic aridity is one of the main preconditions for the development of endorrheic basins. The deserts in Iran are mostly located in vast basins in the central parts of the country. They are often surrounded by mountains. Thus, the whole area can be subdivided into several parts (Fig. 1). The central Persian desert basins (Fig. 1: 1A,1B), are connected with smaller basins (Fig. 1: 2A, 2B). Seperated by the lower east Persian Mountains, the basin of Seistan leads to the south Afghan deserts (e.g. Seistan).

The mountain regions in central Afghanistan (Fig. 1: 4E, 4F) are mostly semidesertic or desertic. However, the easterly mountains (Fig. 1: 4G) are more humid and free of sabkha formation. The northern slopes (3E) are semiarid, but some interior cold valleys are very arid (2G).

Barth & Böer (eds.), SABKHA ECOSYSTEMS, 109-122

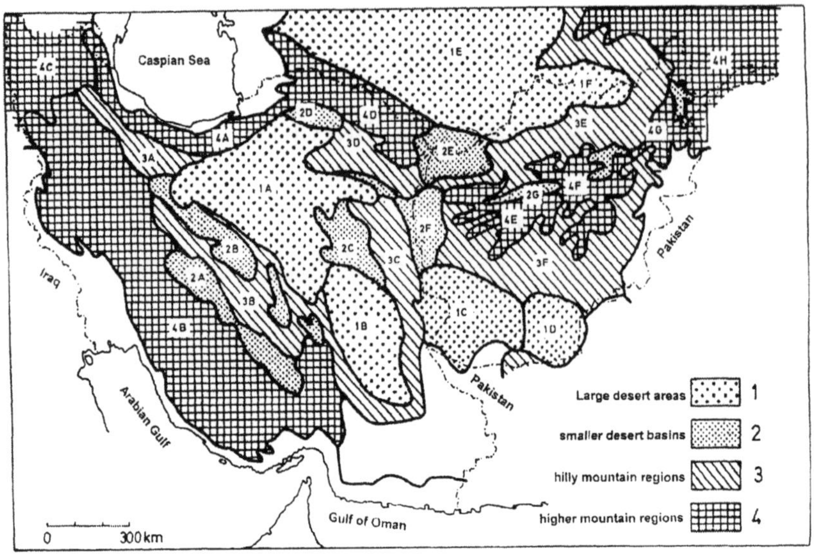

Figure 1. The main regions of the Iran-Afghanistan area (in part from Scharlau 1969, Schweizer 1975, Breckle 1983). Desert regions: 1A: = the Great Kavir and other central Persian desert basins; 1B = the Dasht-e-Lut Desert basin; 1C = Seistan and Dasht-e-Margo (S-, SW-Afghanistan-desert); 1D = Registan, SE Afghanistan sandy desert; 1E = Karakum and adjacent desert regions (Kyzylkum, Aralkum); 1F = Oxus basin (Amu Darja). Smaller marginal deserts and semideserts: 2A = basins within the eastern Zagros Ranges; 2B = deserts in the basins of Ardestan, Na'in etc.; 2C = deserts in the Chusf-Shur-region; 2D= deserts in the Esfarayen basin; 2E = semi-desert in western Afghanistan (Khorassan); 2G = desertic valleys and basins within Afghan mountains. Transitional hill regions, semi-deserts and lower mountains: 3A = plains and hills between the Zagros and Elburz mountains; 3B = Kuh-Rud mountains in C-Iran; 3C = Kuhistan and Zabolistan hill region in eastern Iran; 3D = Khorasan mountain region; 3E = northern slopes of Afghan mountains and circum Oxian-hill region; 3F = southern slopes and hills of Afghan mountains. Mountain regions: 4A = Elburz range; 4B = Zagros range; 4C = eastern Anatolian mountains; 4D = Turkmenian and Khorasan ranges; 4E = Ghorat mountain region in western central Afghanistan; 4F = Hazarajat mountain region in central Afghanistan; 4G = Hindu Kush high ranges; 4H = the Pamir high ranges.

GEOLOGY AND PALAEOHISTORY

The geological mosaic of the area is very complex. However, it is important to keep in mind, that most of the mountain ridges (Hindu Kush, Elburz and some other ridges) were folded only during Tertiary. The mountains consist of a big variety of Paleozoic and Mesozoic metamorphic series, including limestones. Towards the margins of the mountains, the rocks dip under talus and gravel, sand dune and loess deposits. Silt and clay deposits characterise the various basins and surrounding deserts.

Glaciation in the Quaternary was not a prominent feature for interior Iran. There is also no good evidence for assuming higher humidity (Zohary 1973). Some of the basins might have been somewhat less desertic than today (Scharlau 1958). The slight differences in water budget might have been enough for the existence of inland lakes where sandy or clay regions often with high percentage of salinity occur today. Vast areas were filled with gravels, sand, and clay materials during glacial and interglacial periods. Because of the endorrheic situation, salts accumulated and are yet accumulating. The lowlands, therefore, are often covered by salt deserts on Quaternary deposits. Alluvial fans, takyrs, clay pans and sand dunes with local loess coverings are mixed together. More details on geological history and the periglacial situation are given by Bobek (1961) and Pias (1977).

Climatic aridity is the key factor in developing endorrheic geomorphological features. But the climatic conditions in transitional regions may have changed tremendously in the past. An example from the easternmost parts of the Irano-Turanian plains and highlands concerns the aridity in East-Afghanistan. The monsoonal influence seems to have occurred rather further to the west of the Kabul area during the Late Tertiary and the beginning Quaternary than today (Breckle 1967). This can be concluded from fossils of tree leaves. The case-study on a small Iranian playa (Breckle 1981) indicates more humid conditions in late glacial times. Van Zeist (1967) provided data on the much greater extension of trees in Persia during the Pleistocene, but there are some indications that the climate during the pleniglacial was more arid than today.

HYDROLOGY

Rivers of the region are untamed mountain streams which discharge highly variable amounts of water. With few exceptions, they are inland rivers, disappearing in the deserts or steppes or flowing into inland lakes. The Zagros Mountains, the Baluchistan and the Makran areas discharge water into the Persian Gulf and the Gulf of Oman. All of central Iran, however, is characterised by several small and some large endorrheic systems (vast basins, called kavirs; like the North African shotts with clay and salt swamps, or the American playas) (cf. Krinsley 1970, Breckle 1981). In all endorrheic regions smaller or larger playa formation can be observed.

The northern parts of Iran are drained to the Caspian Sea, the greatest inland lake on earth. The northern part of Afghanistan is drained by several rivers which partly join the Amu-Dar'ya and thus discharge to the diminishing Aral Sea. The Hari Rud and the Murgab Rud disappear in desert oases and feed large groundwater reservoirs in the Karakum. West, south and central Afghanistan is drained by the Farah Rud, Khash Rud, Helmand and Arghandab Rivers, which discharge into the swampy inland lakes at the Afghan-Iranian border (Hamun-e-Sabori, Hamun-e-Puzak). Only a small section of eastern Afghanistan is drained to the Indian Ocean via the Kabul and the Indus Rivers.

CLIMATE

The entire Irano-Turanian area (including most of Afghanistan) has an extremely continental and arid climate. The highly accentuated surface relief, the great variability from year to year and the short period of meteorological records impart an incomplete perspective of the climatic conditions. Some of the data available are used for representative climatic diagrams (Fig. 2). Additional sources for climatic data are Bobek (1952), Adle (1960), Ganji (1968), Sabeti (1969), Herman et al. (1971) and Lalande et al. (1974).

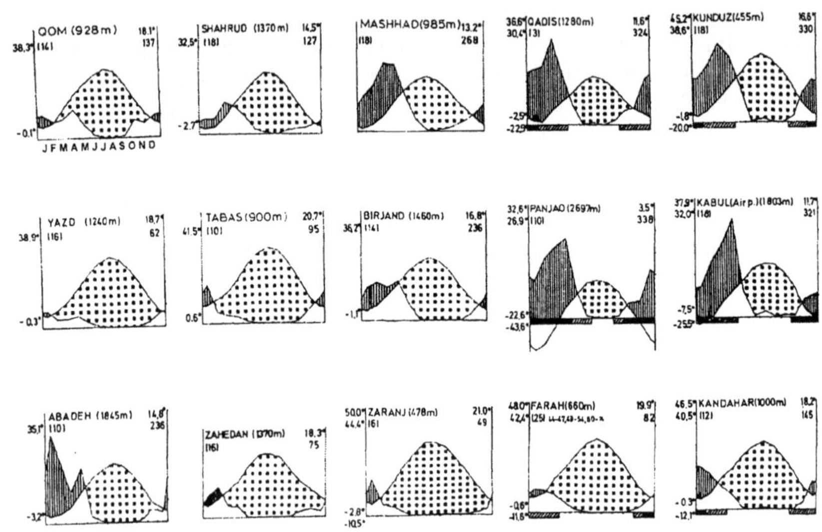

Figure 2. Selected climatic diagrams of the desert and semi-desert areas of Iran and Afghanistan (Data from Afghan Air Authority, Walter & Lieth 1960-67, Walter 1974).

In general, the climate of the area resembles the arid climate of the adjacent Karakum with very high annual fluctuations of temperature. The absolute temperature fluctuations in some places reach 70 K. The area is characterised by the regular occurrence of frost days (Fig. 3). Only a few transitional regions of both countries (e.g., Dasht-e-Lut, or Sistan) exhibit less than thirty days with minimum temperatures below 0°C per year. Iran is thus a typical transitional desert country between hot deserts (zonobiome III) and cool deserts with cold winters (zonobiome VII/III) in the sense of Walter & Breckle (1984, 1999).

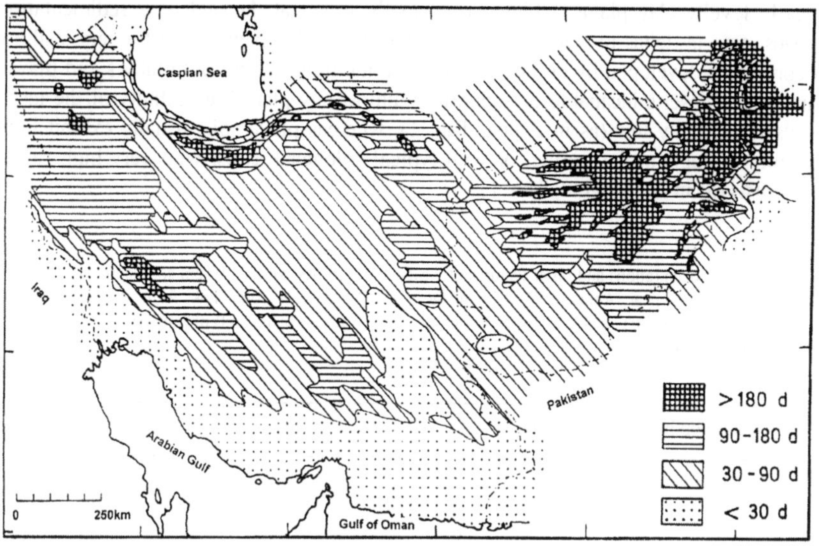

Figure 3. Number of days per year with temperatures equal or below 0°C in the Iran-Afghanistan area, shown by an isoline map (Breckle 1983).

On the other hand, from Dasht-e-Lut and from Sistan, there are reported more than five days per year with an average temperature exceeding 45°C and absolute maxima around 51°C, similar to the conditions shown by the climatic diagram of Zaranj. Precipitation varies greatly depending on the elevation and geographic position of the area. Vast areas have total annual precipitation of less than 200 mm. In the mountains and at their foothills there are zones with annual precipitation (rain and snow) from 200 to 500 mm. The main period of precipitation is during winter and spring in the north, ceasing some weeks later than in the south. The summers are dry and hot; autumn also is an arid season. The global solar radiation in all the desertic areas is very high. The relative humidity of the air is low, with annual means around 50% or lower. Extremely low values (<5%) have been recorded. It is obvious that evaporation is very high. Stenz (1941, 1957) has reported an annual evaporation of 2305 mm for Kabul.

SOILS

Vast areas are covered by immature soils, derived from unconsolidated materials, such as gravel, rocks or sand. The most common categories (Zohary 1973) in the sense of geomorphology and desert-types are hammadas, serirs, regs, ergs, nebkhas, takyrs, sabkhat and playas - each in a variety of forms. The term takyr is applied to large, nearly levelly clay plains, which are normally found at the end of a sedimentation series, e.g., in deltas of wadis and rivers. Playas are clay pans in mostly very shallow depressions developed by deflation or by tectonic movements. Both takyrs and playas are barren deserts incapable of supporting almost any plant life. Salinity of the takyr soils is normally low (as in Aqina, Tab. 1), whereas playa soils usually exhibit high salinity throughout the profile, or with typical solonchak soils, with very high saline saltcrusts (as in Ali Moghul, Tab. 1).

Table 1. Ion concentrations and other features of typical soil profiles within 110 cm soil depth. Aqina = soil from takyr plains near Aqina (NW-Afghanistan); Ali Moghul = solonchak from Ali Moghul (N-Afghanistan). Na^+, K^+, Cl^-, and SO_4^{2-} (mMol×kg^{-1} dry soil). $CaCO_3$ (% dry soil) pH and electric conductivity (EC, S×cm) (Breckle 1983).

soil depth (cm)	0-1	1-5	10-20	30-40	60-70	100-110
Aqina/NW-Afgh						
Na^+	70	125	360	410	190	105
K^+	5	6	2	3	3	5
Cl^-	8	17	95	520	140	30
SO_4^{2-}	40	100	220	260	110	80
$CaCO_3$ (%)	23	27	22	24	20	14
pH	8,3	8,3	8,3	8,0	8,5	9,1
EC (S×cm)	4	6	16	24	6	2
Ali Moghul/N-Afgh						
Na^+	1750	240	110	120	170	220
K^+	6	5	9	10	6	11
Cl^-	1270	780	100	110	410	670
SO_4^{2-}	1750	160	90	70	60	50
$CaCO_3$ (%)	15	22	21	23	19	12
pH	8,6	8,2	8,3	8,3	7,9	8,0
EC (S×cm)	115	13	6	7	11	10

Cemented (petrocalcic horizons often develop in the soil profile (Salem & Hole 1969). The cementing compounds are usually calcium carbonate, calcium sulfate, but also salts of sodium, which can form thick hard layers (Fig. 4). Halomorphic soils may occur on various substrates, such as loams, loess, and sands, but often they are limited to alluvial lowlands and basins with very fine-textured parent materials. The amount of soluble salts tolerated by plants ranges between 0,3 and 3% in the majority of saline soils (Zohary 1973). In some cases, higher percentages of salts have been found, but such habitats usually lack plants. Saline soils may however, harbour quite a number of plant communities and a rather rich flora, even with some endemic species. This depends not only on the amount of salts, but even more on the grain size conditions of the soil and the availability of water, and probably also of calcium. In fine-textured soils evaporation causes losses of water from rather deep parts in the profile. Thus, deficiency of moisture turns salines into a very adverse habitat of plant life. The plants living in them, the xerohalophytes, have to overcome both drought and excess of salts (Fig. 4). Only under more moderate climatic conditions (Breckle 1976) is plant life possible.

Figure 4. The north eastern margin of the Great Kavir, Iran, marked by the last occurring plants, mainly scattered *Halocnemum strobilaceum* (photo: Breckle, Aug. 1977).

Buringh (1960) divides the solonchaks of Iraq into the following groups (applicable also to Iran and Afghanistan):

a) Flooded solonchaks in shallow depressions, flooded in winter, with a thick salt crust on the surface.
b) Internal solonchaks with some salt concentrations in deeper parts of the soil (Tab. 1); annuals and shallow-rooted dwarf-shrubs may grow successfully.
c) External solonchaks, more saline in upper layers and with a salt crust on the surface, but saline throughout the soil profile. These habitats support several kinds of halophytes varying in degree of salt tolerance.
d) Solonchaks with a puffed of fluffy surface layer, almost bare of vegetation in our areas (Tab. 1), often very alkaline.

e) Sabkha and playa soils with a rather high salinity and with some magnesium sulfate in the permanently moist surface layer. These soils are often densely vegetated by meso-halophytes reaching fresh water table (Fig. 5), such as *Tamarix* and *Halocnemum*.

Alkali soils of solonez are rare in the area being considered. They are alkaline, but normally do not exceed a pH of 9,0. The vegetation is very sparse (e.g., *Camphorosma lessingii*). Ecological factors such as aeration, seasonality of flooding, organic matter balance, water table movement and movement of salts within the soil profile are probably important, but need more specific investigation.

FLORA AND VEGETATION

Vast areas of inner Iran are almost bare of vegetation. These are mostly saline plains with soils of high salt content, (Fig. 4) and fine-textured. Despite the high number of halophytic species, the associations have low diversity, are often monotonous and repeat themselves over vast expanses. In the deserts of Iran and Afghanistan the following groups can be distinguished and we include here also for better comparison the low-salinity vegetation types:

I. *Halophytic vegetation*
   1. saline flats (almost bare of vegetation, soil with very high salt content)
   2. euhalophytic vegetation (halo-hammada, on gravel, sand, but mainly clayey soils)
   3. mesohalophytic vegetation (less salt accumulation within soil profile)

II. *Gypsophilous vegetation* (desert and semi-desert, often with dwarf-shrubs, on gypsiferous soils)

III. *Psammophytic vegetation* (erg; on sand dunes and other sandy areas)

IV. *Vegetation on loess* (predominantly covered by semi-desert vegetation rich in ephemerals)

V. *Vegetation on scree, gravel, rocks* (hammada, serir and reg)
   1. desertic vegetation, non-saline (hammada)
   2. semi-desert regions (Artemisia semi-desert, very common in the foothills and lower mountain regions)
   3. shrubby semi-desert (only in few, rather inaccessible areas)
   4. open woodlands (semi-desert with single, small trees)

VI. *Mountain vegetation* (in higher areas, normally above 3000 m)
   1. sub-alpine cushion shrublands and semi-deserts
   2. alpine semi-desert and steppe

VII. *Vegetation of oases and river margins* (with additional water supply).

The playas are most saline in the center of the basin. *Halocnemum strobilaceum*, together with *Seidlitzia rosmarinus*, dominate on the most saline sites in these areas as well as in the Kavirs, in eastern Iran (Fig. 4, 5), Sistan and in northern Afghanistan. These species grow on the most extrem habitats - for instance, in Dasht-e-Lut and at the margins of numerous salt pans of the Kavirs. The vegetation of these regions is clearly arranged according to grain size and distribution of salt and gypsum layers within the profile. Enormous areas occur in the northern Kavirs where *Halocnemum strobilaceum*, sometimes with *Seidlitzia rosmarimus*, is surrounded by a second belt with *Haloxylon aphyllum* (Fig. 6). *Anabasis setifera* also occurs in larger patches in places on the fringe of the depressions. *Aeluropus lagopoides, Atriplex verrucifera, Limonium iranicum, Lycium ruthenicum, Fortuynia garcinii, Alhagi* and *Salsola spp.* form a mosaic of communities on sites of intermediate salinity and intergrade to semi-deserts with *Zygophyllum eurypterum* or *Artemisia herba-alba s.l.* (Fig. 6).

Figure 5. *Halocnemum strobilaceum, Halostachys caspica* and *Seidlitzia rosmarinus*, forming a rather dense mosaic of halophytic shrubs where high water tables occur at the north eastern margin of the Kavir, Iran. The very saline soil surface is puffy and rugose (photo: Breckle, Aug. 1977).

Enormous portions of the saline flats are bare of higher vegetation. *Halostachys caspica* and several *Tamarix spp.* sometimes dominate near water courses or on alluvial fans with lower water tables and often moist soils. *Reaumuria fruticosa* occupies dry, salty ridges. *Seidlitzia rosmarinus, Salsola dendroides, Anabasis setifera, Halocharis spp., Haloxylon salicornicum, Reaumuria persica, Aellenia glauca, Alhagi maurorum* were reported from other parts of the kavirs on saline flats (Zohary 1963). *Tamarix spp., Prosopis farcta* and a salt-tolerant subspecies of *Artemisia herba-alba* are found on disturbed sites (Frey & Probst 1979). Mobayen (1975) mentions *T. aphylla* and *T. macrocarpa*, as occurring with *Calligonum sp.* on the margins of the more sandy and gravelly parts of the Dasht-e-Lut. The center of Dasht-e-Lut seems to be bare of any vegetation and was called a truly abiotic region (Hallier 1975, 1978).

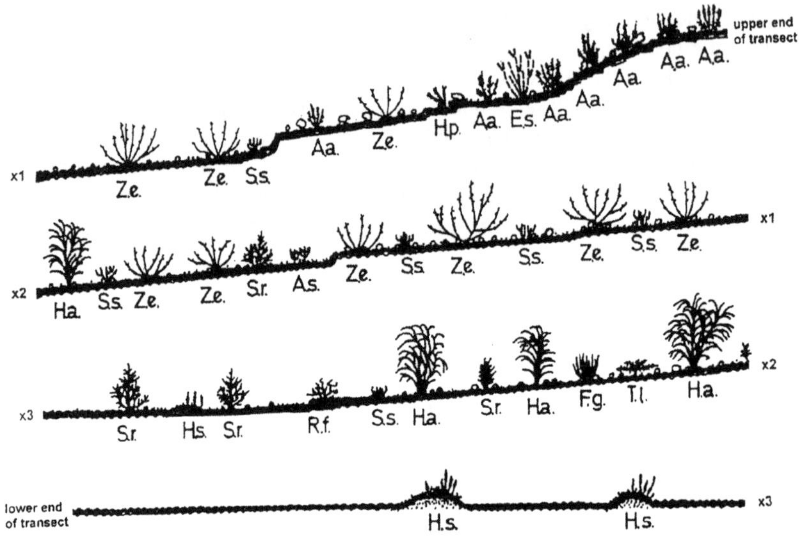

Figure 6. Schematic vegetation profile through the north eastern margin of the Great Kavir with a typical sequence of halophytic vegetation types along most large playa margins in Iran (Breckle 1983). Plant species: A.a.: *Artemisia herba-alba* (s.l.); A.s.: *Anabasis setifera*; E.s.: *Ephedra strobilacea*; F.g.: *Fortuynia garcinii*; H.a.: *Haloxylon aphyllum* (up to 3 m high); H.p.: *Hulthemia persica*; H.s.: *Halocnemum strobilaceum*; R.f.: *Reaumuria fruticosa*; S.r.: *Seidlitzia rosmarinus*; S.s. *Salsola spec.* (*S. tomentosa* etc.); T.l.: *Tribulus longipetalus ssp. macropterus*; Z.e.: *Zygophyllum eurypterum*.

Zohary (1963) found vegetal cover of less than 0,1% around Dasht-e-Lut. It consisted of only a few *Anabasis setifera, Salsola baryosma, S. dendroides, Cornulaca leucantha* and *Tribulus longipetalus*. More to the north, he found these plus *Suaeda dendroides, Seidlitzia rosmarinus* and *Haloxylon aphyllum*. A more detailed description and a vegetation map of the Dasht-e-Lut can be found in Mobayen (1976). *Halocnemum strobilaceum*, the most salt-tolerant and most abundant halophytic plant in northern Iran, Afghanistan and whole Central Asia, is less common in southern Iran and the Arabian Peninsula.

The Iranian Kavir is a vast system of plains. Parts of it have been under protection since 1964. This constitutes one of Iran's National Parks (Firouz 1964). The Kavir National Park covers an area of 6094 km^2 and has deserts, semi-deserts, dry steppes and semi-desertic mountains, all within a mosaic of plains and rocky areas (Vahedi

1976). 283 species of higher plants have been found in that area, more than 50% of which are annuals (Rechinger & Wendelbo 1976). The family with the greatest representation is the Chenopodiaceae (35 species). For comparison, on the dry sea floor of the Aral Sea, in the newly formed Aralkum desert with an area of about 40.000 km^2 since 1960 about 270 species have invaded that area, with 70 species of Chenopodiaceae (Wucherer et al. 1999). Chenopods dominate most communities on all non-sandy areas with annual precipitation less than 150 mm. They exhibit the peculiarity of flowering during the arid season and being active during most of the dry summer.

SALT GRADIENTS AND ZONATION

Playas and sabkhat in the Middle East comprise a wide range of geomorphological features, according to their geology and history. Nevertheless, the often concentric arrangement around these salt flats and their salinity lead to a typical zonation of halophytic communities; some eco-zones are very narrow, and at a range of a few metres one can encounter a whole gradient of salinities. On the other hand, there are areas where changes in salinity are rather gradual, and each of the vegetation types stretches over several kilometres. Furthermore, the reduction in salinity may be linear with distance from the centre, but more often it changes rather abruptly.

The following halophyte-types according to morphology and ecophysiology can be distinguished: stem-succulents (**S**); leaf-succulents (**L**); recreto-halophytes (**X**); pseudo-halophytes (**P**); and non-halophytes (**N**). It is clear that these types are arranged according to salinity-zones, as was checked in various playas in Iran and Afghanistan. Rather similar results were obtained in other regions, though the salinity-gradient varied, being lower or higher, or other features, such as gypsum-crusts or alkaline soils, had modified the general picture. On the average calculated for all 8 regions in Iran and Afghanistan it can be shown that, within the salinity gradient with a salinity variation of 1000 times, the dominant stem-succulents are gradually replaced by leaf-succulents, those by recreters, those by pseudo-halophytes, and on almost salt-free soil non-halophytes predominate (Tab. 2).

The biogeographical distribution influences the occurrence within vegetation types of a distinct area more than salinity levels. Dependent normally on phytogeographical areas, there is a somewhat N-S gradual change with increasing percentages of C4-plants in more southern areas. According to Frey & Kürschner (1983), C4-plants are dominant mainly among xerohalophytes in zones of the transect which are transitional biotopes in respect of edaphic factors.

Table 2. Percentage values of calculated abundancies of halophytic types (S, L, X, P, N, %) arranged according to salinity levels (logarithmic intervals, % NaCl of dry soil), averages from 8 halophytic catenas in Iran and Afghanistan (Breckle 1986).

salt content in soil (%):
15 - 10 - 5 - 3 - 2 - 1 - 0.5 - 0.2 - 0.1 - 0.05 - 0.02 - 0.01 - <0.01%

S	**71**	**63**	**51**	18	3	12	7	4	11	10	0	0
L	29	32	30	**72**	30	**56**	**51**	18	**35**	18	10	0
X	0	5	19	0	**60**	27	28	**41**	4	9	0	2
P	0	0	0	5	7	5	15	23	24	**45**	**68**	10
N	0	0	0	0	0	0	0	14	26	24	22	**88**

A typical halo-catena is characterised by adaptation-types of plants to salinity. Even if in some individual cases or in disturbed areas one member of the catena might only be weakly developed, under natural conditions such a halo-catena can be drawn as a general rule along salt-gradients. The S-L-X-P-N-sequence is apparently a characteristic way to describe the adaptation advantages of halophytes under ecological conditions where competition narrows the physiological amplitude. It remains open which ecophysiological or physiological adaptation in the various halophyte types is governing such competition abilities within the whole ontogenesis of a species.

ECOLOGY OF HALOPHYTES

Usually saline habitats are poorer in species as compared with neighbouring low-salt or salt-free areas. Along seashores all over the world, there is a distinct zonation of such halophytic communities (Chapman 1974, Walter & Breckle 1985). However, in inland areas, only semiarid and arid regions exhibit large saline areas, where various types of halophytic communities occur. Semideserts and desert areas are thus often a mosaic of saline and non-saline stands, according to the typical catenas which develop with time (Walter & Breckle 1984, 1999). Some halophytic localities, especially in endorrheic playas, are in chemical aspects unique. They exhibit an ion pattern, which was accumulated during geological and climatic processes according to the adjacent rocks and mountains with their specific "chemistry". Thus, in some areas, more rare soluble cations as e.g. Li, Rb or anions, as e.g. boric acid, nitrate (as in parts of the Atacama) may play a role. Sulphate or nitrate have other physiological effects on halophytes than NaCl (Freitas & Breckle 1993). Inland salines are, however, mainly characterised by the anion pattern, by the chloride or the sulphate salinity. This determines to a great extent the vegetation types. Especially in Central Asia under continental climates the sulphate salinity is important. the whole mineral metabolism of the halophytic species is influenced and nutrient supply, as well as trace elements, their uptake, their accumulation and physiological effects in the plant tissues may be altered (Joshi et al. 1986).

In studying halophytes all complexity levels (molecular biology with genes and cell membranes, ecophysiology with photosynthesis, respiration and ion compartmentation, ecology with adaptation and formative effects of single plants, population biology and phytosociology with succession and regeneration processes in halophytic plant communities, landscape ecology and global ecology with salt cycling in ecosystems) are important for revealing a better unterstanding of NaCl as an overall ecological factor. Intact sabkha and playa areas with their typical zonation and gradients are best suited for such studies.

ACKNOWLEDGEMENTS

These studies were supported by a joint research programme of the Universities of Bonn and Kabul, by a Schimper Fellowship for extra-European ecological studies (Hohenheim) and by the Department of Environment, Iran. Especially we have to thank Rasik Bhadresa/London for his assistance during field-work. Often we were guests of many of the local people in Iran and Afghanistan; for their kind hospitality and for their help we are very grateful.

References

Adle, A.M. 1960. Régions climatiques et végétations en Iran. Publ. Univ. Teheran No. 626, Teheran.
Bobek, H. 1952. Beiträge zur klimaökologischen Gliederung Irans. *Erdkunde* 6: 65-84.
Bobek, H. 1961. Die Salzwüsten Irans als Klimazeugen. Anz. Österr. Akad. Wiss. Philos.- Hist. Kl. 2: 7-19.
Breckle, S.-W. 1967. Fossile Pflanzenreste am Latahbandpass in Afghanistan. *Science (Kabul)* 4: 16-21.
Breckle, S.-W. 1976. Zur Ökologie und den Mineralstoffverhältnissen absalzender und nicht absalzender Xerohalophyten (in Utah/USA). Diss. Bot. 35: 169 pp.
Breckle, S.-W. 1981. The time scale of salt-accumulation in a desertic hydrotope in north-central Iran. Beih. Tübinger Atlas des Vorderen Orients, Reihe A 8: 51-60.
Breckle, S.-W. 1983. Temperate deserts and semi-deserts of Afghanistan and Iran. In: Goodall, D.W. & West, N. (eds.), Ecosystems of the world, Vol. 5: 271-319. Elsevier, Amsterdam.
Breckle, S.-W. 1986. Studies on halophytes from Iran and Afghanistan. II. Ecology of halophytes along salt-gradients. *Proceed. Roy. Bot. Soc. Edinb.* 89B: 203-215
Buringh, P. 1960. Soils and soil conditions in Iraq. Rep. Iraq Ministry of Agric., Baghdad, 322 pp.
Chapman, V.J. 1974. Salt marshes and salt deserts of the world. 2nd reprint edn. Cramer, Lehre.
Firouz, E. 1964. Environmental and nature conservation in Iran. *Environ. Conserv.* 3: 33-42.
Freitas, H. & Breckle, S.-W. 1993. Accumulation of nitrate in bladder hairs of *Atriplex* species. *Plant Physiol. Biochem.* 31: 887-892.
Frey, W. & Kürschner, H. 1983. Photosyntheseweg und Zonierung von Halophyten an Salzseen in der Türkei, in Jordanien und im Iran. *Flora* 173: 293-310.
Frey, W. & Probst, W. 1979. Flora und Vegetation von Iran. Encyclopedia Persica
Ganji, H.M. 1968. Climate. In: Fisher, W.B. (ed.), The land of Iran. pp. 212-249. Cambridge Univ. Press.
Hallier, U.W. 1975. Die abiotische Zone der Lut/Zentraliran. Unp. Manusc. 15 pp.

Hallier, U.W. 1978. Die Wüste Lut. *Bild d. Wiss.* pp. 46-56.
Herman, N.M., Zillhardt, J. & Lalande, P. 1971. Recueil de données des stations météorologiques de l'Afghanistan. *Publ. Inst. Météorol. Kaboul* 2: 1-58.
Joshi, A., Engenhart, M., Wickern, M. & Breckle, S.-W. 1986. Seasonal changes in the trace metals in salt marsh angiosperms. *J. Plant Physiol.* 128: 173-177.
Krinsley, D.B. 1970. A geomorphological and palaeoclimatological study of the playas of Iran. U.S. Government Printing Office, Washington, D.C.
Lalande, P., Hermann, N.M. & Zillhardt, J. 1974. Cartes climatiques de l'Afghanistan. Publ. Inst. Météorol., Kaboul, 4, 47 pp.
Mobayen, S. 1975. Aspects phytogéographiques de l'Iran et particularités de végétation désertique du Lut. *Envir. Stud. (Teheran)* 4: 181-190.
Mobayen, S. 1976. Structure géobotanique du Loute. Acta Ecol. Iran. 1: 73-86.
Pias, J. 1977. Différentes aspects de l'évolution périglaciale en Afghanistan. *Mém. Soc. Géol. Fr.* 8: 169-178.
Rechinger, K.H. & Wendelbo, P. 1976. Plants of the Kavir Protected Region. *Iran. J. Bot.* 1: 23-56.
Sabeti, H. 1969. Les études bioclimatiques de l'Iran. Teheran No.1231. 266 pp.
Salem, M.Z. & Hole, F.D. 1969. Soil geography and factors of soil formation in Afghanistan. *Soil Sci.* 107: 289-295.
Scharlau, K. 1958. Zum Problem der Pluvialzeiten in Nordost-Iran. Z. Geomorphol. 2: 258-277.
Scharlau, K. 1969. Iran. In: Westermann Lexikon der Geographie. Westermann, Braunschweig.
Schweizer, G. 1975. Der Naturraum. In: Gehrke, U. & Mehner, H. (eds.), Iran. Erdmann Ländermonogr. 5: 13-39.
Stenz, E. 1941. On the evaporation capacity in Kabul. Min. Public works, Afghan Meteor. Serv. Publ. 2, 8 pp.
Stenz, E. 1957. Precipitation, evaporation and aridity in Afghanistan. Acta Geophys. Polon. 244-263.
Vahedi, M. 1976. Ecological data of the Kavir protected region. pp. 23. Department of Environment, Teheran.
Van Zeist, W. 1967. Late quaternary vegetation history of western Iran. Rev. Palaeobot. Palynol. 2: 301-311.
Walter, H. 1974. Die Vegetation Osteuropas, Nord- und Zentralasiens. Vegetationsmonographien VII. Fischer, Stuttgart.
Walter, H. & Breckle, S.-W. 1984. Ökologie der Erde. Bd. 2 Spezielle Ökologie der Tropischen und Subtropischen Zonen. UTB Große Reihe, 2. Aufl. (1990) Fischer, Stuttgart.
Walter, H. & Breckle, S.-W. 1985. Ecological systems of the geobiosphere, vol. 1. Ecological principles in global perspective. Springer, Berlin.
Walter,H. & Breckle, S.-W. 1999. Vegetation und Klimazonen. Ulmer, Stuttgart.
Walter, H. & Lieth, H. 1960-1967. Klimadiagramm-Weltatlas. Fischer, Jena.
Wucherer, W., Breckle, S.-W. & Dimeyeva, L. 2000. Flora of the dry sea floor of the Aral Sea. In: Breckle, S.-W., Veste, M. & Wucherer, W. (eds.), Sustainable Land use in deserts. Springer, Heidelberg.
Zohary, M. 1963. On the geobotanical structure of Iran. Bull. Res. Counc. Isr., Sect. D, Bot. Suppl. 1: 1-113.
Zohary, M. 1973. Geobotanical foundations of the Middle East. Fischer, Stuttgart.

SALT TOLERANT PLANTS OF COASTAL SABKHAT OF PAKISTAN

M. AJMAL KHAN & BILQUEES GUL

Department of Botany, University of Karachi, Karachi-75270, Pakistan

Abstract

Sabkhat of subtropical maritime deserts of coastal Pakistan are dry for most of the year, seasonally wet, and generally poor in species. About 108 halophytic species belonging to 36 families are found in this region. About half of them are hydrohalophytes and a quarter belongs to psammophytes. Dominant life forms are herbaceous perennials, perennial grasses, and shrubs. *Suaeda fruticosa, Arthrocnemum macrostachyum, and Halopyrum mucronatum* are the most common and dominant species of the sabkhat. *Arthrocnemum macrostachyum, Suaeda fruticosa*, and *Cressa cretica* were most abundant in the region of frequent tidal inundation; this region has correspondingly lower soil salinity and pH. Areas with infrequent inundation were dominated by *Zygophyllum simplex* and *Salsola imbricata*. Coastal dunes were dominated by *Halopyrum mucronatum*.

INTRODUCTION

The coastline of Pakistan (ca. 900 km) can be divided into two areas: the Balochistan coast, which contributes the largest segment (640 km) beginning at Bella (including Makran) and extending westward to the Iranian border, and the Sindh coast which contributes about 240 km and ranges from Karachi eastward to the Indian border (Fig. 1). Almost the entire Sindh coast is dominated by the mangrove *Avicennia marina* and the area covered by mangrove is estimated to be about 410.000 ha (Ansari 1987). The Balochistan coast, reported to be one of the most arid coastlines in the world (Chapman 1974), has little or no mangrove development (the exception being the Miani Hor estuary). Ahmed (1951) divided the coastal area of

Barth & Böer (eds.), SABKHA ECOSYSTEMS, 123-139

Pakistan into two regions: Lower Sindh coast and Makran-Lasbella coast. Annual rainfall decreases from 220 mm to around 25 mm from west to east. The eastern part of the coast receives summer rains while the western part receives winter rains. Rain patterns throughout the coastal area are generally erratic. The mean annual temperature ranges from 32-40°C, with extreme maximum values reaching as high as 50°C. Strong coastal winds are characteristic of the region resulting in highly dynamic dune areas and sand deposition in urban areas. Humidity is high with only occasional dew formation. Overall, the area can be characterised as extremely arid, with very low plant diversity. Coastal sabkhat are present all along Balochistan and the coastal areas in Karachi and its vicinity. Succulent perennials or perennial grasses dominate the vegetation (Chaudhri & Qadir 1958, Chaudhri 1961, Chaudhri et al. 1966, Chaudhri & Chuttar 1966, Qadir et al. 1966, Karim & Qadir 1979). The present study was undertaken to document the halophytic species found in the coastal sabkhat of Pakistan and their ecological relationships.

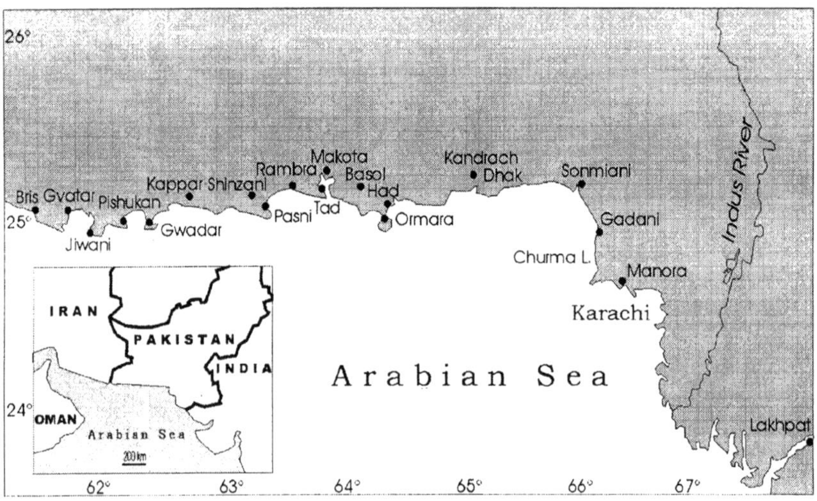

Figure 1. Location of the area under investigation.

METHODS

A reconnaissance of the entire coast was conducted to prepare the list of halophytes found in the area. The Flora of Pakistan (Ali et al. 1999) has provided the basis for taxonomic nomenclature and other relevant details. There are numerous ways of defining halophytes. For our purpose the best definition seems to be "plants that complete their life cycle in saline habitats" (Ungar 1991) where salt concentration of soil solution is about 5 g/l total dissolved solids (7-8 mS cm^{-1}, Aronson et al. 1988). This list is organised alphabetically by botanical family, and within each family, by genus and species.

Life form: Only one life form is given per species, even though many species show a certain amount of plasticity in this regard. A = annual, AQ = aquatic, CH = chamaephyte (small shrub, under 0.5 m average height), CHN = nano-chamaephyte (dwarf shrub, under 0.25 m average height), H = hemicryptophyte, HP = herbaceous perennial, PG = perennial grass, SH = shrub, T2 = tree, average height 2 m, T4 = tree, average height 4 m, and T(x) = tree, average height (x) m.

Plant Type: This category bases on the habitats in which the taxon is distributed. Hyphal = hydrohalophytes (present in salt marshes), Xeorh = Xerohalophyte = salt desert species, Psamm. = Psammophytes (sand loving plants found on littoral or inland sand dunes), Xero = xerophytes (desert species suspected as halophytes), Chasm = Chasmophytes (cliff-dwelling species), Weedy = fugitive species, Phrea = phreatophytes.

Max mS cm^{-1} : This heading gives the maximum reported salinity tolerance of a taxon as cited in the references cited.

RESULTS

The coastal sabkhat of Pakistan are represented with about 108 halophytic species distributed in 36 families (Fig. 2). Grasses dominated (23) the area followed by the sedges (14) and chenopods (13) while the other 33 families are represented by five or less than five species (Fig. 2, Appendix). About 380 halophytic species are found in Pakistan and the halophytes from coastal sabkhat constitute about 30% of this halophytic flora. In addition among the total 107 halophytes about 46 of them could be classified as hydrohalophytes, followed by psammohalophytes (24), xerophytes (18), xerohalophytes (10), weedy (5) and phreatohytes (2,). Life forms of these plants also showed a high degree of variation (Appendix). Herbaceous perennial (32) is the dominant life forms of Pakistani coastal halophytes, followed by trees (21), perennial grasses (18), shrubs (15), annuals (13), and others like aquatic, chaemaephytes, geophyte, hemi-cryptophytes and small trees are represented by less than five individuals (Appendix).

The diversity of plants along the coastal sabkhat is very low. Vegetation on the entire Sindh coast is dominated by *Avicennia marina.* While less common species include *Ceriops tagal, Rhizophora mucronata, Agiceras corniculata, Salicornia brachiata, Arthrocnemum macrostachyum, Suaeda fruticosa, Cressa cretica,* and *Halopyrum mucronatum*, other species are rare in occurrence.

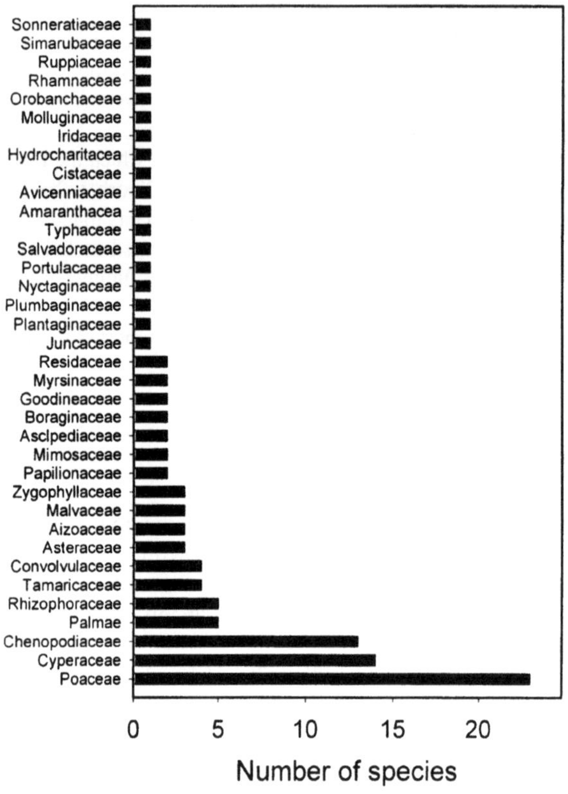

Figure 2. Distribution of halophytes among various families.

Only three species *Suaeda fruticosa*, *Arthrocnemum macrostachyum* and *Halopyrum mucronatum* could be considered the dominant species of the Balochistan region of the Pakistani coast. The species with secondary dominance would include *Cressa cretica,* and *Salsola imbricata*. The remaining 102 species, can be considered uncommon. Coastal areas near Karachi are primarily dominated by *Arthrocnemum macrostachyum, Suaeda fruticosa, Halopyrum mucronatum,* and

Cressa cretica. In addition to these common species other less abundant species like *Aeluropus lagopoides, Atriplex stocksii, Prosopis juliflora, Ipomaea pescaprae, Limonium stocksii, Avicennia marina, Salvadora persica,* and *Arthrocnemum indicum* can be found in pure stands. Gadani and Sonmiani coastal areas (Fig. 1) consist of hills, sand dunes and marshes. Marshes are dominated by either *Arthrocnemum macrostachyum* or *Arthrocnemum indicum*. Sandy shores primarily inhabited by *Suaeda fruticosa*. Ormara (Fig. 1) is characterised by high ambient temperature and high wind velocity causing frequent soil erosion. Large sand dunes were dominated by *Halopyrum mucronatum, Paganum hermala* and *Suaeda fruticosa*.

Areas around Pasni (Fig. 1) are also characterised by high wind and dust storms. This area has both permanent and embryonic sand dunes. Western dunes have soil red in color and dominated by *Halopyrum mucronatum* and *Indigofera argentia*. Eastern parts of the Pasni area have large stable dunes with small embryonic dunes and coastal marshes. Marshy areas are dominated by *Arthrocnemum macrostachyum* and *Cressa cretica*. Whereas, sand dunes are inhabited by *Suaeda fruticosa* and *Halopyrum mucronatum*. The next vegetation zone studied is near Gwadar (Fig. 1). This area is also very hot and has high winds. Soil erosion is very common. The dominant vegetation of this area was *Halopyrum mucronatum, Suaeda fruticosa,* and *Cyperus sp.* Jiwani is the driest area with annual rainfall less than 25 mm/yr and very high-sustained humidity. This study area had a flat plain, a very low water table, and was relatively less warm compared to other Balochistan localities. The average wind velocity is also relatively low. The dominant vegetation of the area consists of pure stands of *Halopeplis perfoliata, Arthrocnemum macrostachyum,* and *Halopyrum mucronatum*.

DISCUSSION

The coastline of Pakistan can be divided into two socio-political and physiographic provinces, geographically divided by the city of Karachi. To the east lies the province of Sindh, with vegetation dominated primarily by mangrove species. To the west lies the province of Balochistan, which is characterised by halophytic vegetation. In addition to vegetation differences, the two provinces differ markedly in their climate. The average rainfall on the western edge of the Sindh coast is regionally high (about 220 mm/yr). Proceeding west along the Balochistan coast, rainfall drops to only 25 mm/yr, making it one of the most arid coastlines in the world. Thus, the Balochistan coast is extremely dry and consists of halophytic vegetation distributed quite sparsely.

Tolerance of salinity by halophyte seeds may be expressed either as the ability of un-germinated seeds to tolerate high salinity without loosing viability or the ability of seeds to germinate at high salinities (Khan & Ungar 1995). Seeds of halophytes do not only germinate at higher salinities but also remain viable for long periods of time when immersed in saline water (Khan & Ungar 1996). Halophytes vary in their upper limit of salt tolerance and increase in salinity usually delays their germination (Ungar 1995). Seeds of salt marsh species like *Atriplex stocksii,* and *Zygophyllum simplex* show little germination above 7300 mg/l NaCl (Khan & Rizvi 1994, Khan & Ungar 1996, Khan et al. 2000a), however, species like *Aeluropus lagopoides, Haloxylon stocksii, Sporobolus iocladus, Suaeda fruticosa, Limonium stocksii,* and *Urochondra setulosa* could germinate in up to 29200 mg/l NaCl (Gulzar & Khan 2001, Gulzar et al. 2000, Khan & Rizvi 1994, Khan & Ungar 1997a, 1998, Khan, unpublished data). A third group of species like *Arthrocnemum macrostachyum, Cressa cretica, Salicornia brachiata, Salicornia bigelovii, Salsola iberica,* and *Tamarix pentandra* could germinate at 46720 mg/l or higher NaCl concentrations (Khan & Gul 1998, Khan 1991, 1999, Ungar 1995). Species like *A. stocksii, H. stocksii,* and *S. fruticosa* could be classified as moderately salt tolerant and *A. macrostachyum* and *C. cretica* as highly salt tolerant. Sharma & Sen (1989) observed an extremely fast germination in the seeds of *Haloxylon stocksii* and *H. salicornicum,* occurring within an hour. An ecophysiological adaptive role is assigned to such a phenomenon of germination, which appears like uncoiling of the young embryo out of the testa immediately after contact with water with an unusually high rate of cell elongation, soon after imbibition. Such quick seed germination indicates the adaptive strategy by the plants as the availability of water with reduced NaCl content in soil during the rainy season is for a short duration. Increase in salinity leads to dormancy of seeds in halophytes and glycophytes. More investigations with halophytes (Ungar 1995) have demonstrated that seeds of several species, including *Arthrocnemum macrostachyum, Salicornia bigelovii, Salicornia brachiata, Cressa cretica, Tamarix pentandra, Aeluropus lagopoides, Atriplex stocksii, Haloxylon stocksii, Sporobolus iocladus, Suaeda fruticosa, Limonium stocksii* and *Urochondra setulosa* remained dormant at high salinity and these will germinate when returned to distilled water (Khan 1999, Khan & Gul 1998, Khan & Ungar 1996, Khan & Ungar 1997abc, Khan & Ungar 1998, Ungar 1995).

Salt tolerance of species varies with the stage of their development. Some species like *Suaeda fruticosa, Haloxylon stocksii, Atriplex stocksii* and *Zygophyllum simplex* were not very highly salt tolerant at germination but showed a high salinity tolerance at growth stage (Khan 1999, Khan & Ungar 1996, Khan & Ungar 1997abc, Khan & Ungar 1998, Ungar 1995, Khan et al. 2000abc). While other species like *Arthrocnemum macrostachyum,* and *Cressa cretica* showed a higher degree of salt tolerance both at germination and growth stages (Khan 1991, Khan & Aziz 1998, Khan & Gul 1998, Khan & Ungar, unpublished data).

A number of different mechanisms are used by halophytes to achieve osmotic adjustment, including inorganic ion accumulation, synthesis or accumulation of organic compounds and water loss (Ungar 1991). Classification schemes have been constructed which attempts to match morphological and physiological characters to specific halophyte habitats and growth strategies (Breckle 1983). However, all these classifications have little predictive value. Under physical or physiological stress conditions, the leaves of saline plants play an important role and develop certain xeromorphic adaptive characteristics like succulence, reduction in surface area, thick cuticle, or a cover of waxy layers on epidermis, hairs on stem and leaves, sunken stomata, and salt glands, etc. The succulents often lack the ability to secrete salts but they thwart the rise of salt concentration by an increase of their water content, and they become more and more succulent during their development. *Arthrocnemum, Haloxylon, Heliotropium, Salicornia, Suaeda, Salsola*, and *Zygophyllum* are prominent succulent halophytic genera. *Suaeda fruticosa, Salsola imbricata, Haloxylon stocksii, Cressa cretica, Sporobolus iocladus, Urochondra setulosa*, and *Aeluropus lagopoides* are characterised by thick cuticle and a cover of waxy layers, while stem and leaves of last five species remain covered with hairs. Only a small number of halophytes are able to excrete salts through glandular cells. Liphschitz and Waisel (1982) listed active secreting glands in *Avicennia, Aeluropus, Aegiceras, Limonium, Bruguiera, Tamarix* and *Reaumaria*. A similar function of salt recretion is ascribed to bladder trichomes of some Chenopodiaceae, e.g. *Atriplex* species. The basic role of bladders is the protection of young developing shoots and leaves from toxic salt levels first in the apoplast and subsequently in the symplast.

It has been assumed that survival of plants in saline environments depends upon the altered biochemical relations and on the quantitative ratio between toxic and protective compounds like betaine. Khan et al. (1998) while studying *Halopyrum mucronatum, Atriplex stocksii, Haloxylon stocksii* and *Suaeda fruticosa* found high accumulation of betaine with the corresponding increase in salinity. The betaine is said to function as a source of solute for intracellular osmotic adjustment. Betaine accumulation occurs in the tissues of plants exposed to a saline substrate and there is a positive correlation between betaine content and the amount of Na^+ and Cl^- in the cell sap. It is also estimated that about 11.680 mg/l plant water or more betaine concentration is needed to successfully achieve osmotic adjustment under saline conditions and most of the Pakistani species tested have betaine concentration higher then this.

ECONOMIC POTENTIALS OF LOCAL HALOPHYTES

Promising species with economic utilisation could be selected from the coastal sabkhat flora according to their structure, chemical content or other useful properties. Potential utilisation of halophytes grown with saline irrigation, which include crops for food, feed, fodder, wood, chemical, landscaping and ornamentals. Fruits of *Salvadora oleoides* and *Avicennia marina* are used as food among the coastal populations in Pakistan. The young leaves and shoots of *Salicornia bigellovi, S. brachiata, Sesuvium portulacastrum* and *Portulaca oleracea* have also been used for vegetables, salads and pickles. *Suaeda fruticosa* is used to prepare a kind of baking soda,, which is used in preparation of food. Radicles of *Rhizophora*, *Bruguiera* and *Ceriops*; tender leaves of *Thespesia populnea*, and fruits or karnals of littoral species such as *Terminalia catappa* are consumed by the local populations.

The foliage of *Aegiceras corniculata*, *Avicennia marina*, *Sonneratia alba*, *Ceriops tagal*, and *Rhizophora mucronata* is a high quality forage for livestocks and camels in particular. Many species of *Salicornia*, *Chenopodium*, *Atriplex*, *Salsola*, *Suaeda* and *Kochia* are common fodder shrubs. Among grasses *Aeluropus lagopoides*, *Dactyloctenim sindicum*, *Cynodon dactylon*, *Paspalum vaginatum*, *Sporobolus marginatus*, *Chloris gayana*, *C. virgata*, *Echinochloa turnerana*, *E. colonum* and *Puccinellia distans* are common species found in saline and alkaline areas and used as forages.

Production of vegetable oil from seed-bearing halophytes appears promising. Seeds of various halophytes like *Suaeda fruticosa*, *Arthrocnemum macrostachyum*, *Salicornia bigelovii*, *S. brachiata*, and *Haloxylon stocksii* contain sufficient quantity of high quality edible oil with un-saturation ranging from 70 – 80% (Weber et al. 2001, Khan, unpublished data). Seeds of *Salvadora oleoides* and *S. persica* contain 40-50% fat and are good source of lauric acid. Purified fat is used for soap and candle making and is a potential substitute for coconut oil. Quite often fuel wood is obtained from salt tolerant trees and shrubs, which may include species of *Prosopis*, *Tamarix*, *Salsola*, *Suaeda*, *Kochia*, *Capparis* and *Salvadora* (Dagar 1995). In coastal areas the mangroves are used frequently for fuel and timber, which has contributed a lot to deforestation of these habitats. Species of *Rhizophora*, *Ceriops*, *Avicennia*, and *Sonneratia*, are excellent fuel woods and also contribute to form charcoal. Some of these halophytes are conventionally used as medicinal plants such as to relieve itches (*Salsola imbricata*); cough (*Salvadora oleoides*); stomach upset (*Thespesia populnea*); to cure eyes ailments (*Zygophyllum simplex*) etc.

Fresh water resources are becoming increasingly limited and water for agricultural irrigation systems will steadily increase in salinity in the near future. It is about time to develop sustainable biological production system which could use seawater for irrigation of halophytic crops in coastal saline lands. The coastal area of Balochistan has about 108 halophytic species and therefore a promising biological potential.

RECOMMENDATIONS

- Investigation of local halophytic crops on their suitability as feed, food, fibre, forage, turf, and oil-seed (see economic potentials above).
- Research on selected species regarding their suitability of seawater irrigation for the specific area (for species see economic potentials above).
- Establishments of demonstration sites in order to introduce the research results and new techniques to the farmers, administrators and the public.

- Governmental support to encourage farmers to introduce non-conventional halophytic crops using seawater for irrigation, which could lead to considerable economic improvement of the region.

References

1. Abbas, J.A. & El-Oqlah, A.A. 1992. Distribution and communities of halophytic plants in Bahrain. *Journal of Arid Environments* 22: 205-218.
2. Abdel-Razik, M.S. & Ismail, A.M.A. 1990. Vegetation composition of a maritime salt marsh in Qatar in relation to edaphic features. *Journal of Vegetation Science* 1: 85-88.
3. Ahmed, K.S. 1951. Climate regions of Pakistan. *Geography Review* 6: 1-35.
4. Ahmed, N. & Chaudhary, G.R. 1988. Irrigated Agriculture of Pakistan. Shahzad Nazir, Lahore.
5. Ahmed, R. 1987. Saline agriculture at coastal sandy belt. University of Karachi, Pakistan.
6. Ali, S.I., Nasir, E. And Qaiser, M. (eds.) 1999. Flora of Pakistan, University of Karachi, Karachi.
7. Ansari, T.A. 1987. Mangroves of the Asia and the Pacific: Status and Management. In: Tech. Rep. UNDP/UNESCO Res. and Training. Pilot Prog. on Mangrove Ecosystem in Asia and Pacific (RAS/79/002), pp. 151-173.
8. Aronson, J. 1989. *HALOPHYTES;* Salt Tolerant Plants for the World - A Computerised Global Data Base of Halophytes with Emphasis on their Economic Uses. University of Arizona Press. Tucson, USA.
9. Aronson, J.A., Pasternak, A. & Danon, A. 1988. Introduction and first evaluation of 120 halophytes under seawater irrigation. In: Whitehead, E.E., Hutchinson, C.F., Timmermann, B.N. & Varady R.G. (eds.), Arid lands: Today and tomorrow. Westview, Boulder, Colorado. pp. 737-746.
10. Aziz, S. 1994. Population biology of a desert perennial halophyte *Cressa cretica*. M. Phil thesis. University of Karachi, Pakistan.
11. Aziz, I. & M.A. Khan. 2000. Physiological adaptations to seawater concentration in *Avicennia marina* from Indus delta, Pakistan. *Pakistan Journal of Botany* 32: 151-170.
12. Baro, D. & Medrano, F.G. 1985. Halophytic flora and vegetation of Tamaulipas, Mexico. Unpublished manuscript.
13. Barros, M. 1960. Las ciperáceas del Estado de Santa Catalina. *Sellowia* 12:182-450.
14. Batanouny, K H. & Zaki, M. 1974. Edaphic factors and the distribution of plant associations in a sector in the coastal Mediterranean zone in Egypt. *Phyton* 15: 193-202.
15. Batanouny, K.H. 1981. Ecology and flora of Qatar. University of Qatar, Doha.
16. Baum, B. 1978. The genus *Tamarix*. Israel Academy of Sciences and Humanities, Jerusalem.
17. Bernstein, L., Francois, L.E. & Clark, R.A. 1972. Salt tolerance of ornamental shrubs and ground covers. *Journal American Horticultural Science* 97: 550-556.
18. Boelcke, O. 1981. Plantas vasculares de la Argentina nativas y exóticas. Hemisferio Sur, Buenos Aires.
19. Boyko, H. 1964. Principles and experiments regarding irrigation with highly saline and seawater without desalinization. In: Transactions of the New York Academy of Sciences Series II New York Academy of Sciences, New York. pp. 1087-1102.
20. Bramwell, D. & Bramwell, Z. 1974. Wildflowers of the Canary Islands. Stanley Thorne Ltd, London.
21. Breckle, S.-W. 1983. Temperate deserts and semi-deserts of Afghanistan and Iran. In: Goodall, D.-W. & West, N. (eds.), Ecosystems of the world 5: Elsevier, Amsterdam. pp. 271-319.
22. Bustanai, M., Waisel, Y., Kuller, Z., Guggenheim, J. & Tillman, A.G. 1971. The use of saline water for growing Rhodes grass on coastal sand dunes. *Hassadeh* 51: 603-605.

23. Chapman, V.J. 1974. Salt marshes and salt deserts of the world. Stechert-MacMillan, Pennsauken, NJ.
24. Chaudhri, I.I. 1961. The vegetation of Karachi. *Vegetatio* 10: 229-246.
25. Chaudhri, I.I. & Chuttar, M.S. 1966. The vegetation and range flora of Thar desert. The West Pakistan Forest Department, Hyderabad, 165 pp.
26. Chaudhri, I.I., Hussain, S.I. & Shah, B.H. 1966. The vegetation of riverian tract of Indus around Ghulam Mohammad Barrage, West Pakistan. *Vegetatio* 10: 310-338.
27. Chaudhri, I.I. & Qadir, S.A. 1958. Sand dune vegetation of coastal regions of Karachi. *Pak. J. For.* 8: 337-341.
28. Choukr-Allah, R. 1996. The potential of halophytes in the development and rehabilitation of arid and semi-arid zones. In: Choukr-Allah, R., Malcolm, C.V., Hamdy, A. (eds.), Halophytes and biosaline agriculture. Marcel Dekker, Inc. New York. pp 3-16.
29. Cooper, A. 1982. The effects of salinity and waterlogging on the growth and cation uptake of salt marsh plants. *New Phytologist* 90: 263-275.
30. Corre, J.J. 1985. Institut de Botanique, Université de Montpellier, France. Personal communication to Pasternak, D., in Pasternak, D. 1989.
31. Dafni, A. 1975. Comparative biology of *Prosopis farcata* in four different habitats in Israel. Hebrew University, Jerusalem. Ph.D. diss.
32. Dagar, J.C. 1995. Characteristics of halophytic vegetation in India. In: Khan, M.A. & Ungar, I.A. (eds.): Biology of salt tolerant plants. University of Karachi, Karachi. pp. 255-276.
33. Daoud, H.S. 1985. Flora of Kuwait. In: Dicotyledoneae Vol.: 1.
34. Eshel, A. 1985. Response of *Suaeda aegyptiaca* to KCl, NaCl and Na_2SO_4 treatments. *Physiologia Plantarum* 64: 308-315.
35. Fakhary, A.M. 1994. Species richness and diversity in the vegetation of the Western Mediterranean Desert of Egypt. Ph. D. Thesis, University of Alexandria, Alexandria.
36. Farquhar, G.D., Ball, M.C., Von Caemmerer, S. & Roksandic, Z. 1982. Effect of salinity and humidity on $d^{13}C$ value of halophytes - evidence for diffusional isotope fractionation determined by the ratio of intercellular atmospheric partial pressure of CO_2 under different environmental conditions. *Oecologia* 52:121-124.
37. Feinbrun, N. 1978. Flora Palaestina. Part III. Israel Academy of Science, Jerusalem.
38. Feinbrun, N. 1986. Flora Palaestina. Part IV. Israel Academy of Sciences, Jerusalem.
39. Felger, R.S. & Moser, M.B. 1973. Eelgrass (*Zostera maritima* L.) in the Gulf of California: Discovery of its nutritional value by the Seri Indians. *Science* 181: 355-356.
40. Felker, P., Clark, P.R. Lagg, A.E. & Pratt, P.F. 1981. Salinity tolerance of the tree legumes: Mesquite *(Prosopis glandulosa* var. *torreyana, P. velutina* and *P. articulata*), Algarrobo (*P. chilensis*), Kaiwe (*P. pallida*) and Tamarugo (*P. tamarugo*) grown in sand cultures on nitrogen free media. *Plant and Soil* 61: 311- 317.
41. Ferguson, C.R. 1952. Salt-tolerant plants for south Florida. Florida State Horticultural Society. pp. 306-313.
42. Glenn, E.P. Brown, J.J. & Blumwald, E. 1999. Salt tolerance of crop potential of halophytes. *Critical Reviews in Plant Sciences* 18: 227-255.
43. Glenn, E.P., Tanner, R., Miyamoto, M., Fitzsimmons, K., & Boyer, J. 1998. Water use productivity and forage quality of the halophyte *Atriplex nummularia* grown on saline waste water in a desert enviroment. *Journal of Arid Environments* 38: 45-62.
44. Gulzar, S. & M.A. Khan. 2001. Seed germination of a halophytic grass *Aeluropus lagopoides*. *Annals of Botany* 87: 319-324.
45. Gulzar, S., M.A. Khan & I.A. Ungar. 2001. Effect of temperature and salinity on the germination of *Urochondra setulosa*. *Seed Science & Technology* 29: 21-29.
46. Hoffman, G.J. & Shannon, M.C. 1986. Relating plant performance and salinity. *Rehabilitation & Revegetation Research* 5: 211-225.

47. Joshi, G.V., B.B. Jamale & Bhosale, L.J. 1975. Ion regulation in mangroves. In: Walsh, G.E., Snedaker, S.C. & Teas, H.J. (eds.): Proc. Int. Symp. Biol. & Management of Mangroves. Vol.2. Int. Food Agric. Sci. Univ. Florida, Gainesville. pp. 595-607.
48. Karim, A. & Qadir, S.A. 1979. Phytosociological studies of the coastal swamps near Karachi. *Bangladesh Journal of Botany* 8: 20-29.
49. Kartawinata, K, Adisumarto, S., Soemodihardjo, S. & Tantra, I.G.M. 1979. Status pengetahuan hutan bakau Indonesia. (The state of knowledge in the Indonesian Mangrove forest.).
50. Kassas, M. & Zahran, M.A. 1967. On the ecology of the Red Sea littoral salt marsh, Egypt. *Ecological Monograph* 37: 297 - 316.
51. Khan, M.A. 1991. Studies on germination of *Cressa cretica* L. seeds. *Pakistan Journal of Weed Science Research* 4: 89-98.
52. Khan, M.A. 1999. Comparative influence of salinity and temperature on the germination of subtropical halophytes. Halophyte Uses in different climates I: Ecological and Ecophysiological Studies. *Progress in Biometeriology* 13: 77-88. Edited by H. Lieth, M. Moschenko, M. Lohman, H. -W. Koyro & A. Hamdy. Backhuys Publishers, The Netherlands.
53. Khan, M.A. 2001. An ecological overview of halophytes from Pakistan. In: Lieth, H. & Moschenko, M. (eds.), Cash crop halophytes. Kluwer Academic Press, Netherlands.
54. Khan, M.A. & Aziz, S. 1998. Some aspects of salinity, plant density, and nutrient effects on *Cressa cretica* L. *Journal of Plant Nutrition* 21: 769-784.
55. Khan, M.A & Aziz, I. 2001. Salinity tolerance of some mangroves from Pakistan. *Wetland Ecology and Management* 9: 228-332.
56. Khan, M.A. & Gul, B. 1998. High salt tolerance in the germinating dimorphic seeds of *Arthrocnemum indicum*. *International Journal of Plant Sciences* 159: 826-832.
57. Khan, M.A. & Rizvi, Y. 1994. The effect of salinity, temperature and growth regulators on the germination and early seedling growth of *Atriplex griffithii* Moq. var. *stocksii* Boiss. *Canadian Journal of Botany* 72: 475-479.
58. Khan, M.A. & Ungar, I.A. 1984. Effects of salinity and temperature on the germination and growth of *Atriplex triangularis* Willd. *American Journal of Botany* 71: 481-489.
59. Khan, M.A. & Ungar, I.A. 1995. Biology of Salt Tolerant Plants. Department of Botany, University of Karachi, Pakistan. 424 pp.
60. Khan, M.A. & Ungar, I.A. 1996. Influence of salinity and temperature on the germination of *Haloxylon recurvum*. *Annals of Botany* 78: 547-551.
61. Khan, M.A. & Ungar, I.A. 1997a. Effect of light, salinity, and thermopcriod on the seed germination of halophytes. *Canadian Journal of Botany* 75: 835-841.
62. Khan, M.A. & Ungar, I.A. 1997b. Effect of thermoperiod on recovery of seed germination of halophytes from saline conditions. *American Journal of Botany* 84: 279-283.
63. Khan, M.A. & Ungar, I.A. 1997c. Germination responses of the subtropical annual halophyte *Zygophyllum simplex*. *Seed Science & Technology* 25: 83-91.
64. Khan, M.A. & Ungar, I.A. 1998. Germination of salt tolerant shrub *Suaeda fruticosa* from Pakistan: Salinity and temperature responses. *Seed Science & Technology* 26: 657-667.
65. Khan, M.A., I.A. Ungar & Showalter, A.M. 1999. The effect of salinity on growth, ion content, and osmotic relations in *Halopyrum mucronatum* (L.) Stapf. *Journal of Plant Nutrition* 22: 191-204.
66. Khan, M.A., I.A. Ungar & Showalter, A.M. 2000a. Effects of sodium chloride treatments on growth and ion accumulation of the halophyte *Haloxylon recurvum*. *Communication of Soil Science and Plant Nutrition* 31: 2763-2774.
67. Khan, M.A., I.A. Ungar & Showalter, A.M. 2000b. Growth, water, and ion relationships of a leaf succulent perennial halophyte, *Suaeda fruticosa* (L.) Forssk. *Journal of Arid Environments* 45: 73-84.
68. Khan, M.A., I.A. Ungar & Showalter, A.M. 2000c. Salt tolerance in the subtropical perennial halophyte *Atriplex griffithii* Moq. var. *stocksii* Boiss. *Annals of Botany* 85: 225-232.
69. Khan, M.A., I.A. Ungar, A.M. Showalter & Dewald, H. 1998. NaCl – induced accumulation of

glycinebetaine in four subtropical halophytes from Pakistan. *Physiologia Plantarum* 102: 487-492.

70. Lavranos, J. 1986. Bramley, South Africa. Personal communication to Pasternak, D., in Pasternak, D. 1989.
71. Le Houérou, H.N. 1980. Browse in Africa - the current state of knowledge. International Livestock Center of Africa, Addis Ababa, Ethiopia.
72. Le Houérou, H.N. Forage halophytes and salt-tolerant fodder crops in the Mediterranean Basin. In: Squires, V.R. & Ayoub, A.T. (eds.), Halophytes as a resource for livestock and for rehabilitation of degraded lands. Tasks for vegetation science Vol.: 32. Kluwer Academic Publisher, Dordrecht.
73. Lieth, A.F. 1994. Use of sea water for growth and productivity of halophytes in the Gulf Region. United Arab Emirates University, Faculty of Science.
74. Lindacher, R. 1995. PHANART Datenbank der Gefäßpflanzen Mitteleuropas. Vol.: 125. Geobotanisches Institut der ETH, Stiftung Rübel, Zürich. Veröffentlichung des Geobotansichen Institutes.
75. Liphschitz, N. & Y. Waisel. 1982. Adaptation of plants to saline environments: salt excretion and glandular structure. In: Sen, D.N. & RajPurohit, K.S. (eds.), Contribution to the Ecology of Halophytes. Dr. W. Junk Publishers. The Hague. pp. 197-214.
76. McMillan, C. 1959. Salt tolerante within a *Typha* population. *American Journal of Botany* 46: 1521-1526.
77. Menzel, U. & Lieth, H. 1999. Halophyte Database Vers. 2.0. pp. 77-88. In: Lieth, H., Moschenko, M., Lohman, M., Koyro, H.-W. & Hamdy, A. (eds.), Halophyte Uses in different climates I: Ecological and Ecophysiological Studies. Progress in Biometeriology, Vol. 13. Backhuys Publishers, The Netherlands.
78. Mepham R.H. & Mepham, J.S. 1985. The flora of tidal forests - a rationalization of the use of the term "mangrove.". *South African Journal of Botany* 51: 75-99.
79. Mepham R.H. 1983. Mangrove floras of the southern continents. Part 1. *South African Journal of Botany* 2:1-8.
80. Moghaddam, P.R. & Koocheki, A. 1999. A comprehensive survey of halophytes in Khorasan province of Iran. Unpublished manuscript.
81. Morton, J.F. 1982. Wild plants for survival in south Florida. Fairchild Tropical Garden, Miami, Florida.
82. O'Leary, J.W., Glenn E.P., & Watson, M.C. 1986. Agricultural production of halophytes irrigated with seawater. *Plant and Soil* 89: 311-322.
83. Pasternak, D. & Nerd, A. 1996. Research and utilization of halophytes in israel. In: Choukr-Allah, R., Malcolm, C. V. & Hamdy, A. (ed.), Halophytes and biosaline agriculture. Marcel Dekker, Inc. 270 Madison Avenue, New York. pp. 325-348.
84. Peltier, J.P. 1982. La végétation du Bassin Versant de L'Oued Sous (Maroc). Université Scientifique et Medicaie de Grenoble. Unpublished Ph.D. diss.
85. Qadir, S.A., Qureshi, S.Z., & Ahmad, M.A. 1966. A phytosociological survey of Karachi University Campus. *Vegetatio* 9: 339-362.
86. Quézel, P. 1965. La végétation du Sahara. Fischer, Stuttgart.
87. Qureshi, R.H. & Barrett-Lennard, E.G. 1998. Saline agriculture for irrigated land in Parkistan: A handbook. Australian Centre for International Agricultural Research Canberra, Australia.
88. Ragonese, A.E. & Covas, G. 1947. La flora halofile del sur de la provincia de Santa Fe. *Darwiniana* 7/3:401-495.
89. Saxena, S.K. & Gupta, R.K. 1972. Vegetation of Pachpadra Salt Basin in Western Rajasthan. *Journal of Bombay Natural History* 70/1:104-127.
90. Saxena, S.K., Squire, V.R., & Ayoub, A.T. 1994. Halophytes of the Indian desert-communities on the various 'ranns'. In: Squires, V.R. & Ayoub, A.T. (eds.), Tasks for vegetation science 32: 211-215. Kluwer Academic Publisher, Dordrecht.
91. Schulz, A.G. 1963. Plantas y frutos comestibles de la región chaquena. *Rev. Agronómica Noroeste Argeatino* 4: 57-83.

92. Sharma, S.K. & Gupta, R.K. 1971. Effects of salts on seed germination of some desert grasses. *Annals of Arid Zone* 10: 33-36.
93. Sharma, T.P. & D.N. Sen. 1989. A new report on abnormally fast germinating seeds of *Haloxylon* spp. - an ecological adaptation to saline habitat. *Current Science* 58: 382-385.
94. Shomer-Ilan, A. 1983. Photosynthetic pathways in the flora of Israel and its ecological significance. B. The Gramineae and the Chenopodiaceae in Israel. Rotem Vol. 6: 5-17.
95. Shomer-Ilan, A. 1985. Tel Aviv University, Israel. Unpublished data.
96. St. John, H. 1970. Classification and distribution of the *Ipomoea pescaprae* group (Convolvulaceae). Botanische Jahrbücher für Systematik, Pflanzengeschichte und Pflanzen-geograpbie 89/4: 563-583.
97. Szabolcs I. 1994. Salt affected soils as the ecosystem for halophytes. In: Squires, V.R. & Ayoub, A.T. (eds.), Tasks for vegetation science 32: 19- 24. Kluwer Academic Publisher, Dordrecht.
98. The Royal Botanic Garden. 1993. Index Kewensis. Computerized data base.
99. Toutain, G. 1977. Éléments d'agronomie Saharienne. Institut National de Recherche Agronomique, Marrakech, Maroc.
100. Ungar, I.A. 1991. Ecophysiology of Vascular Halophytes. CRC Press, Boca Raton, 209 pp.
101. Ungar, I.A. 1995. Seed germination and seed-bank ecology of halophytes. In: J. Kigel & Galili, G. (eds.), Seed Development and Germination. Marcel and Dekker Inc. New York. pp 599-627.
102. Vareschi, V. 1979. Plantas entre el may e la tierra. Armitano, Caracas.
103. Weber, D.J., Gul, B., Khan, M.A., Williams, T., Wayman, P. & Warner, S. 2001. Composition of vegetable oil from seeds of native halophytic shrubs. In: McArthur, E. Durant; Fairbanks, Daniel J., comps. 2000. Proceedings: Shrubland Ecosystem Genetics and Biodiversity; 2000 June 13-15; Provo, UT. Proceedings RMRS-P-000. Ogden, UT: U.S. Department of Agriculture, Forest Service Rocky Mountain Research Station.
104. Zahran, M.A. 1973. Comparative eco-physiological studies on *Puccinellia maritima* and *Festuca rubra*. Bulletin Faculty of Science Manura University 1: 43-66.
105. Zahran, M.A. 1977. Africa Wet formations of the African Red Sea coast. In Chapman, V.J. (ed.), Ecosystem of the world. 1. Wet coastal ecosystems. New York: Elsevier. 215 pp.
106. Zohary, M. 1966. Flora Palaestina. Part I. Israel Academy of Sciences, Jerusalem.
107. Zohary, M. 1972. Geobotanical foundations of the Middle East. Fischer, Stuttgart.

Appendix

Alphabetical listing of halophytes from coastal sabkhat of Pakistan.

Genus, Species and Author	Plant Type	Life form	Max mS/cm	References
Aizoaceae				
Aizoon canariense L.	Xero	A		15
Sesuvium sesuvioides (Fenzl.) Verdc.	Psamm.	HP		87
Trianthema triquetra Rottl.ex Willd.	Xero	A		29, 41
Amaranthaceae				
Aerva javanica (Brum. f.) Juss. ex J.A. Schultes	Xero	SH		96
Asclepediaceae				
Leptadenia pyrotechnica (Forssk.) Dene.	Xero	SH		8
Pergularia tomentosa L.	Xero	V		25
Asteraceae				
Launea sarmantosa (Willd.) Alsoton	Psamm.	HP		46
Pulicaria hookeri Jafri	Psamm.	HP		46
Sonchus tenerrimus L.	Xero	HP		84
Avicenniaceae				
Avicennia marina (Forssk.) Vierh.	Hyphal	T8	58	11, 52
Boraginaceae				
Heliotropium bacciferum Forssk.	Xero	HP		79
Heliotropium currassavicum L.	Weedy	HP		12, 34
Chenopodiaceae				
Arthrocnemum indicum (Willd.) Moq.	Hyphal	CHN	90	38, 77
Arthrocnemum macrostachyum (Moric.) C. Koch	Hyphal	CHN	90	54
Atriplex leucoclada Boiss.	Xero	SH	79	8, 82
Atriplex stocksii Boiss.	Xerohal.	SH	36	55, 67
Bienertia cycloptera (Bunge ex Trautv.) B.ex Boiss.	Hyphal	A		15, 79
Cornulaca monocantha Del.	Xerohal.	A		70, 71
Halocnemum strobilaceum (Pall.) M. Bieb.	Xero	CH		15
Halopeplis perfoliata (Forssk.) Bunge ex Schweinf.	Hyphal	HP	10	15, 71
Halostachys belangerana (Moq.) Botsch.	Hyphal	SH		93, 98
Halothamnus iranicus Botsch.	Xero.	CH		6
Haloxylon stocksii (Boiss.) Benth. & Hook.	Psamm.	SH	35	65
Salicornia brachiata Roxb.	Hyphal	CH	59	8, 79
Salsola imbricata Forssk.	Xerohal.	SH		6
Suaeda fruticosa (L.) Forssk.	Xerohal.	HP	97	62, 66
Cistaceae				
Helianthemum lippii (L.) Pers.	Xero	SH		6
Convolvulaceae				
Cressa cretica L.	Hyphal	HP	97	49, 64

Evolvulus alsinoides (L.) L.	Hyphal	HP		81
Ipomoea alba L.	Psamm.	HP		12
Ipomoea pes-caprae (L.) R. Br.	Psamm.	HP		94
Cyperaceae				
Bolboschoenus affinis (Roth.) Drobov	Hyphal	HP		92
Carex stenophylla ssp. *Interrupta* (V. Krecz.) Egor	Hyphal	HP		96
Cyperus arenarius Retz.	Hyphal	HP		6
Cyperus conglomeratus ssp. *Conglomeratus* Rottb.	Hyphal	HP		15, 79
Cyperus conglomeratus ssp. *culvulus* Rottb.	Hyphal	HP		15, 79
Cyperus corymbosis Rottb.	Hyphal	HP		13, 86
Cyperus laevigatus L.	Hyphal	HP		13
Cyperus malaccensis Lam.	Hyphal	HP		47, 80
Cyperus pachyrhizhus Nees ex Boeck	Hyphal	HP		6
Cyperus rotundus L.	Hyphal	HP		46
Cyperus stoloniferous Retz.	Hyphal	HP		38
Fimbristylis cymosa R. Br.	Hyphal	HP		46
Pycerus polystachyos (Rottb.) P. Beauv	Hyphal	HP		46
Schoenus nigricans L.	Hyphal	HP		94
Goodeniaceae				
Scaevola plumieri (L.) Vahl.	Psamm.	SH	60	30, 40
Scaevola taccada (gaertn.) Roxb.	Psamm.	SH	10	82
Iridaceae				
Iris ensata Thunb.	Hyphal	AQ		79
Juncaceae				
Juncus maritimus Lam.	Hyphal	HP	87	73, 82
Malvaceae				
Hibiscus tiliaceus L.	Psamm.	T12	58	30, 77
Thespesia populnea (L) Sol.	Hyphal	T4	58	30, 82
Thespesia populneoides (Roxb.) Kostel.	Hyphal	T6	58	30, 77
Mimosaceae				
Acacia nilotica (L.) Delile	Xero	T6		6
Prosopis juliflora (Swartz)	Xero	T8		36, 39
Molluginaceae				
Glinus lotoides L.	Psamm.	A		6
Moraceae				
Ficus microcarpa L.	Hyphal	T8		6
Myrsinaceae				
Aegiceras corniculata (L.) Blanco	Hyphal	T6	58	30, 34
Ardisia solanacea Roxb.	Hyphal	SH		96
Nyctaginaceae				
Pisonia grandis R. Br.	Psamm.	T4		1

Orobanchaceae				
Cistanche tubulosa (Schrenk) Hook	Hyphal	A		79
Palmae				
Cocos nucifera L.	Hyphal	T8	58	28, 76
Hyphaene thebaica (L.) Mart.	Hyphal	T8		37, 84
Livistonia chinenesis (N. J. Jacquin) R. Brown	Hyphal	T8		82
Pheonix dactylifera Linn.	Hyphal	T8		8, 19
Pheonix canarienses L.	Hyphal	T8		8, 97
Papilionaceae				
Alhagi maurorum Medic.	Hyphal	SH		15, 71
Melilotus indica (L.) All.	Chasm	A		79, 95
Plantaginaceae				
Plantago major L.	Psamm.	HP		98, 103
Plumbaginaceae				
Limonium stocksii (Boiss.) O. Ktze	Hyphal.	HP	50	15, 82
Poaceae				
Aeluropus lagopoides (L.) Trin. ex Thw.	Hyphal	PG	56	15, 44
Cenchrus ciliaris Rich.	Psamm.	PG		41, 88
Crypsis aculeate (L.) Ait.	Weedy	A		46
Crypsis schoenoides (L.) Lam.	Weedy	A		46
Cynodon dactylon (L.) Pers.	Weedy	PG	17	41, 82
Dactyloctenium aegyptium (L.) P. Beauv.	Weedy	PG		17, 41
Dactyloctenium scindicum Boiss.	Xerohal.	PG		20, 52
Desmostachya bipinnata (L.) Stapf.	Xerohal.	PG	5.6	17, 41
Dichantheum annulatum (Forssk.) Stapf.	Xero	PG		90
Digitaria bicornis (Lam.) Roem & Schulte	Psamm.	A		46
Eragrostis superba Peyr.	Psamm.	PG		6
Halopyrum mucronatum (L.) Stapf.	Psamm.	PG	35	15, 69
Imperata cylindrica (L.) Raeuschel.	Psamm.	PG		96
Lasiurus scindicus Forssk.	Psamm.	PG		90
Ochthochloa compressa (Forssk.) Hilu	Hyphal.	PG		6
Oryza coarctata Roxb.	Hyphal.	HP		96
Paspalum pasploides (Michex) Scribner	Hyphal	HP		22, 89
Phragmites karka (Retz.) Trin. ex. Steud.	Hyphal	PG		77
Polypogon monspeliensis (L.) Desf.	Psamm.	PG		83, 84
Spinifex littoreaus (Brum. F.) Merr.	Psamm.	PG		46
Sporobolus iocladus (Nees ex Trin.) Nees	Psamm.	PG	35	15, 79
Sporobolus virginicus (L.) Kunth.	Psamm.	PG	73	15, 82
Urochondra setulosa (Trin.) C.E. Hubb	Xerohal.	PG	30	44
Portulacaceae				
Portulaca oleracea L.	Xero	AP		88
Residaceae				
Ochradenus baccatus Del.	Xero	SH	58	102
Oligomeris linifolia (Vahl) Macbride	Xero	A	58	38

Rhamnaceae				
Zizyphus nummularia (Burm. f.) Wight and Arn.	Xero	T4		9
Rhizophoraceae				
Bruguiera gymnorhiza (L.) Savigny	Hyphal	T8	58	15, 77
Ceriops decandra (Griff.) Ding.	Hyphal	T9	58	72
Ceriops tagal (Perr.) C.B. Robinson	Hyphal	T8	58	52, 77
Rhizophora apiculata Blume, Bijdr.	Hyphal	T8	58	46, 77
Rhizophora mucronata Poir.	Hyphal	T8	59	15, 52
Salvadoraceae				
Salvadora oleoides Dene	Xero	T8		17, 45
Simarubaceae				
Suriana maritima L.	Hyphal	SH	58	101
Sonneratiaceae				
Sonneratia caseolaris (L.) Engl.	Hyphal	TX	58	78, 94
Tamaricaceae				
Reaumaria stocksii Boiss.	Hyphal	CHN		79
Tamarix aphylla (L.) Karst.	Phrea	T8	56	15, 16
Tamarix indica Willd.	Hyphal	T8		16
Tamarix pakistanica Qaiser	Phrea	T2		6
Typhaceae				
Typha domingensis Pers.	Hyphal	HP	26	14, 15
Zygophyllaceae				
Nitraria retusa (Forssk.) Aschers	Xero	SH	90	15, 71
Tribulus terrestris L.	Xero	HP		10
Zygophyllum simplex L.	Xerohal.	A	29	61, 102

THE COASTAL SABKHAT OF YEMEN

ABDUL NASSER AL-GIFRI & SAEED A. GABALI

Department of Biology, University of Aden, P.O. Box 6014, Khormaksar, Aden, Yemen.

Abstract

The coastal belt of Yemen consists of the Gulf of Aden and Indian Ocean coast in the south and the Red Sea coast (Tihama) in the west. The sabkha bordering halophytic vegetation of each of the coastal zones is described shortly as well as the basic climatic and soil characteristics. The lack of existing literature demonstrates the need of a systematic documentation of Yemen's coastal flora in order to allow for a professional environmental management of the Yemeni coastal zone.

INTRODUCTION

Yemen is located at the base of the Arabian Peninsula between latitude 12°40'N and 19°N and longitudes 42°30'E and 53°E with an area of about 500.000 km^2. It is bordered in the north by Saudi Arabia, in the south by the Gulf of Aden and the Indian Ocean, in the east by the Sultanate of Oman and in the west by the Red Sea. The coastal belts of Yemen are about 1950 km in length and extend along the Red Sea for about 550 km (this represents the western coastal belt of Yemen), and along the Gulf of Aden and part of the Indian Ocean for about 900 km and 500 km respectively, representing its southern coastal belt.

The Red Sea coastal plain of Yemen, the Tihama, is a sandy coastal desert about 30-60 km wide sloping gently from the base of the foothills (at altitudes of 300-400 m a.s.l.) of the Red Sea mountains. The Gulf of Aden coast is principally a wide sandy plain backed along most of its length by high mountains which approach the sea in some places with no coastal plain (El-Anbaawy 1985).

Barth & Böer (eds.), SABKHA ECOSYSTEMS, 141-146

CLIMATE

Over the whole area of the Red Sea basin, winter may extend from mid-October to mid-April with summer occupying the rest of the year. The January mean daily temperature ranges from about 20°C in the far north to about 29°C in the far south. The correspondingly July figures being 35°C and 40°C, respectively. In the Yemen section of the Red Sea, the mean daily maximum and minimum temperatures range between 41.1°C and 24.7°C in summer (July) and 30.1°C and 18.3°C in winter (January) respectively, (Zahran & Al-Kaf 1996).

Along the Red Sea coast, humidity increases from north to south. In the Yemeni section, the monthly means of relative humidity ranges in maximum between 83% and 90% and between 41% and 57% in minimum with rather great daily variations.

Rainfall over the Red Sea coast as a whole is sparse and occurs sporadic and often very localised. A particular location may receive no rain for months or years, then it experiences a brief heavy rainfall, which may not be repeated for a similar lengthy period. This can be observed in available data of the rainfall over the Hodaydah coast. During 1989, the total annual rainfall was 207.3 mm distributed between January (31.4 mm), February (10.5 mm), March (56.4 mm) and April (89.0 mm). In 1988, the total annual rainfall was 127 mm distributed between April (39.0 mm), August (61.4 mm) and September (27.5 mm). These two years were exceptionally wet. However, the mean annual precipitation at the Hodaydah coast usually ranges between 35-60 mm, mainly during the October-April period (Anonymous 1993).

The southern coastal belt of Yemen (Gulf of Aden coast) is influenced by two distinct monsoon seasons: the SW monsoon of the Indian Ocean that occurs between May and September and the NE monsoon, which occurs between October and April. Although the term "monsoon" generally implies heavy seasonal rains, both monsoons of the Yemeni southern coastal belts are dry. The means of the annual maximum and minimum temperatures are 32°C and 25°C. The highest temperatures usually occur in June (37°C) and the lowest temperatures in January (19°C). Relative humidity in this coastal belt varies within a narrow range: between 56-58% in December, as the lowest values, and 73-75% in September, as the highest values. The mean annual relative humidity ranges between 65 and 66%. Precipitation, generally low, may occur in all months of the year except in June. The total annual rainfall ranges between 39 mm and 63 mm.

SOIL PROPERTIES

The climatic aridity of the Yemeni coastal belts have pronounced effects on their edaphic characteristics. High temperature causes an increase of water loss from soil and as precipitation is low there is insufficient leaching and salts accumulate particularly on the surface layers. Surface salt crusts are common features in the sabkhat of these coastal belts.

Representative soil samples were collected from different habitats of Yemeni coastal belts. The results indicate that these soils are generally sandy in texture, salt affected, alkaline in reaction and contain a considerable amount of carbonates (estimated as calcium carbonates).

The sabkhat contain the highest amounts of soluble salts ranging between 10.3 and 16.0% on the surface layers and 4.2-8.9 in the subsurface layers. On the sandy soils of palm groves, the soil salinity ranges between 3.4 and 2.3%. In the surface and subsurface layers, respectively. The main bulk of the soluble salts in all soil samples is formed of chlorides and sulphates. Bicarbonates, however, occur in smaller amounts whereas carbonates are nil.

Except for the mangrove muddy soil, the sand fraction forms the major constituent of the soils of the Yemeni coastal belts with values ranging between 74-85% and 72-74% in the surface and subsurface layers (for the sabkha soils) and up to 88% and 87% on the palm groves soils. The pH values indicate that the soils of the mangroves and palm groves are slightly alkaline (pH = 7.4 and 7.5, respectively). The soils of the salt marshes are moderate to highly alkaline (pH = 8.2-9.2) (Zahran & Al-Kaf 1996).

THE COASTAL VEGETATION

The vegetation of the coastal part of Yemen including the sabkhat can be mentioned below as explained by Ulrich Deil (1998).

Tihamah coast

The vegetation of the Tihamah coast has been described by Al Hubaishi & Müller-Hohenstein (1984), Scholte et al. (1991), Deil & Müller-Hohenistein (1983), Dubaie & Al-Khulidi (1993) and Wood (1983).

Along the Yemeni Red Sea coast northwards from Wadi Siham the following plant community zonation occurs: An *Avicennia marina* zone followed by *a Limonium cylindrifolium - Suaeda fruticosa - Limonium axillare* community, forming hummocks. A sterile sabkha is present after which raised beaches above the high tide level are covered by *Atriplex farinosa, Zygophyllum hamiense, Aeluropus lagopoides and Halopyrum mucronatum.* Sand dunes towards the seaward side are colonised by *Suaeda monoica* and *Salsola spinescens,* and the inland dunes by *Odyssea mucronata, jatrapha pelargoniifolia* and *Leptadenia pyrotechnica.* Further inland, away from the sea, species such as *Indigofera argentea, Blyttia spiralis* and *Jatropha glandulosa,* perennial grasses like *Panicum turgidum* and *Lasiurus scindicus,* and tall shrubs such as *Cadaba rotundifolia, C. glandulosa, Ebretia obtusifolia,* and succulents such as *Euphorbia triaculeata* and *E. fractiflexa* replace the halophytic coastal species. The dum palm *Hyphaene thebaica* characterises silty and sandy depressions. Coastal salt marshes are characterised and dominated by *Suaeda monoica, Aeluropus lagopoides, A. Littoralis, Sporobolus spicatus, S. virginicus* and *Cressa cretica.* The *Suaeda monoica* community can also be found in saline depressions further inland.

Gulf of Aden Coast

The coastal vegetation of the Gulf of Aden coast between Aden and Mukalla has been studied by Al-Gifri & Hussein (1993), Al-Gifri & Al-Subai (1994), Zahran & Al-Kaf (1996) and Kurschner et al. (1998). The southwestern corner of the Arabian Peninsula is characterised by the occurrence of a coastal species, *Odyssea mucronata,* endemic to this part of Arabia. *O. mucronata* is a clump forming, spiny, rhizomatous perennial which colonises semi-mobile dunes and flat sandy areas. Depending on the depth of the sand, an *Odyssca mucronata - Suaeda monoica* community can be distinguished on flat sandy layers overlying saline silts, and an *Odyssea mucronata - Panicum turgidum* community on deeper sand. Other species in the first community are *Sporobolus spicatus, S. consimilis* and *Halopyrum mucronatum,* and in the second, *Salsola spinescens* and *S. vermiculata.* Lagoons on this coast are fringed with *Typha domingensis, Cyperus laevigatus* and *Bacopa monnieri,* and along wadis *Salvadora persica* and *Tamarix aphylla* are the dominant species. Towards the coast, coastal dunes are covered with *Ipomoea pes-caprae* and the endemic, *Conyza cylindrica. I. pes-caprae* is found up to the central part of the eastern coast of Oman, which delimits its northernmost distribution.

Hadhramaut Coast

The Hadhramaut coast has been studied by Gabali & Al-Gifri (1991) and Kurschner et al. (1998). The vegetation is seen from the *Cyperus conglomerates* associations, where the Omano - Makranian element, *Coclachyrum piercei* and the Eritreo - Arabian element *Odyssea mucronata* are common members. The coastal vegetation is also rich in endemics and shows a strong phytogeographical relationship with the coasts of northeast Africa. The species zones are:

(1) Coastal dunes colonised by sedges and grasses *(Cyperus conglomeratus, Halopyrum mucronatum, Odyssea mucronata, Coelachyrum piercei* and *Panicum turgidum).*

(2) Sandy-salty depressions colonised by the endemic *Urochondra setulosa* association, with the co-dominant *Arthrocnemum macrostachyum, Limonium cylindrifolium* and *Crotalaria saltiaina.*

(3) Clayey-salty, relatively wet areas colonised by monospecific stand of *Arthrocnemum macrostachyum.*

(4) Sandy coastal plains colonised by the endemic *Anabasis ehrenbergii-Pulicaria hadramautica-Zygopllum hamiense* association (Deil 1998).

THE SABKHAT

The aridity of Yemen has pronounced effects on the edaphic characteristics of the coastal belt, causing an increase of water loss from soil, and as rainfall is low, salts accumulate on the surface layers. Polyhaline soils (solonchaks) and salt crusts therefore are a common feature in most of the low lying and seasonally inundated salt flats of the coast of Yemen.

Much of the Yemeni sea littoral belts is low lying and is seasonally inundated leading to the formation of sabkhat. These sabkhat dry out during dry periods (seasonally) to form biotic salt pans. Some of these sabkhat may partly reflect a previous sea level rise of up to 2 metres about 4000-6000 years ago (IUCN 1986). Sabkhat usually extend just into the lower intertidal to the point where salinity stress is minimised by regular tidal flushing. Here, it may be slightly overlapped with intertidal mangroves in soft-bottom habitats (Zahran & Al-Kaf 1996).

According to Al-Gifri (1991), Gabali & Al-Gifri (1991), Scholte et al. (1991), Thulin et al. (1992), Al-Gifri & Hussein (1993) and Kurschner et al. (1998) the littoral salt marshes of Yemen are located at Al Lyhaiah on the Red sea coast, Aden, Daufas at Aden- Abyan road and the Hadramaut coastal area along the Gulf of Aden. They are mainly white plains with vegetation surrounding them. It comprises communities dominated by: *Suaeda monoica, S. vermiculata, Tamarix nilotica, Sporobolus spicatus, Arthrocnemum macrostachyum, Salicornia fruticosa, Halopeplis perfoliata* and *Aeluropus lagopoides.*

CONCLUSION

It is clearly shown, that the coastal plains, sabkhat and their vegetation in Yemen received to-date only very limited attention (see number of published literature). We therefore recommend:

- The systematic documentation of Yemen's coastal flora as well as sabkhat vegetation in detail is urgently required in order to allow for professional environmental management of the Yemeni coastal zone.
- Conservation of some coastal locations near to the centres of endemism in Yemen, as the coastal part in front of Jebal Aryes (Abyan), Jebal Rimah (Tihamah), Hadramout Jols (Hadramout), Ras Fartak and Jatheb (Al- Mahrah). Al-Gifri & Gabali (1996)

References

Al-Gifri, A.N. 1991. Salt tolerant plants in Aden (Yemen). *Frgm. Flor. Geobot.* 36(2): 289-292.

Al-Gifri, A.N. & Hussein, M.A. 1993. Plant communities along the road from Aden to Sheikh Salem (Abyan). Yemen. *Feddes Rep.* 104(93-4): 267-270.

Al-Gifri, A.N. & Al-Subai, M.Y. 1994. Vegetation between Abyan and Modia (Abyan). Yemen. *Feddes Rep.* 105: 229-234.

Al-Gifri, A.N. & Gabali, S.A. 1996. The present status of floristic knowledge in the Republic of Yemen. Proc. APSG Workshop, Riyadh, pp. 78-85.

Al-Hubaishi, A. & Müller-Hohenstein, K. 1994. An introduction to the vegetation of Yemen, ecological bases, floristic composition and human influence. Eschborn GTZ: 206 pp.

Anonymous 1993. Climatic normals of the northern section of the Yemen Republic. *Civil Aviation and Meteorological Authorities.* Computer Department, Sheets of Al-Hodaydah, means of 1980-1992.

Deil, U. 1998. Coastal and sabkha vegetation. In: Ghazanfar, S.A. & Fisher, M. (eds.), Vegetation of the Arabian Peninsula Chapter, Kluwer, Dordrecht, pp. 209-228.

Deil, U. & Müller-Hohenstein, K. 1983. Pflanzenwelt des Yemen am Beispiel Sukkulenter, Euphorbies Yemen. Report. *Mitt. Der Deutsch Jemenitischen Ges.*, 14(2/5): 12-16.

Dubaie, A.S. & Al-Khulaidi, A. 1993. Studies on the flora of Tihamah (Y.R.) *Feddes Rep.* 104: 256-265.

El-Anbaawy, M.I. 1985. Geology of Yemen Republic. Dar El-Kotob and El-Wathaiq El-Kawmiah, Sanaa Yemen, 337 pp.

Gabali, A.S. & Al-Gifri, A.N. 1991. A survey of the vegetation of Hadramout, Republic of Yemen. *Fragm. Flor. Geobot.* 36: 127-134.

IUCN 1986. Marine conservation survey of the Yemen Arab Republic. Red Sea Coast, 84 pp.

Kürschner, H., Al-Gifri, A.N., Al-Subai, M.Y. & Rowaishad, A.K 1998. Vegetational patterns within coastal salines in southern Yemen. *Feddes Rep.* 109: 147-159.

Scholte, P.T., Al-Khuleidi, A.W. & Kessler, J.J. 1991. The vegetation of the Republic of Yemen. Western Part. *RLIP. OHV Consultants BV*: 56 pp.

Thulin, M., Eriksson, M., Langstrom, E. & Al-Gifri, A.N. 1992. Report on the flora within the Masila Block performed from CANOXY. Yemen. In: Environmental Impact Assessment, Canadian Accidental Petroleum Ltd., 9 pp.

Wood, J.R.I. 1983. An outline of the vegetation of Yemen Arab Republic. In: King, J.W. (ed.), Soil survey of the Yemen Arab Republic. Cornell University, Washington, pp. 29-89.

Zahran, M.A. & Al-Kaf, H.F. 1996. Introduction to the ecology of littoral halophytes of Yemen. *Arab Gulf J. Sci. Res.* 14: 691-703.

SABKHAT AND HALOPHYTES IN THE HORN OF AFRICA (DJIBOUTI, ERITREA, ETHIOPIA, SOMALIA, SUDAN)

BENNO BÖER[1], HANS-JÖRG BARTH[2]
& OMYMA OSMAN[3]

[1]UNESCO Regional Office in the Arab States of the Gulf, P.O. Box 3945, Doha, Qatar,
[2]University of Regensburg, Department of Physical Geography, Universitätsstr. 31, D-93040 Regensburg, Germany,
[3]University of Juba, P.O. Box 12311, Khartoum, Sudan

Abstract

Literature dealing exclusively with aspects of halophytes, saline deserts, and saline water agro-ecosystems in the Horn of Africa region is limited. This contribution is a first attempt to summarise the information available on the halophyte flora, and the sabkha resources of the region. We try to provide useful and realistic recommendations on the next steps towards the sustainable development of the region's environmental geo-biological resources.

INTRODUCTION

The southern Red Sea and Gulf of Aden region's main income sources are based on the production and trade with fish, livestock, and agricultural products. The geo-biological resources are heavily utilised, and subject to a variety of environmental problems. Ground water exploitation and pollution, tree cutting for charcoal production, rangeland grazing, deforestation, as well as fishing beyond the ecological carrying capacities, soil erosion, and desertification, all posing a potential threat to the agricultural and fishing industry. The resource exploitation accelerated due to the population pressure. New, sustainable land use systems are urgently required, in order to redress adverse environmental impacts, which threaten the main

Barth & Böer (eds.), SABKHA ECOSYSTEMS, 147-159

conventional income sources. On the other hand, it is likely that the existing fresh groundwater resources, which are re-charged by the run-off water from the Abessinnian highlands, are not even fully exploited, and available for future agricultural land use schemes.

The information provided is not complete, however, it is a first review on the available information on saline irrigation, together with an introduction to the regional halophyte and sabkha resources in the region.

PHYSICAL SETTING

The coastal structure of the region of interest is greatly determined by the Tertiary geological history and the tectonic activity, which continues until today. Present rifting in the Red Sea depression is about 2 cm per year (Stanley 1994). The Red Sea is the result of a large rift system which extends from the Gulf of Aqaba and the Dead Sea Rift in the North to The Afar triangle in Ethiopia, a so called "triple junction". There it diverges through the Gulf of Aden into the Arabian Sea and down the African mainland as the large African Rift Valley System in the South. The separation of the Arabian plate as a fragment of the large African plate started about 25 Ma ago. The tensional forces led to the formation of a graben structure (Red Sea depression) with a pronounced topographic relief between the plateau and the floor of the rift. Erosion in the fault-block mountains provided clastic sediments such as conglomerates and sandstones which were shed from the steep valley walls and accumulated in the rift. Magma rising up the faults formed basaltic lava, dikes and sills. Continued subsidence as the rift widened led to the incursion of sea water with associated marine sedimentation.

The geomorphology of the coastal areas bordering the Red Sea and the Gulf of Aden is a direct result of the tectonic history in the past. The western shores from Egypt in the north to Sudan, Eritrea, Djibouti and Somalia in the south are quite similar to the opposite side in Saudi Arabia and Yemen. The Red Sea is bordered by a narrow coastal lowland (it's width ranging between only two and some tens of kilometres). It is a narrow band of alluvial gravel and sands covering pediments cut on the crystalline and Tertiary rocks of the mountains to the west. The eastern part of the coastal plain is basically a depositional surface on coralline rocks. Sandy and muddy coastal plains are scattered between these carbonaceous rocks, beach terraces and wadi sediments. Most of them show high salinity and are therefore coastal sabkhat. They occur all along the western coast of the Red Sea and Gulf of Aden, but are always restricted to flat and sheltered depositional environments.

The climate of the coastal lowlands bordering the southern shores of the Red Sea and Gulf of Aden is a dry desert climate. Annual average precipitation rates lie between 50 and 150 mm, and average annual temperatures vary from 28 to 32°C. These conditions lead to potential evapo-transpiration rates of 1800 to 3000 mm yr^{-1}

(Griffiths & Hemming 1966, Griffiths 1972). The area is free of frost events. The Red Sea is highly saline with salinities from 36.5 - 40.5 ppt (Sheppard et al. 1992) and offshore surface temperatures lie according to the time of the year between 22 and 32°C (Sheppard et al. 1992). The Temperature and salinity variations of the Gulf of Aden are less extreme. The daily tidal range of the Red Sea varies between almost no difference in the central parts to 0.6 - 0.9 m in the southern parts. Additionally there is a basically wind driven seasonal tide superimposed on the daily cycle. In winter the mean sea level is about 0.5 m higher than in summer. The change from summer low to winter high occurs over less than a month (Sheppard et al. 1992).

The coastal plains support mainly halophytic vegetation. The hydrology is characterised by permanent water-deficit. Numerous wadis, which cross the coastal plains are dry most time of the year. Only during the rainy season they carry water from the highlands to the coastal plains and into the sea.

The main geomorphic and soil units between Sudan and Somalia are:

- Exposed coral shelves with low to medium salinity.
- Raised coral beaches, 1-3 m above sea level, salinity may be moderate to high.
- Quartz sand-sheets of a moderate to light salinity.
- Sandy beaches and coral sand-dunes, occasionally connected to a lagoonal mangrove.
- Mangrove swamps, consisting mainly of *Avicennia marina* and *Rhizophora mucronata*.
- Sabkhas, often with thin layers of crystallised salt on the surface, and barren of any vegetation.
- Alluvial fans with low to medium salinity.
- Pediments and desert pavements at the foothills of the escarpments, occasionally reaching almost the shoreline. They are often strewn with pebbles and are usually referred to a "Serir" or "Reg" throughout the Sahara and Arabia, both words meaning desert pavement in Arabic.
- Recent lava flows.

THE HALOPHYTES

Numerous authors provided information on the flora and vegetation of the Sudan-Somalia Red Sea and Gulf of Aden coast region (Adams 1967, Andrews 1945, 1948, Bally 1968, 1976, Bari 1968, Beals 1968, Begué 1958, Boaler & Hodge 1962, 1964, Bruneau de Miré 1960, Bunting & Lea 1962, Chaffey 1979, Chedeville 1972, Chevalier 1939, Chipp 1929, 1930, Ciferri 1939, Collenette 1931, Cufodontis 1940, Engler 1904, 1906, Eyre et al. 1953, Gay Berry 1959, Gillet 1941, Gilliland 1952, Good 1924, Halwagy 1961, 1962a, b, 1963, Hancock 1944, Harrison & Jackson

1958, Hedberg 1971, 1975, Hemming 1961, 1965, 1966, 1968, Hunting Technical Services 1958, 1964, 1968, Jackson 1950, 1951, 1956, Jenkin et al. 1977, Jonglei Investigation Team 1954, Kassas 1956a, b, 1957, Klötzli 1975, Lamprey 1975, Logan 1946, Macfadyen 1950, Mahmoud & Obeid 1971, Migahid 1947, Mooney 1963, Morison et al. 1948, Negri 1913, Obeid & Mahmoud 1971, Obeid & Seif el Din 1971a, b, Pichi-Sermolli 1938, 1939, 1940, Posnett & Reilly 1977, Quezel 1969, 1970, Radwanski & Wickens 1967, Ramsay 1958, Ruxton & Berry 1960, Schweinfurth 1868, Scott 1952a, b, 1955, 1958, Seebald 1972, Senni 1935, Smith 1949, Verdcourt 1969, Wickens 1977, Wickens & Collier 1971, Willimot 1957, Worall 1959, 1960a, b). However, the most relevant documents might be Hemming (1961, 1966, 1998), Audru et al. (1987), and White (1983).

Characteristic features of the region's flora are:

- A large number of glycophytic succulent perennials
- The presence of many CAM species
- A large number of perennial C4 grasses
- The expansion of the Burseraceae family, particularly the frankincense *Boswellia sacra*
- The occurrence of over 40 species of halophytes.

Over 40 species of halophytes in the region have been recorded (Tab. 1), and a number of those is endemic to the region. The number of halophytes of course depends on the definition of the term "halophyte". For the purpose of this paper we are only interested in those halophytes, which can grow, reproduce and germinate in the presence of either periodic sea water inundation, or the clear presence of a salt surface crust.

The halophytes occur in brackish *reed swamps*, sea grass beds, mangroves, salt marshes, and beach communities (different beach types with different substrates).

Table 1. Halophytes, that have been recorded within the study area.

Family	Genus and species
Aizoaceae	*Mesembrianthemum nodiflorum* L.
	Sesuvium sesuviodes (Fenzl) Verdc.
Arecaceae	*Phoenix caespitosa* Chiov.
	Phoenix dactylifea L.
	Phoenix reclinata Jacq.non Schwartz
Asclepiadaceae	*Calotropis procera* (Ait.) Ait. F.
Avicenniaceae	*Avicennia marina* (Forssk.) Vierh.
Caryophyllaceae	*Sphaerocoma hookeri* T. Anderson

Chenopodiaceae	*Anabasis setifera* Moq. (syn. A. ehrenbergeri Boiss.)
	Arthrochnemum macrostachuym (Moric.) Moris
	Atriplex coriacea Forssk.
	Atriplex farinosa Forssk.
	Atriplex griffithii (Moq.) Boulos
	Atriplex halimus L.
	Atriplex nogalensis Friis & Gilbert
	Cornulaca ehrenbergii Asch.
	Cornulaca monacantha Delile
	Fadenia zygophylloides Allen & Townsend
	Halocnemum strobilaceum (Pall.) M.- Bieb.
	Halopeplis perfoliata (Forssk.) Asch.
	Halothamnus beckettii Botsch.
	Halothamnus somalensis (N.E. Br.) Botsch.
	Haloxylon persicum Bunge
	Haloxylon salicornicum (Moq) Bunge
	Lagenantha cycloptera (Stapf) Gilb. & Friis
	Salsola cyclophylla Baker
	Salsola imbricata Forssk.
	Salsola foetida Del. ex Spreng.
	Salsola rubescens Franch.
	Salsola spinescens
	Salsola villosa Schult in Roem.
	Sarcocornia fruticosa (L.) A.J. Scott
	Seidlitzia rosmarinus Bunge ex Boiss.
	Sevada schimperi Moq
	Suaeda aegyptiaca (Hasselq.) Zohary
	Suaeda monoica Forssk. ex Gmel
	Suaeda pruinosa Lange
	Suaeda vermiculata Forssk. ex Gmel
Combretaceae	*Conocarpus lancifolius* Engl. & Diels
	Terminalia brownii Fresen
Compositae	*Pluchea* sp.
Convolvulaceae	*Cressa cretica* L.
	Ipomoea donaldsonii Rendle
	Ipomoea pes-caprae (L.) R. Br.
	Ipomoea sultani Chiov.
Cymodoceaceae	*Halodule uninervis* (Forssk.) Asch.
	Syringodium isoetifolium (Asch.) Dandy
	Thalassodendron ciliatum (Forssk.) Hartog.
Cyperaceae	*Cyperus laevigatus* L.
	Schoenoplectus lacustris (L.) Palla
	Schoenoplectus littoralis (Schrad.) Palla
	Schoenoplectus mucronatus (L.) Palla
Hydrocharitaceae	*Halophila ovalis* (R.Br.) Hook
	Halophila stipulacea (Forssk.) Asch.
	Thalassia hemprechii (Solms.) Asch.

Mimosaceae	*Prosopis cineraria* (L.) Druce
	Prosopis juliflora (Swartz) DC.
	Prosopis koelziana Burk.
Orobanchaceae	*Cistanche tubulosa* (Schrenk) Hook
Pandanaceae	*Pandanus odoriferus* (Forssk.) Chiov.
Plumbaginaceae	*Limonium axillare* (Forssk.) O. Kuntze
	Limonium cyndrifolium (Forssk.) Verdc. Ex Cuf.
	Limonium somalorum Chiov. Hemming
	Limoniastrum migiurtinorum Hemming
Poaceae	*Aeluropus lagopoides* (L.) Trin. ex Twaites
	Aeluropus littoralis (Gouan) Parl.
	Halopyrum mucronatum (Nees) Stapf
	Ochthochloa compressa (Forssk.) Hiln.
	Paspalidium desertorum (A. Rich.) Stapf
	Paspalidium geminatum (Forssk.) Stapf
	Paspalum distichum L.
	Paspalum vaginatum Sw
	Phragmites autralis (Cav.) Trin & Steud.
	Sporobolus helvolus (Trin.) Dur. & Schinz
	Sporobolus ioclados (Nees ex Trin) Nees
	Sporobolus robustus Kunth.
	Sporobolus ruspolianus Chiov.
	Sporobolus spicatus (Vahl) Kunth
	Sporobolus pungens (Shreb.) Kunth
Rhiziphoraceae	*Bruguiera gymnorrhiza* (L.) Lam.
	Ceriops tagal (Perr.) C.B. Robinson
	Rhizophora mucronata Lam.
Ruppiaceae	*Ruppia maritima* L.
Salvadoraceae	*Salvadora persica* L.
Tamaricaceae	*Reaumuria hirtella* Jaub. & Spach
	Tamarix aphylla (L.) H. Karst
	Tamarix macrocarpa (Ehrenb.) Bunge
	Tamarix nilotica (Ehrenb.) Bunge
	Tamarix stricta Boiss.
Typhaceae	*Typha domingensis* Pers.
	Typha aff. *latifolia* L.
Zygophyllaceae	*Kelleronia quadricornuta* Chiov.
	Nitraria retusa (Forssk.) Asch.
	Zygophyllum album L.
	Zygophyllum coccineum L.
	Zygophyllum hildebrandtii Engl.
	Zygophyllum simplex L.

SALINE IRRIGATION

Comprehensive environmental management will contribute to the sustainable utilisation of the geo-biological resources. One discipline in the field of environmental management, which gains steadily the interest of the Arab region's decision makers, as well as of the scientific community, is the production of sea water irrigated cash crop halophytes, which can potentially offer a new source for jobs and income (Böer & Lieth 1999). In order to develop such systems on a sustainable basis, assessments of the available resource of land, which can be irrigated with saline water under the avoidance of soil salinisation is essential. This is technically possible, but saline soils, such as found in sabkhat, need to be identified, and suitable irrigation techniques, and draining techniques need to be applied. It is also of importance to identify the indigenous halophyte resources, since these are best adapted to the local environmental conditions, and as such, need the least maintenance. This is best to be done by first delineating the coastal zone (Böer & Gliddon 1998), and then to carry out a detailed survey of the entire coastline (Böer & Saenger in press.). Once the resources are known, feasibility studies towards sustainable man-made agro-ecosystems should be carried out, together with socio-economic impact studies, in order to determine the suitability and capacity of cash crop halophyte production in the Sudan - Somalia region. It is also of great importance, to carry out comprehensive environmental impact assessments, in order to avoid adverse impacts, which can potentially be caused by coastal halophyte production systems, such as habitat loss, loss of biological diversity, the discharge of polluted effluents, soil salinisation, and others.

The best known project in the region is probably the Seaphire project in Eritrea, north of the coastal city of Massawa. Information was provided at the EXPO 2000 in Hannover, and more information is available on the net (http://www.seawaterfarms.com/). Published information on that project has not been obtained by the authors. However, the project is based on sea water irrigation, and mainly focuses on the production of indigenous mangroves and the annual salt marsh species *Salicornia europaea* L.. The agricultural potential is to produce and market the products based on this species, such as edible oil, pickles, salad, vegetable, as well as a livestock feed. Furthermore, the project aims at the production of fish and shrimp.

The achievements, current research, and prospects for saline agriculture and coastal restoration projects in the Arabian peninsula have been documented by Böer & Lieth (1999), and Böer (in press.). Under the consideration of the geo-biological resources, such as hyper-saline coastal land, sea water, numerous indigenous halophytes, limited freshwater availability, as well as harsh climatic conditions, it seems that the region has a realistic potential for saline irrigation based agriculture.

CONCLUSIONS

The available information on halophytes and saline agro-ecosystems in the horn of Africa region is incomplete. People of the region are highly dependant on the utilisation of the natural resources, as sources of income, and for the maintenance of the day-to-day life. The current utilisation of the resources is not sustainable, and therefore an urgent need for land-use planning systems towards the sustainable development and utilisation of the resources is required. New jobs and income may be generated via the development of sustainable land use methods. Man-made agro-ecosystems, based on seawater irrigation and halophytes might be developed as an alternative source of income for farmers, aquaculturalists, and labourers.

RECOMMENDATIONS

- Coastal resources maps are required, considering all biotic and abiotic resources, as well as the human infrastructure in the region.
- Pilot projects towards the establishment of seawater based agro-ecosystems should be carried out, under consideration of comprehensive environmental impact assessments, social impact assessments, as well as feasibility studies.
- The success and failure of existing pilot projects in the region should be carefully investigated, and possibly adopted and improved, towards their applicability in the region.

ACKNOWLEDGEMENTS

We offer our thanks to Chris Hemming (London), Johannes Schroeder (Berlin), and Mats Thulin (Uppsala), for their kind support with information on the flora and geo-ecology of the Sudan-Somalia region.

References

Adams, M.E. 1967. A study of the ecology of Acacia mellifera, A. seyal, and balanites aegyptiaca in relation to land clearing. *J. appl. Ecol.* 4: 221-237.

Andrews, F.W. 1945. Water plants in the Gezira canals. *Ann. appl. Biol.* 32: 1-14.

Andrews, F.W. 1948. The vegetation of the Sudan. In: Toothill, J.D. (ed.), Agriculture in the Sudan, pp. 32-61. London, Oxford University Press.

Audru, J., César J., Forgiarini G & Lebrun, J.P. 1987. La végétation et les potentialités pastorales de la République de Djibouti. 384 pp., l Carte Coul. h.t, 2 feuilles 1 / 100.000. IEMVT, Maisons-Alfort, & CIRAD, Montpellier.

Audru, J., Cesar, J. & Lebrun, J.P. 1994. Les plantes vasculaires de la République de Djibouti. 3 vols, 965 pp.., CIRAD-EMVT, Montpellier.

Ayyad, M.A. & Ghabbour, S.I. 1986. Hot deserts of Egypt and Sudan. In: M. Evenari, I. Noy-Meir & Goodall, D.W. (eds.), Hot deserts and arid shrublands, Ecosystems of the World. Vol. 12 B: 149-202. Elsevier, Amsterdam.

Bally, P.R.O. 1968. Somali Republic South. In: Hedberg, I. & Hedberg, O. (eds.), Conservation of vegetation in Africa south of the Sahara. *Acta Phytogeogr. Suec.* 54: 145-148.

Bally, P.R.O. 1976. The vegetation of Somalia. *Boissiera* 24: 447-450.

Bari, E.A. 1968. Sudan. In: Hedberg, I. & Hedberg, O. (eds.), Conservation of vegetation in Africa south of the Sahara. *Acta Phytogeogr. Suec.* 54: 59-64.

Beals, E.W. 1968. Ethiopia. In: Hedberg, I. & Hedberg, O. (eds.), Conservation of vegetation in Africa south of the Sahara. *Acta Phytogeogr. Suec.* 54: 137-140.

Begué, L. 1958. Les forêts de la République du Soudan. *Bois Forêts Trop.* 62: 3-19.

Boaler, S.B. & Hodge, C.A.H. 1962. Vegetation stripes in Somaliland. *J. Ecol.* 50: 465-474.

Boaler, S.B. & Hodge, C.A.H. 1964. Observations on vegetation arcs in the northern region, Somali Republic. *J. Ecol.* 52: 511-544.

Böer, B. in press. Halophyte development in the Arabian peninsula - current research and prospects. Proceedings of the symposium on saline agriculture, International Center of Biosaline Agriculture, Dubai, March 18-20, 2001.

Böer, B. & Gliddon, D. 1998. Mapping of coastal ecosystems and halophytes (case study of Abu Dhabi, United Arab Emirates). *Marine & Freshwater Research* 49: 297-301.

Böer, B. & H. Lieth 1999. Halophytes for seawater irrigation in the Arabian peninsula - a review. In: Hamdy et al. (eds.), Halophyte uses in different climates II. 33-55. Blackhuys Publishers, Leiden, The Netherlands.

Böer, B. & P. Saenger, in press. The bio-geography of the coastal vegetation of the Abu Dhabi Gulf Coast. *Marine Pollution Bulletin.*

Bruneau de Miré, P. 1960. Note préliminaire sur l'étage culminal du Djebel Marra (Republic of the Sudan) et ses affinités avec les hauts sommets du Tibesti. *Cr. Somm. Séanc. Biogéogr.* 37: 11-18.

Bunting, A.H. & Lea, J.D. 1962. The soils and vegetation of the Fung, east central Sudan. *J. Ecol.* 50: 529-558.

Chaffey, D.R. 1979. South-west Ethiopia forest inventory project. A reconnaissance inventory of forest in south-west Ethiopia. Tolworth Tpwer, Surbiton, Surrey, Directory of Overseas. 316 pp.

Chedeville F. 1972. La végétation du Territoire Français des Afars et des Issas. *Webbia* 26: 243-266.

Chevalier, A. 1939. La Somalie française. Sa flore et ses production végétals. *Rev. Bot. appl. Agric. trop.* 19: 663-687.

Chipp, T.F. 1929. The Imatong Mountains, Sudan. *Kew Bull.* 177-197.

Chipp., T.F. 1930. Forest and plants of the Anglo-Egyptian Sudan. *Geogr. J.* 75: 123-143.

Ciferri, R. 1939. Les associazione del litorale marino della Somalia meridionale. *Riv. Biol. colon.* 2: 5-42.

Collenette, C.L. 1931. North-east British Somaliland. *Kew Bull.* 401-414.

Corra, M. 1983. Contribution a l'étude des paturages extensifs du plateau Harar-Ogaden (Harargue, Ethiopie): Flore, végétation, écologie. multigr. Ph.D. thesis Université des Sciences et Techniques du Languedoc (USTL). 261 pp., Montpellier.

Cufodontis, G. 1940. La vegetazione. In: Zavaratti, R. & Cufodontis, G. (eds.), Missione biologica nel Paese dei Borana. I. Condizioni biografiche e antropiche. 141-255. Roma, R. Acad. Ital.Centro Studi AOI.

Engler, A. 1904. Über die Vegetationsverhältnisse des Somalilandes. Sitzber. *Preuss. Akad. Wiss.* 10: 355-416.

Engler, A. 1906. Über die Vegetationsverhältnisse von Harar und des Galla-Hochlandes auf Grund der Expedition von Freiherrn von Erlanger und Herrn Oscar Neumann. Sitzber. *Preuss. Akad. Wiss.* 10: 726-747.

Eyre, V.E.F., Ramsay, D.M. & Jewitt, T.N. 1953 Agriculture, forests and soils of the Jur ironstone country of the Bahr el Ghazal Province, Sudan. *Bull. Minist. Agric. Sudan* 9: 1-40.

Gay, P.A. & Berry, L. 1959. The water hyacynth. A new problem on the Nile. *Geogr. J.* 125: 89-91.

Gillet, J.B. 1941. The plant formations of Western British Somaliland and the Harar Province of Abessynia. *Kew Bull.* 37-199.

Gilliland, H.B. 1952. The vegetation of Eastern British Somaliland. *J. Ecol.* 40: 91-124.

Good, R. 1924. The geographical affinities of the flora of Jebel Marra. *New Phytol.* 23: 266-281.

Griffith, J.F. (ed.) 1972. Climates of Africa. In: Landsberg, H.E. (ed.), World survey of climatology, 10. 604 pp., Amsterdam, London, New York, Elsevier.

Griffith, J.F. & Hemming, C.F. 1963. A rainfall map of Eastern Africa and Southern Arabia. Nairobi, Meteorological Dep., E.A. Common Saervices Organizations, 42 pp. (Mimeo.).

Halwagy, R. 1961. The vegetation of the semi-desert north-east of Khartoum, Sudan. *Oikos* 12: 87-110.

Halwagy, R. 1962a. The impact of man on semi-desert vegetation in the Sudan. *J. Ecol.* 50: 263-273.

Halwagy, R. 1962b. The incidence of the biotic factor in northern Sudan. *Oikos* 13: 97-117.

Halwagy, R. 1963. Studies on the succession of vegetation on some islands and sand banks in the Nile near Khartoum, Suda. *Vegetatio* 11: 217-234.

Hancock, G.M. 1944. The grass-Acacia cycle in the Blue Nile Province, south of Sennar, and proposals for its management. Sudan Govt Rep. Soil Conserv. Comm. 78-86.

Harrison, M.N. & Jackson, J.K. 1958. Ecological classification of the vegetation of the Sudan. *Forests Bull. Sudan* 2: 1-45.

Hassan, H.M. 1974. An illustrated guide to the plants of Erkowit. 106 pp., Khartoum University Press, Khartoum.

Hedberg, O. 1971. The high mountain flora of the Galama Mountain in Arussi Province, Ethiopia. *Webbia* 26: 101-128.

Hedberg, O. 1975. Studies of adaptation and speciation in the Afroalpine flora of Ethiopia. *Boissiera* 24: 71-74.

Hemming C.F.1961. The ecology of the coastal area of northern Eritrea. *Journal of Ecology* 49: 55-78.

Hemming, C.F. 1965. Vegetation arcs in Somaliland. *J. Ecol.* 53: 57-67.

Hemming, C. F. 1966. The vegetation of the northern region of the Somali Republic. *Proc. Linn. Soc. Lond.* 177: 173-249.

Hemming, C.F. 1968. Somali Republic North. In: Hedberg, I. & Hedberg, O. (eds.), Conservation of vegetation in Africa south of the Sahara. *Acta Phytogeogr. Suec.* 54: 141-145.

Hemming C.F. 1973. An ecological classification of the Bosaso region. Working Paper, AGP SF SOM 70/512, 67 pp. FAO, Rome.

Hemming C.F. 1972. Ecological and grazing survey of the Mudugh region.. Working Paper, AGPC, DP/SOM/70/512, 21 pp. FAO, Rome.

Hemming, C.F. 1998. The vegetation and soils of Somalia. In: Ash, J.S. & Miskell, J.E (eds.), Birds of Somalia. Pica Press.

Hunting Technical Services 1958. Jebel Marra investigations. Report on phase I. Studies. London, Hunting Technical Services Ltd. 112 pp.

Hunting Technical Services 1964. Land and water use survey in Kordofan Province of the Republic of Sudan. Report on the survey of geology, geomorphology and soils, vegetation and present land use. Athens, Dixiades Associates. 349 pp.

Hunting Technical Services 1968. Land and water resources survey of the Jebel Marra area. Reconnaissance vegetation survey. Rome, FAO.

Jackson, J.K. 1950. The Dongotona Hills, Sudan. Emp. For. Rev. 29: 139-142.

Jackson, J.K. 1951. Mount Lotuke, Didinga Hils. Sudan Notes Rec. 32: 339-341.

Jackson, J.K. 1956. The vegetation of the Imatong Mountains, Sudan. *J. Ecol.* 44: 341-374.

Jenkin, R.N., Howard, W.J., Thomas, P., Abell T.M.B. & G.C. Deane, G.C. 1977. Vegetation. In: Forestry development prospects in the Imatong Central Forest Reserve, Southern Sudan. Tolworth Tower, Surbiton, Surrey, Directorate of Overseas Surveys, 61-78.

Jonglei Investigation Team 1954. The Equatorial Nile Project and its effects in the Anglo-Egyptian Sudan. Khartoum, Govt Printer.

Kassas, M. 1956a. Land forms and plant cover in the Omdurman Desert, Sudan. Bull. Soc. Géogr. *Égypte* 29: 43-58.

Kassas, M. 1956b. The Mist Oasis of Erkowit, Sudan. J. Ecol. 44: 180-194.

Kassas, M. 1957. On the ecology of the Red Sea coastal land. Sudan. *J. Ecol.* 45: 187-203.

Klötzli, F. 1975. Zur Waldfähigkeit der Gebirgssteppen Hoch-Semiens (Nordäthiopien). *Beitr. Naturk. Forsch. Südw. Dtl.* 34: 131-147.

Lamprey, H. 1975. Report on the Desert Ecroachment Reconaissance in Northern Sudan, November 1975. UNESCO/UNEP. 16 pp.

Lebrun J.-P., Audru J. & César, J 1989. Catalogue des plantes vasculaires de la République de Djibouti. Etudes & Synthèses n° 34, IEMVT, 277 pp. Maisons-Alfort.

Le Houérou H.N. 1972 Rangeland improvement and related problems in Northern Somalia. Report AGP/SF/SOM/12, Plant Production and Protection Division, 18 pp. FAO, Rome.

Le Houérou H.N. 1984. Aperçu écologique de l'Ethiopie. *Journal d'Agriculture Traditionelle et de Botanique Appliquée* (JATBA) 31: 2&é-223.

Le Houérou H.N. & Haywood M. 1978. Survey of the Eastern Rangelands of Ethiopia (Northern Ogaden and Jijigga). International Livestock Center for Africa. (ILCA) Livestock and Meat Board, Addis Ababa.. 250 pp. mimeo, 1 colour map 1 : 200,000 Bioclimatology, 5 colour maps 1: 50,000 : Land-Use, Vegetation, Geology, Soils, Hydrology and Erosion. 30,000 km^2.ILCA, Addis Ababa.

Logan, W.E.M. 1946. An introduction to the forests of central and southern Ethiopia. Inst. Pap. Imp. For. Inst. 24: 1-65.

MacFayden, W.A. 1950. Vegetation patterns in the semi-desert plains of British Somaliland. *Geogrl. J.* 116: 199-211.

Mahmoud, A. & Obeid, M. 1971. Ecological studies in the vegetation of the Sudan. I: General features of the vegetation of Khartoum Province. *Vegetation* 23: 153-176.

Migahid, A.M. 1947. An ecological study of the 'sudd' swamps of the Upper Nile. *Proc. Égypt. Acad. Sci.* 3: 57-86.

Mooney, H. 1963 An account of two journeys to the Araenna Mountains in Balé Province (south-east Ethiopia), 1958 and 1959-60. *Proc. Linn. Soc. Lond.* 174: 127-152.

Morison, C.G.T., Hoyle, A.C. & Hope-Simpson, J.F. 1948. Tropical soil-vegetation catenas and mosaics. *J. Ecol.* 36: 1-84.

Negri, G. 1913: Appunti di una escursione botanica nell'Etiopia meridonale (marzo-agosto 1909). Rome, Ministero delle Colonie. 177 pp.

Obeid, M. & Mahmoud, A. 1971. Ecological studies in the vegetation of the Sudan. II: The ecological relationships of the Khartoum Province. *Vegetatio* 23: 177-198.

Obeid, M. & Seif El Din, A. 1971a. Ecological studies of the Sudan. I: *Acacia senegal* (L.) Willd. and its natural regeneration. *J. appl. Ecol.* 7: 507-518.

Obeid, M. & Seif El Din, A. 1971b. Ecological studies of the Sudan. III: The effect of simulated rainfall distribution at different isohyets on the regeneration of *Acacia senegal* (L.) Willd. on clay and sandy soils. *J. appl. Ecol.* 8: 203-209.

Pichi-Sermolli, R.E.G. 1938. Ricerche botaniche nella regione del Lago Tano e nel Semièn. In: Danielli, G. et al. (eds.), Missione di Studio al Lago Tano, 1. Relazioni preliminari, 77-103. Rome, R. Accad. Ital. Centro Studi AOI.

Pichi-Sermolli, R.E.G. 1939. Aspetti del paesaggio vegetale nell'alto Semièn (Africa Orientale Italiana). *Nuovo G. bot. ital.* 45: 515-524.

Pichi-Sermolli, R.E.G. 1940. Osservazioni sulla vegetazione del versante occidentale dell' Altipiano Etiopico. *Nuovo G. bot. ital.* 47: 609-623.

Pichi-Sermolli R.E.G. 1957. Una carta Geobotanica dell'Africa Orientale (Eritrea, Etiopia, Somalia). Con una carta al 1 / 5 000.000 *Webbia* 13 (1): 14-132.

Pignati S, Moggi G. & Raimondo, F.M. 1993. Dry coastal ecosystems of Somalia. In: E. Van der Maarel (ed.), Dry Coastal Ecosystems: Africa, America, Asia and Oceania., 31-36, Ecosystems of the World, Vol. 2 B, Elsevier, Amsterdam.

Posnett, N.W. & Reilly, P.M. 1977. Ethiopia. Tolworth Tower, Surbiton, Surrey, Land Resources Division, Ministry of Overseas Development. (Land resource bibliography 9), 194 pp..

Quezel, P. 1969. Flor et végétation des plateaux du Darfur Nord-occidental et du jebel Gourgeil (Rép. Du Soudan). CNRS. 146 pp.

Quezel, P. 1970. A preliminary description of the vegetation in the Sahel region of the North Darfur. Sudan Notes Rec. 51: 119-125.

Radwanski, S.A. & Wickens, G.E. 1967. The ecology of *Acacia albida* on mantle soils in Zalingei, Jebel Marra, Sudan. *J. appl. Ecol.* 4: 569-579.

Ramsay, D. McC. 1958. The forest ecology of central Darfur. *Sudan Govt. For. Bull.* 1: 1-80.

Ruxton, B.P. & Berry, L. 1960. The Butana grass patterns. *J. Soil Sci.* 11: 61-62.

Schroeder J.H. 1983. Die Saumriffe von Port Sudan. II Gefärdung, Schutz, Entwicklungshilfe. *Essener Geogr. Arb.* Bd. 6: 45-57.

Schroeder, J.H., Kachholz, K.D. & Lange, H. 1984. Eolian dust in coastal lagoon and reef sediments in the Sudanese Red sea. In: Massoud A.H. Saad, (ed.), Proceedings of the Symposium on coral reef environment of the Red Sea, Jeddah.

Schroeder, J.H. & Mansour, N. 1995. Sedimentary Environments in the Coastal plain of the Red Sea / NE Sudan. Thematic map 1 / 50,000 Jebel Lagagat. Institut für Geologie und Paläontologie, Technische Universität, Berlin.

Schroeder, J.H. & Mansour. N. 1993. Sedimentary environments in the coastal plain of the Red Sea / NE Sudan Thematic map 1.50,000 Sheet Marsa Arakyai. Institut für Geologie und Paläontologie, Technische Universität, Berlin.

Schroeder, J.H. & Mansour, N. 1994. Sedimentary Environments in the Coastal plain of the Red Sea / NE Sudan. Thematic map 1:50,000, Sheet Marsa Fijja. Institut für Geologie und Paläontologie, Technische Universität, Berlin.

Schroeder. J.H. & Nasr. D.H. 1983. The fringing reefs of Port Sudan, Sudan I : Morphology, Sedimentology and Zonation. *Essener Geogr. Arb.*, Bd. 6: 19-44.

Schweinfurth, G. 1968. Pflanzengeographische Skizze des gesamten Nilgebiets und der Uferländer des Roten Meeres. *Petermanns Geogr. Mitt.* 14: 113-129, 155-169, 244-248.

Scott, H. 1952a. Journey to the Gughé Highlands (Southern Ethiopia) 1948-1949. *Proc. Linn. Soc. Lond.* 163: 85-189.

Scott, H. 1952b. La végétation de la haute Éthiopie centrale et méridionale. *Lejeuna* 16: 67-80.

Scott, H. 1955. Journey to the high Semien District, Northern Ethiopia 1952-1953. *Webbia* 11: 425-450.

Scott, H. 1958. Biogeographical research in High Semien (Northern Ethiopia), 1952-1953. *Proc. Linn. Soc. Lond.* 170: 1-91.

Sebald, O. 1972. Bericht über botanische Studien und Sammlungen bei Lalibela, am Tana-See und im Awash-Tal (Äthiopien). *Stuttgarter Beitr. Naturk.* 236: 1-32.

Senni, L. 1935. Gli alberi e le formazione legnose della Somalia. Florence, Inst. Agric. Colon. Ital. 305 pp.

Sheppard, C., Price, A. & Roberts, C. 1992. Marine ecology of the Arabian region. Academic Press, London, 360 pp.

Smith, J. 1949. Distribution of tree species in the Sudan in relation to rainfall and soil texture. *Bull. Minist. Agric. Sudan* 4: 1-63.

Stanley, S. M. 1994. Historische Geologie. Eine Einführung in die Geschichte des Lebens. Heidelberg, Berlin, Oxford. 634 pp.

Thulin, M. (ed.) 1993. Flora of Somalia. Vol. 1, 493 pp., Royal Kew Botanical Gardens, Kew, Richmond, UK.

Thulin M. (ed.) 1995. Flora of Somalia. Vol. 4, 298 pp. Royal Botanic Gardens, Kew, Richmond, UK.

Thulin M. (ed.) 1999. Flora of Somalia. Vol. 2, 303 pp. Royal Botanic Gardens, Kew, Richmond, UK.

Verdcourt, B. 1969. The arid corridor between the north-east and south-west areas of Africa. *Palaeoecology of Africa*. 4: 14-144.

White, F. 1983. The vegetation of Africa. A descriptive memoir to accompany the UNESCO/AETFAT/UNSO vegetation map of Africa. UNESCO.

Wickens, G.E. 1977. The flora of Jebel Marra (Sudan Republic) and its geographical affinities. *Kew Bull. Add. Ser.* 5: 1-368.

Wickens, G.E. & Collier, F.W. 1971. Some vegetation patterns in the Republic of Sudan. *Geoderma* 6: 43-59.

Willimot, S.G. 1957. Soils and vegetaion of the Boma plateau and eastern District, Equatoria. *Sudan Notes Rec.* 38: 10-20.

Wilson, R.T. 1977. The vegetation of Central Tigre, in relation to its Land-Use. *Webbia*, 32 (1) : 235-270.

Worall, G.A. 1959. The Butana grass patterns. *J. Soil Sci.* 10: 34-53.

Worall, G.A. 1960a. Patchiness in vegetation in the Northern Sudan. *J. Ecol.* 48: 1-7-115.

Worall., G.A. 1960b. Tree patterns in the Sudan. *J. Soil. Sci.* 11: 63-67.

SHORT COMMUNICATION

FLORA AND PHYTOGEOGRAPHICAL ASPECTS OF THE RED SEA SABKHAT IN EGYPT

AHMED G. FAHMY[1] & BENNO BÖER[2]

[1]*Department of Botany, Faculty of Science, University of Helwan, Cairo, Egypt.*

[2]*UNESCO Regional Office in the Arab States of the Gulf, P.O. Box 3945, Doha, Qatar*

INTRODUCTION

Egypt occupies a geographical position of special interest regarding the location towards the Red Sea, where it possesses two coasts: the western African and the eastern Asian. The western Red Sea coast of Egypt represents the northeast corner of Africa. It extends from Suez to Marsa Halaib covering a distance of ca. 1100 km. The eastern coast is located along the Sinai Peninsula. It includes the eastern shores of the Gulf of Suez and western coasts of the Gulf of Aqaba. The eastern coast of the Gulf of Suez extends from El Shatt in the north to Ras Mohammed in the south for about 340 km. The western coast of the Gulf of Aqaba extends for about 235 km from Taba to Ras Mohamed, where it meets the eastern coast of the Gulf of Suez.

The Red Sea coast of Egypt belongs to the subtropical high pressure belt with an extremely arid climate. Available meteorological data obtained from Zahran & Willis (1992) show that the mean annual rainfall ranges between 25 mm in Suez, 4 mm in Hurghada and 3.4 mm in Quseir. Rain usually occurs during the winter months and the summer is dry (i.e. Mediterranean affininty). The temperatures are generally high and range between 14°C and 21.7°C in winter and 23.1°C and 46.1°C in summer. Average values of relative humidity range between 43% in summer and 65% in winter. The Pich evaporation is in summer is between 13.7 and 21.5 mm/day and in winter between 5.2 and 10.4 mm/day. Orographic precipitation, especially on

Barth & Böer (eds.), SABKHA ECOSYSTEMS, 161-169

slopes of high mountains, is an appreciated source of water which supports plant life along the Red Sea coast (Kassas 1956).

Since the 19th century, botanists and geographers have focused their studies on plant life at the Red Sea coast in Egypt. In this regard floristic studies were undertaken by Ruprecht (1849), Schweinfurth (1896-1899), Ascherson & Schweinfurth (1889), Drar (1936). Habitat types of the Hurghada district and a Hurghada-Qussier-Qena transect were described by Montasir (1938). A baseline survey of plant life in the Red Sea region of Egypt was done by Hassib (1951).

The studies of Kassas (1956, 1957 and 1960) have made significant headway in our understanding of plant ecology of the Red Sea coastal land and give detailed analyses of the plant community types, their vegetation and ecological relationships. These studies also have formed a corner stone for further detailed account of the vegetation types of the Red Sea coastal land of Egypt by Kassas & Zahran (1962, 1967, 1971) and Zahran (1962, 1964, 1965, 1977). These studies show that the coastal land is organised into three principal ecosystems: coastal salt marsh, coastal desert plain and coastal mountains. The type and extent of the plant cover in each ecosystem is influenced by various environmental conditions like tidal and wave movement, sea spray, salt water seepage, character of surface sediments and substratum, land relief and local climate as well as microclimate (Zahran & Willis 1992).

METHODS

A list of plant taxa of the study area has been prepared based on herbarium specimens kept at Cairo University Herbarium as well as on records from the scientific literature dealing with the subject. For each taxon, valid Latin name is checked in the up to date literature (Boulos 1995, 1999, 2000). Floristic categories are according to Wickens (1976).

FLORA AND VEGETATION

The present study concludes that flora of the Red Sea sabkhat in Egypt includes at least 104 plant taxa belonging to 94 genera and 38 families (Tab. 1). This figure is relatively high when compared to the harsh environmental conditions that prevail in littoral salt marshes (high salinity and temperature). This could be attributed to the capability of many xerophytes to grow on outskirts of the salt marshes. This is clear at localities close to the wadi outlets at the Red Sea. These taxa are characterised by a wide ecological amplitude and immense power of expansion. The following xerophytes have been recorded from sabkhat of the Red Sea in Egypt: *Alhagi graecorum, Anabasis articulata, A. setifera, Astragalus vogelii, Atriplex lecuoclada,*

Calligonum polygonoides, Calotrois procera, Capparis deciduas, Capparis sinaica, cornulaca monacantha, Crotalaria aegyptiaca, Ephedra alata, Fagonia glutinosa, Gymnocarpus decander, Heliotropium arbainense, Hyoscyamus muticus, Leptadinia pyrotechnica, Monsonia nivea, Neurada procumbens, Ochradenus baccatus, Opophytum forsskaolii, Panicum turgidum, Pergularia tomentosa, Pulicaria incisa, Retama raetam, Salsola schweinfurthii, Salvadora persica, Schowuia purpurea, Zilla spinosa, Zygophyllum album, Z. coccineum, Z. decumbens.

74 species have been recorded from the African coast while 31 species from the eastern coast of the Gulf of Suez and 34 species from the western coasts of the Aqaba Gulf.

Zahran & Willis (1992) and Kassas (1993) note that vegetation of the littoral belt of the Red Sea in Egypt comprises of mangal vegetation and coastal salt marshes. Lagoons, bays, coral and sand bars parallel to the shore provide niches for successful growth of *Avicennia marina* seedlings (Zahran 1977). In Egypt, mangrove vegetation is usually monotonous and consists of *Avicennia marina.* On the other hand, *Rhizophora mucronata* may grow in form of pure stands or mixed with *Avicennia marina* as a co-dominant or as a common species forming a tree layer above the thickets of *Avicennia marina* (i.e. Mersa Abu Fissi, Lat. 23° N to 22°40' N). Kassas (1993) notes that the ground layer of the mangal vegetation includes submerged phanerogams: *Cymodocea ciliata, C. serrulata, Halophila stipulacea, H. ovalis* and *Halodule uninervis.* The mangal vegetation is restricted to the Red Sea, i.e. it does not penetrate northward into the Gulf of Suez (Kassas 1960).

According to Zahran & Willis (1992) the coastal salt marsh vegetation comprises the following community types: *Halocnmum strobilaceum* community type, *Anthrocnemum macrostachyum* community type, *Halopeplis perfoliata* community type, *Limonium pruinosum* community type, *Limonium axillare* community type, *Aeluropus lagopoides* community type, *Zygophyllum album* community type, *Nitraria retusa* community type, *Halopyrum mucronatum* community type and *Tamarix nilotica* community type. Kassas & Zahran (1967) note that these community types are organised into zones following the shore-line.

Kassas (1993) notes that a number of halophytic community types are limited in their distribution: *Tamarix passerinoides* community type (El Mallaha, 260 km south of Suez), *Juncus rigidus* community type (Ain Sokhna salt marsh, 50 km south of Suez), *Cressa cretica* community type (Wadi Hommath, 30 km of Suez and Wadi Araba, 120 km south of Suez). It was very interesting to record a *Sarcocornia fruticosa* community type at Ras Ghareb, 240 km south of Suez which is common in the Mediterranean littoral.

According to Zahran & Willis (1992), the following community types were recorded from eastern coast of the Gulf of Suez: *Halocnemum strobilaceum, Aeluropus sp., Zygophyllum album, Nitraria retusa, Halopeplis perfoliata, Limonium pruinosum* and *Cressa cretica.*

The littoral salt marsh vegetation of the Gulf of Aqaba western coast is dominated by *Limonium axillare* community type. *Avicennia marina, Nitraria retusa* and *Zygophyllum album* have been recorded as common associates (Zahran & Willis, 1992). On the other hand, the coastal desert and wadis of the same area bears xerophytic vegetation (Tab. 1).

Table 1. Flora of the Red Sea sabkhat in Egypt.

Taxon	**African Coast**	**Gulf of Suez**	**Gulf of Aqaba**
Abutilon fruticosum Guill. & Perr.			■
Acacia tortilis (Forssk.) Hayne	■		■
Achillea fragrantissima Sch. Bip.		■	
Aeluropus lagopoides (L.) Trin ex Thwaites	■		
Aerva javanica (Burm. F.) Juss. ex Schult.	■		■
Agathophora alopecuroides Bunge		■	
Azioon canariense L.	■		
Alhagi graecorum Boiss.	■	■	
Amaranthus graecizans L.	■		
Anabasis articulata (Forssk.) Moq.		■	
Anabasis setifera Moq.	■		
Aristida adscensionis L.	■		
Aristida funiculata Trin. & Rupr.	■		
Arnebia hispidissima (Lehm.) DC.	■		
Arthrocnemum macrostachyum (Moric.) K. Koch	■	■	
Asphodelus tenuifolius Cav.	■		
Astragalus eremophilus Boiss.	■		
Astragalus vogelii (Webb) Bornm.	■		
Atriplex farinosa Forssk.	■		
Atriplex leucoclada Boiss.		■	
Avicennia marina (Forssk.) Vierh.	■	■	
Balanites aegyptiaca (L.) Delile	■		
Blepharis ciliaris (L.) B. L. Burtt			■
Calligonum polygonoides L.	■		
Calotropis procera (Aiton) W. T. Aiton	■		■
Capparis decidua (Forssk.) Edgew.			■
Capparis sinaica Veill. In Duh.			■
Caylusea hexagyna (Forssk.) M. L. Green	■		
Cleome chrysantha Decne.			■
Convolvulus lanatus Vahl	■		
Cornulaca monacantha Delile	■		
Cressa cretica L.	■	■	
Crotalaria aegyptiaca Benth.			■
Crotalaria microphylla Vahl	■		
Cymbopogon schoenanthus (L.) Spreng.			■
Cymodocea rotundata Asch.	■		
Cymodocea serrulata (R. Br.) Asch. & Magn.	■		
Cyperus conglomeratus Rottb.	■		■
Ephedra alata Decne.	■		
Fagonia glutinosa Delile		■	
Gymnocarpos decandrus Forssk.		■	■
Halocnemum strobilaceum (Pall.) M. Bieb.	■	■	

Taxon	African Coast	Gulf of Suez	Gulf of Aqaba
Halodule uninervis (Forssk.) Asch.	■		
Halopeplis perfoliata (Forssk.) Asch.	■	■	
Halophila ovalis (R. Br.) Hook. f.	■		
Halophila stipulacea (Forssk.) Asch.	■		
Halopyrum mucronatum (L.) Stapf	■		
Haloxylon elegans Bunge	■	■	■
Heliotropium arbaiense Fresen			■
Heliotropium pterocarpum (DC.) Bunge	■		
Heliotropium ramossissimum A. DC.	■		
Hibiscus micranthus L. f.			■
Hyoscyamus muticus L.	■	■	
Hyphaene thebaica (L.) Mart.	■		
Imperata cylindrica (L.) Raeusch	■		
Indigofera argentea Burm. F.	■		
Juncus rigidus Desf.	■	■	
Lasiurus scindicus Henrard			■
Launaea capitata (Spreng.) Dandy	■		
Launaea cassiniana (Boiss.) Kuntze	■		
Launaea spinosa (Forssk.) Sch. Bip. ex Kuntze	■	■	■
Leptadenia pyrotechnica (Forssk.) Decne.	■		■
Limonium axillare (Forssk.) Kuntze	■		■
Limonium pruinosum (L.) Chaz	■	■	
Lindenbergia sinaica (Decne.) Benth.			■
Lotononis platycarpos (Viv.) Pic. Serm.	■		
Lotus glinoides Delile	■		
Monsonia nivea (Decne.) Webb	■		
Neurada procumbens L.	■	■	
Nitraria retusa (Forssk.) Asch.	■	■	■
Ochradenus baccatus Delile	■		
Opophytum forsskaolii (Boiss.) N. E. Br.		■	
Panicum turgidum Forssk.	■		■
Pergularia tomentosa L.		■	
Polycarpaea repens (Forssk.) Asch.	■	■	
Pterogaillonia calycoptera (Decne.) Lincz.			■
Pulicaria incisa (Lam.) DC. Subsp. *candolleana* E. Gamal-Eldin			■
Reaumuria hirtella Jaub. & Spach.		■	
Retama raetam (Forssk.) Webb & Berthel	■	■	
Rhizophora mucronata Lam	■		
Salsola cyclophyla Baker			■
Salsola imbricata Forssk.	■		
Salsola schweinfurthii Solms			■
Salsola tetrandra Forssk.		■	
Salsola villosa Schult.	■		
Salvadora persica L.			■
Schouwia purpurea (Forssk.) Schweinf.			■
Seidlitzia rosmarinus Bunge ex Boiss.			■
Sevada schimperi Moq.	■		
Sphaercoma hookeri T. Anderson	■		
Sporobolus spicatus (Vahl) Kunth	■		
Suaeda aegyptiaca (Hasselq.) Zohary		■	
Suaeda monoica Forssk.	■		
Suaeda vermiculata Forssk. ex J. F. Gmel	■	■	
Tamarix aphylla (L.) H. Karst	■		■

Taxon	African Coast	Gulf of Suez	Gulf of Aqaba
Tamarix nilotica (Ehrenb.) Runge	■	■	■
Tamarix passerinoides Delile ex Desv.	■		
Taverniera aegyptiaca Boiss.	■		■
Thalassodendron ciliatum (Forssk.) Hartog	■		
Zilla spinosa (L.) Prantl		■	■
Zygophyllum album L. f.	■	■	■
Zygophyllum coccineum L.	■	■	
Zygophyllum decumbens Delile	■		
Zygophyllum simplex L.	■		

PHYTOGEOGRAPHY

Phytogeographical analysis reveals that flora of the Red Sea sabkhat is characterised by a homogenous structure (Figs. 1-3). Table 2 shows in detail that phytogeographical elements influence the area under investigation. The highest percentage (35%) of monoregionals, i.e. plant species confined to a single phytogeographical region, is known from the African coast (Fig. 1). The Saharo-Sindian species are more dominant when compared with the Sudano-Zambesian ones (Tab. 2).

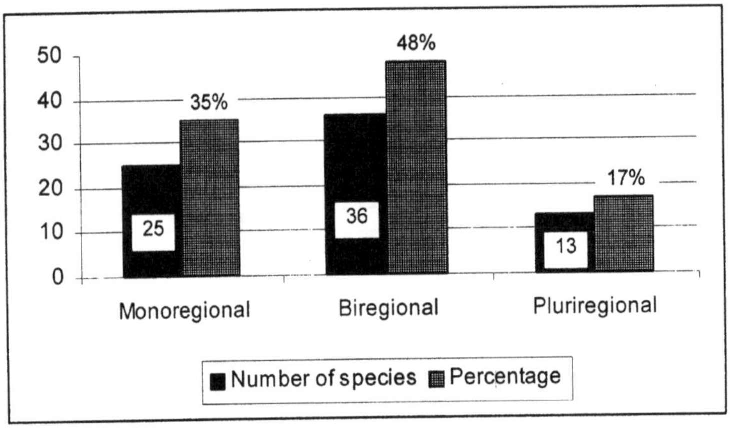

Figure 1. Number and percentage of plant species recorded from the Red Sea sabkhat of the African coast, Egypt.

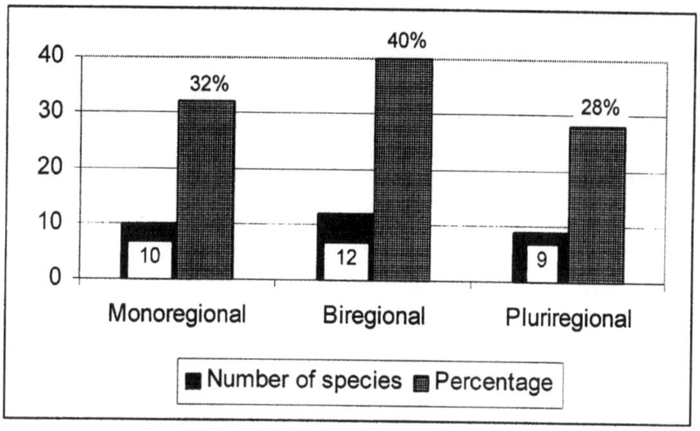

Figure 2. Number and percentage of plant species recorded from sabkhat of the Gulf of Suez, Egypt.

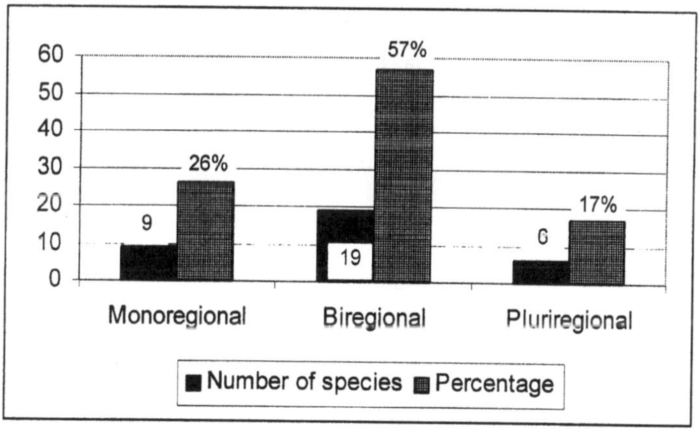

Figure 3. Number and percentage of plant species recorded from sabkhat of the Gulf of Aqaba, Egypt.

The Gulf of Aqaba is characterised by the highest record of biregionals (57%). Saharo-Sindian / Sudano-Zambesian species which are very common across the study area (Fig. 3, Tab. 2). The highest record of pluriregionals (28%) is known from the eastern coast of the Gulf of Suez, Sinai. The pluriregionals are dominated by Saharo-Sindian / Sudano-Zambesian / Irano-Turanian and Saharo-Sindian / Mediterranean / Irano-Turanian taxa.

Table 2. Phytogeographical elements of the study area.

Monoregionals		
African coast	Sa - Si	16 species
	S - Z	5 species
	Tropics and Sub tropics	4 species
Gulf of Aqaba	Sa - Si	9 species
Gulf of Suez, Sinai	Sa - Si	7 species
	S - Z	1 species
	Tropics and Sub tropics	2 species
Biregionals		
African coast	Sa - Si / S - Z	29 species
	Sa - Si / M	3 species
Gulf of Aqaba	Sa - Si / S - Z	18 species
	Sa - Si / Ir - Tr	1 species
Gulf of Suez, Sinai	Sa - Si / S - Z	4 species
	Sa - Si / Ir - Tr	5 species
	Sa - Si / M	3 species
Pluriregionals		
African coast	Sa - Si / S - Z / Ir - Tr	7 species
	Sa - Si / M / Ir - Tr	4 species
	Sa - Si / M / S - Z	2 species
Gulf of Aqaba	Sa - Si / S - Z / Ir - Tr	3 species
	Sa - Si / M / Ir - Tr	1 species
	Sa - Si / M / S - Z	2 species
Gulf of Suez, Sinai	Sa - Si / S - Z / Ir - Tr	1 species
	Sa - Si / M / Ir - Tr	5 species
	Sa - Si / M / S - Z	3 species

Sa - Si : Saharo - Sindian
S - Z : Sudano - Zambesian
M : Mediterranean
Ir - Tr : Irano - Turanian

The present analysis shows that Saharo-Sindian and Sudano-Zambesian elements have major influence on flora of the Red Sea Sabkhat in Egypt. A minor influence of Mediterranean and Irano-Turanian elements has been recorded along the African coast of the Red Sea and western coast of the Gulf of Suez in Sinai. The western coast of the Gulf of Aqaba acts as a migratory track for the Sudano-Zambesian and tropical taxa. Kassas (1993) mentions that the warm strip of the Aqaba area is protected against the cold northerly winds of the Sinai. In this regard the growth of *Avicennia marina* in Ras Mohamed and further north until Nabaq form its most northerly limit in the world.

This short communication presents important phytogeographical data, as well as botanical data, which are important for coastal management practises, and in particular for the establishment of coastal nature reserves, but also for the site selection of halophytic agro-ecosystems.

References

Ascherson, P. & Schweinfurth, G. 1889. Illustration de la Flora d' Egypte. Mem. Inst. Egypte, 3 (1): 25-260.

Boulos, L. 1995. Flora of Egypt: checklist. Al Hadara publishing, Cairo. 283 pp.

Boulos, L. 1999. Flora of Egypt, vol. I. Al Hadara publishing, Cairo. 419 pp.

Boulos, L. 2000. Flora of Egypt, vol II. Al Hadara publishing, Cairo. 352 pp.

Drar, M. 1936 Enumeration of the plants collected at Gebel Elba during two expeditions. Fouad I Agricultural Museum Tech. And Sci. Ser., Ministry of Agriculture, Egypt, No 149, 123 pp. Cairo.

Hassib, M 1951. Distribution of plant communities in Egypt. *Bull. Fac. Sci. Univ. Fouad I*, Cairo, Egypt, 29: 59-261.

Kassas, M. 1956. The mist oasis of Erkwit. *Sudan. J. Ecol.* 44: 180-194.

Kassas, M. 1957. On the ecology of the Red Sea coastal land. *J. Ecol.* 45: 187-203.

Kassas, M. 1960. Certain aspects of landform effects on plant-water resources. *Bull. Soc. Geogr. Egypte* 33: 45-52.

Kassas, M. 1993. Habitat Diversity: Egypt. Publication of National Biodiversity Unit, Egyptian Environment Affair Agency (EEAA), Cairo, 62 pp.

Kassas, M. & Zahran, M 1962. Studies on the ecology of the Red Sea coastal land, I. The district of Gebel Ataqa and El Galala El Baharriya. *Bull. Soc. Geogr. Egypte* 35: 129-175.

Kassas, M. & Zaran, M. 1967. On the ecology of the Red Sea littoral salt marsh, *Egypt. Ecol. Monogr.* 37(4): 297-315.

Kassas, M. & Zahran, M. 1971. Plant life on the coastal mountains of the Red Sea, *Egypt. J. Ind. Bot. Soc.* Golden Jubilee Volume, 50A: 571-589.

Montasir, M. 1938. Egyptian soil structure in relation to plants. *Bull. Fac. Sci., Egyptian Univ.*, Cairo, 15: 1-47.

Ruprecht, F.J. 1849. Die Vegetation des Roten Meeres. *Mem. Soc. Sci. Nat. Petersburg* 6: 71-84.

Schweinfurth, G.A. 1896-1899 Sammlung Arabisch-Aethiopischer Pflanzen. Bull. Inst. De l'Herbier Boissier, App. No II, vols 4-7., Geneve.

Zahran, M. 1962. Studies on the ecology of the Red Sea coastal land. MSc Thesis, Fac. Sci., University of Cairo.

Zahran, M. 1964. Contributions to the studies on the ecology of the Red Sea coast. PhD Thesis, Fac. Sci., University of Cairo.

Zahran, M. 1965. Distribution of mangrove vegetation in Egypt. *Bull. Inst. Desert Egypte* 15(2): 7-12.

Zahran, M. 1977. Africa. A. Wet formations of the African Red Sea coast. In: Chapman, V.J. (ed.), Coastal Ecosystems. Elsevier, Amsterdam, pp. 215-231.

Zahran, M. & Willis, A.J. 1992. The vegetation of Egypt. Chapman and Hall. Pub. London, 424 pp.

PART II

SABKHA ECOLOGY

BENNO BÖER & HANS-JÖRG BARTH

INTRODUCTION

Sabkhat are distinctive geomorphological features because of their flatness and total absence of vegetation. The latter leads to the impression that they are of ecologically limited value. The scientific work in the region has focused on sabkha geology and mineralogy for economic reasons and started in the early 1960s in Qatar and the UAE. Studies related to other aspects were scarce. Therefore up to now no comprehensive picture on their ecological importance can be drawn.

Since we have such limited knowledge about sabkha ecosystems, other than that they are saline flats with shallow ground water, and how they develop, it is important to conduct research into the different ecosystem components. Ecology means the pattern of relationships between organisms and their environment. Therefore, if we want to describe the ecology of sabkha environments, we certainly need to know which organisms occur, what the abiotic factors are like, and how the ecosystems function. Questions about biological productivity and interrelations with adjacent ecosystems are only a few out of hundreds that need answering. Only with a broad and interdisciplinary knowledge based on science will it be possible to estimate the ecological importance of sabkhat. Application of this knowledge is relevant for all coastal managers, as well as inland-sabkha land owners, land users, decision makers, scientists, and other stakeholders. Some of these aspects will be covered in part III of this volume.

This section aims to provide an introduction to the sabkhat as ecosystems. Individual contributions cover abiotic aspects, such as hydrology (Wood & Sanford), geology (Alsharhan et al., Al-Farraj), soils (Lieth & Menzel), sedimentation and erosion (Barth), as well as biotic aspects, including micro-organisms, such as biotic crusts (Belnap), and contributions on diatoms (Round), arthropods (Hogarth & Tigar), vertebrates (Loughland & Cunningham), macro-algae (John), fishes (Zajonz et al.), seagrasses (Phillips) as well as vegetation (Böer) and birds (Aspinall).

Barth & Böer (eds.), SABKHA ECOSYSTEMS, 171

HYDROGEOLOGY AND SOLUTE CHEMISTRY OF THE COASTAL-SABKHA AQUIFER IN THE EMIRATE OF ABU DHABI

WARREN W. WOOD[1] & WARD E. SANFORD[2]

[1]U. S. Geological Survey, 430 National Center, Reston, VA 20192, USA
[2]U. S. Geological Survey, 431 National Center, Reston, VA 20192, USA

Abstract

Studies of the hydrogeology of the coastal-sabkha aquifer in the Emirate of Abu Dhabi indicate that rainfall is the dominant source of water, while ascending continental brine is the dominant source of solutes. A hydrologic analysis of a rectilinear volume of the aquifer perpendicular to the coastline (1-m wide, 10-m deep, and 10-km long) indicates that, on the average, less than 1 m^3 of water enters and exits by lateral ground-water flow each year, 44 m^3 of water enter by upward vertical leakage, and 640 m^3 enter by recharge from annual rainfall. Approximately 690 m^3 of water are lost by evaporation. In contrast, 100 kg of solutes enter by lateral ground-water flow, rainwater introduces 15 kg, and ascending vertical leakage introduces 3700 kg each year. Approximately 200 kg of solutes leave annually by lateral ground-water flow. Water in the sabkha-aquifer has a residence time of approximately 50 years, while a non-reactive solute has a residence time of approximately 26,000 years. The sabkha-aquifer is in equilibrium with the water, but not the solutes. Chemical analyses of solutes, sulfur isotopes, and evaporite minerals are consistent with the proposal that most of the solute mass is derived from ascending continental brines. While the details of this model are specific to the sabkhat of Abu Dhabi, the concepts, approach, and methods used to evaluate solute and water transport are universally applicable to other coastal sabkhat.

Barth & Böer (eds.). SABKHA ECOSYSTEMS, 173-185

INTRODUCTION

Coastal sabkhat (Arabic for "salt flat") are ubiquitous features in many arid and semi-arid areas of the world, with their surfaces typically covered with halite and other soluble chloride and nitrate minerals. A sabkha is by definition a two-dimensional surface. Salts observed on the coastal sabkha generally result from the capillary transport of shallow ground water to the surface, the water evaporates, resulting in the deposition of salts that exhibit polygonal forms defined by upturned pressure ridges (Smoot & Lowenstein 1991). This texture contrasts to typically smooth surfaces, like those of the Bonneville Salt Flat in the United States, where salt precipitates from standing water.

Two models in the literature suggest mechanisms for introducing solutes into the sabkhat of the Emirate of Abu Dhabi. In the "sea-water flooding" model, seawater occasionally floods the low-lying supratidal zone of the sabkha. As flood-water evaporates, anhydrite, calcite, and gypsum, are precipitated and the density of the remaining water increases. The dense water sinks through the sabkha-aquifer and ultimately discharges to the sea as ground water (Kinsman 1969, Butler 1969, Butler et al. 1973, and Patterson & Kinsman 1977, 1981, and 1982). In the second "evaporative-pumping model", water is removed by evaporation and is replaced by sea water (Hsü & Siegenthaler 1969, Hsü & Schneider 1973, McKenzie et al. 1980, Muller et al. 1990). The hydrology of the sabkha-aquifer system, however, suggests that neither of these two models is correct (Sanford & Wood 2001). An alternative "ascending-brine" model presented here suggests that water in the sabkha-aquifer is derived largely from local precipitation, while the solutes are derived largely from underlying continental brines that discharge to the sabkha-aquifer.

METHODS

Hydrogeologic framework

The coastal sabkha in the Emirate of Abu Dhabi in the United Arab Emirates is exposed as a strip approximately 300-km long by 15-km wide (Fig. 1). The area is flat, with a topographic and ground-water gradient of approximately 1 in 5000 toward the Gulf and a depth to ground water between 0.5 and 1 m. Polygons with pressure ridges give the surface 2 to 3 cm of local relief. The general geology associated with the sabkha is given in figure 2. The geologic basement consists of a thick (6000 to 8000 m) Paleozoic-Cenozoic-age sedimentary sequence of carbonates and evaporites that accumulated on the stable Arabian Platform (Alsharhan & Nairn 1997). Above this largely carbonate/evaporite sequence is a Miocene-age clastic deposit derived from rising mountains in western Saudi Arabia, Yemen, and Oman, known by the descriptive name, "Miocene clastic" (Alsharhan & Nairn 1997), or the "Baynunah Formation" (Whybrow 1989). In most areas occupied by the sabkha, the

Miocene clastic is between 20 and 30 m thick, but has been completely removed in some areas at the western extent of the sabkha. A few, widely scattered outcrops of the Miocene clastic exist near the proximal edges of the sabkha and are composed of fine quartz sand with carbonate cement and occasional thin, discontinuous layers of gypsum. These sediments exhibit a reddish-pink color on the Munsell soil color chart (2.5YR-7/4).

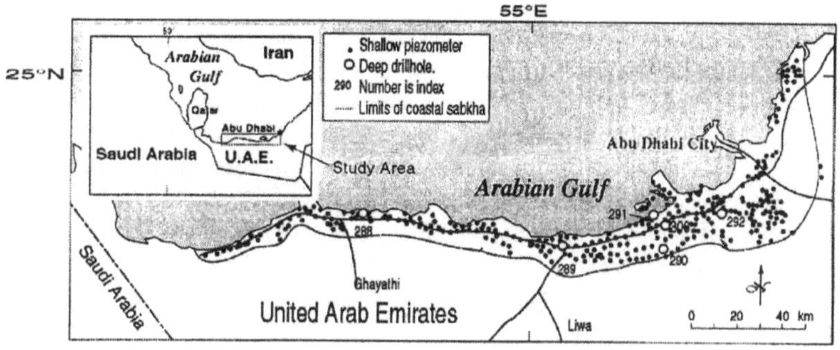

Figure 1. Location of shallow piezometers and deep wells on the coastal sabkha of the Emirate of Abu Dhabi.

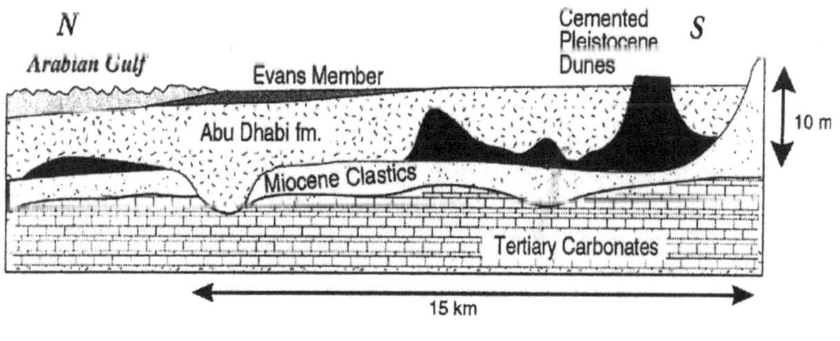

Figure 2. Generalised north-south geologic section of the coastal-sabkha aquifer of Abu Dhabi. Not to scale; representative formation thickness is given in the text. Width of sabkha varies from less than 2 km to more than 20 km.

The Miocene clastics are overlain in some areas by carbonate-cemented sand dunes of Pleistocene age. Remnants of these dunes can be seen on the east-west coastal road and on the roads leading from the coastal highway to both Liwa and Ghayathi (Fig. 1) and have been dated at approximately 26,000 years BP by Optical Stimulated Luminescence (OSL) (Stokes 2000). These carbonate-cemented remnants consist of 60 to 70% carbonate grains with the rest largely quartz and some detrital anhydrite. Their composition, however, becomes more quartz-rich inland from the coast (Hadley et al. 1998). The dune remnants are yellowish-orange (10YR-6/4), exhibit well-developed foreset bedding, and do not contain beds of gypsum. Thus, they are readily distinguishable in the field from outcrops of the Miocene clastic.

Sands of the Holocene-age sabkha-aquifer (Abu Dhabi Formation) are superimposed over Tertiary and Pleistocene sediments. The rise of sea level beginning about 18,000 years BP filled the Gulf to approximately the present level at about 7000 years BP and deposited reworked Holocene, Pleistocene, and Miocene dune sediments to create the aquifer upon which the sabkha later formed. OSL dating of a quartz sample 1.5 m below the surface and 4.5 km inland from the present high tide line suggests that this sample was deposited 8000 years BP (Stokes 2000), confirming the general age of the aquifer. Since the time of formation of the main sand body, a thin, 0.1 to 2 m thick, wedge-shaped sand (Evans member) has been added to the distal half of the sabkha-aquifer by progradation of the sediments into the Gulf. Because of the low topographic gradient and abundant sand source, the progradation has proceeded rapidly at a rate between 1 and 2 km every thousand years (Evans 1995). The low topographic gradient of the sabkha is controlled by the dynamic equilibrium of the ground water elevation, sediment supply, and wind speed. Eolian processes cannot erode the sabkha-aquifer below the capillary fringe 0.5 to 1 m above the water table.

The geometry of the Abu Dhabi Formation profile is variable, but it is generally wedge-shaped, with zero thickness at the proximal edge to approximately 15 m at the distal edge at the coast. The presence of paleo-erosional channels or remnants of cemented Miocene clastic or Pleistocene dunes increases or decreases the thickness locally. Below the water table, the Abu Dhabi Formation consists of uniform, fine sand (0.16-0.22 mm) composed of detrital carbonates (60%) and quartz (35%), with minor amounts of feldspars, anhydrite, and heavy minerals. Several laboratory analyses of repacked sand yield a consistent porosity of 0.38. Because of the uniformity of the sand and its uncemented nature, we believe that this value is also representative of the *in situ* porosity. Field tests indicate a hydraulic conductivity between 1 and 2 m/d (meters per day) (Sanford & Wood 2001). The sand grains are covered with iron and manganese oxide/hydroxide, giving the sediment a light yellowish-orange color (10YR-6/4). X-ray analysis of the grain coatings shows trace amounts of the clay minerals smectite, palygorskite, and illite.

The authigenic chemical sediments are confined to the unsaturated zone approximately 1 m above the water table. In the areas we have investigated, most of the anhydrite consists of clay-sized grains that give it a "toothpaste" consistency. "Chicken wire" anhydrite nodules typically underlie this structureless anhydrite in many areas. Minor amounts of authigenic calcite and dolomite are mixed with the anhydrite. Equidimensional crystals of gypsum up to 0.5 cm in length exist close to the water table. Rubbery mats of cyanobacteria (blue-green algae), typically 5 mm thick, cover the surface of the intertidal zone in many areas, and these have been buried in the prograded sediments.

Anhydrite, calcite, dolomite, and gypsum accumulate in the unsaturated zone rather than on the surface because of their retrograde solubility. Water and solutes in the ground water are at thermodynamic equilibrium or supersaturated with these minerals. As solutes are transported to the surface by capillary forces, anhydrite, calcite, and gypsum precipitate because of increased temperature of the solution during the summer months. The average temperature of ground water in the 10 m thick aquifer is approximately 30°C, while the temperature on the surface in the summer can exceed 65°C, or a gradient of 35°C/m. The gradient is reversed in the winter, when warmer ground water is transported upward, cooled, and becomes thermodynamically undersaturated and able to dissolve some of the previously precipitated minerals. The winter temperature gradient is approximately -15°C/m; thus, the thermodynamic solution capacity of the water is lower. Because the water fluxes are nearly the same in summer and winter (Sanford & Wood 2001), the winter transport of water is unable to dissolve all of the previously precipitated retrograde minerals. This condition results in a net accumulation of these minerals in the unsaturated zone.

Water fluxes

Precipitation on the sabkha decreases from approximately 100 mm/y in the east to less than 50 mm/y in the west. Evaporation from the surface is known to be considerably less than the 3800 mm/y potential evaporation from a free water surface (Bottomley 1996). Measurements were made directly from the sabkha surface using a humidity chamber and a technique described by Stannard (1988). Evaporative flux at a site with approximately 70 mm/y rainfall was found to be 0.24 mm/d (68-mm/y), averaged over four seasons. At a second site, where rainfall is 50 mm/y evaporation was found to average 0.14 mm/d (51 mm/y). Thus, the evaporation rates from the surface of the sabkha were less than 2% of what would be expected from a free water surface.

The water table of the coastal sabkha of Abu Dhabi is flat, with gradients of approximately 1 m for every 4 to 5 km toward the coast. The water table is typically 0.5 to 1 m below the surface. Lateral ground-water flux through the sabkha-aquifer is between 70 and 80 mm/y, calculated using Darcy's Law (equation 1), a hydraulic conductivity of 1 m/d, and a gradient of 0.0002:

$$q = -k\ (dh/dl) \qquad (1)$$

where q is the flux in dimensions of L/T, k is the hydraulic conductivity of the porous medium in units of L/T, and dh/dl is the gradient of the hydraulic head and is dimensionless (L is length and T is time).

Because of the low topographic elevation, the sabkhat act as regional ground-water discharge areas for deeper Paleozoic through Cenozoic aquifers, and hydraulic heads in the piezometers in the Tertiary formation are approximately 10 m above land surface in the area of the coastal sabkha. The hydraulic head for this regional flow system is believed to be provided by the Asir Mountains in western Saudi Arabia and Yemen, and the Hajar Mountains in Oman, where elevations above sea level exceed 3000 m. Upward vertical flux from the underlying formations is approximately 4 mm/y, determined from a vertical gradient of approximately 0.1 and vertical hydraulic conductivity of approximately $1x10^{-4}$ m/d (Sanford & Wood 2001).

The application of these annual water fluxes to a representative rectilinear volume of the aquifer 1-m wide, 10-m deep, and 10-km long results in the following values: less than 1 m^3/y enters and exits by lateral ground-water flow, 45 m^3/y enter by upward vertical leakage, 640 m^3/y enter by recharge of rainfall, and 690 m^3/y are lost to evaporation (Fig. 3a). These values have an error band of ±20 % (Sanford & Wood 2001).

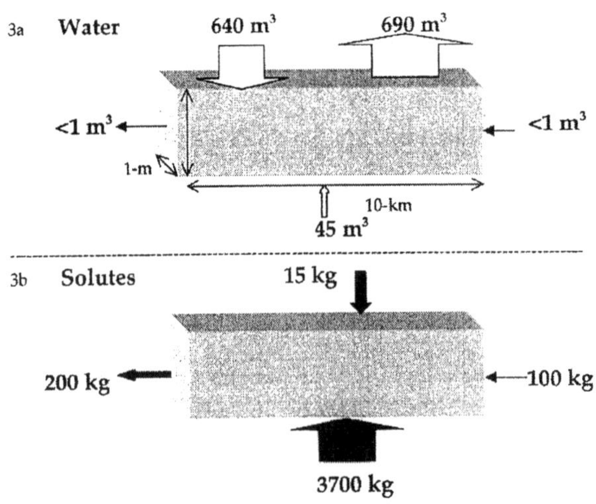

Figure 3. Diagram illustrating annual water (a) and solute (b) flux in a control volume 1-m by 10-m by 10-km representative of the aquifer supporting the coastal-sabkha in the Emirate of Abu Dhabi.

Solute flux and source

Water fluxes determined above multiplied by the average dissolved solids (Tab. 1) result in the following solute-mass flux for the same representative rectilinear volume: less than 100 kg/y enter from lateral flow, approximately 15 kg/y enter by rainfall, and approximately 3700 kg/y enter by upward vertical leakage (Fig. 3b). Approximately 200 kg/y are discharged by lateral flow.

The average steady-state residence time for water in the same rectilinear volume is approximately 50 years, calculated using the following lumped parameter expression:

$$R_w = M_t/M_f \qquad (2)$$

where, R_w is the average residence time of the water (T) in the sabkha-aquifer, M_t is the total mass of water in a representative rectilinear volume, and M_f is net mass flux to the representative rectilinear volume (M/T) (M is mass and T is time).

The lumped parameter approach of dividing total mass by the annual mass flux used above (equation 2) for calculating water residence time is not appropriate for a non-steady-state system, and an alternative method for estimating solute residence time is used by modifying Darcy's Law:

$$q = -k(dh/dl)/\,\theta \qquad (3)$$

and

$$R_s = 2D_t/q \qquad (4)$$

where R_s is the solute residence time (T), θ is porosity (dimensionless), and D_t is total longitudinal flow path distance (L) (L is length).

If the following values are used - horizontal gradient of 0.0002, hydraulic conductivity of 1 m/d, porosity of 0.38, and total distance of 10 km - a residence time in this rectilinear volume would be approximately 26,000 years for a non-reactive solute that is ultimately discharged to the Gulf. The average solute residence time is assumed to be one-half of the total time required for a conservative solute to traverse the full length (10 km) of the rectilinear volume because solutes enter the control volume over its entire length, not just at the proximal end.

The solute flux data and residence time are compelling arguments for the most of the solute mass being derived from the underlying brines, but it must also be shown that the observed solute chemistry in the sabkha-aquifer is consistent with this model. One approach to confirm this consistency is by comparing the concentration of a solute calculated from the water flux data (pore volume) with the analogous concentration ratio (CR) of observed solute data. A pore volume is defined as:

$$PV = v\,\theta \qquad (5)$$

where, PV is the volume of fluid in the pores of a volume of saturated aquifer (L^3), and v is the volume of saturated aquifer (L^3). The porosity θ is defined above.

Hydrologic analysis from vertical gradients and hydraulic conductivity suggests that approximately 7 pore volumes of water from the underlying Tertiary formations have discharged to the original coastal-sabkha aquifer in the 7000 years since its deposition (Sanford & Wood in press).

The concentration ratio is calculated by use of the following relation:

$$CR = C_t/C_i \qquad (6)$$

where, CR is the concentration ratio from solute data (dimensionless), C_t is the average observed concentration in the sabkha aquifer (M/L^3), and C_i is the concentration of average input (M/L^3).

Table 1. Sources of solutes for the coastal sabkhat of Abu Dhabi.

[Analyses in mg/L, isotopes in per mil (o/oo). 2H and 18O are relative to VSMOW, 34S is relative to Canyon Diablo].

Source	Name	Depth, in m	Date sampled	Ca	Mg	Na	K	Sr	SiO_2	Alk as HCO_3	Cl	SO_4	Br	NO_3	Dissolved solids	Density at 20 °C	pH	$\delta^{34}S$ ‰	$\delta^{2}H$ ‰	$\delta^{18}O$ ‰
Sea Water	Gulf of Arabia	surface	Apr-96	625	1750	13000	625	13	4.2	152	24000	3450	72	2	43563	1.0290	7.6		17.7	2.63
Sea Water	Gulf of Arabia	surface	Apr-96	625	1700	13200	625	13	4.3	140	24000	3400	72	1	43656	1.0291	7.6		17.7	2.63
Sea Water	Khor al Bazam	surface	Apr-96	700	2150	16100	750	18	6.0	170	28500	3800	89	1	52130	1.0358	7.9		24.6	3.54
RP 5-35	Local Ground Water	10	May-96	2200	2000	34000	600	50	32	41	61000	6050		40	139697	1.0719	7.2	15.91	15.3	5.63
Tertiary brine	GWP288A	69	Jan-98	1450	950	10500	300	25	36	55	19700	4050	11	33	37078	1.0258	8.0	19.80	-6.2	2.37
Tertiary brine	GWP-289	119	Jan-98	1150	675	5600	120	23	15	49	10300	4000	5	8	21922	1.0146	7.8	19.84	0.7	2.92
Tertiary brine	GWP-290A	88	Feb-98	1500	950	15500	450	27	20	50	25000	6750	15	22	50251	1.0369	7.4	17.72	5.9	4.09
Tertiary brine	GWP-291A	142	Feb-98	2200	1350	31500	950	40	18	39	55100	4650	20	35	95867	1.0686	7.8	18.42	5.0	4.64
Tertiary brine	GWP-292	119	Feb-98	3600	2400	48000	1500	62	23	34	87300	3300	6	33	146239	1.0971	7.4	17.82	7.5	5.19
Tertiary brine	GWP-306A	129	Mar-99	2750	1250	34500	1100	50	16	64	62500	4800	12	50	107056	1.0820	7.7			
Sabkha	Ave. (N =161)	3	97-99	9280	9229	84039	4156	211	53	49	172849	4417	78	2283	286415	1.1858	6.7	15.77	27.6	6.5
Rain	Ave. (N = 4)		95-97	3.73	0.20	0.82	0.52			47	1.46	2.85			39		7.7			

If there are no other sources or sinks of solutes, (i.e., little or no solute discharge), then based on the calculated pore volumes the concentration of a conservative (non-reactive) solute should be approximately seven times the concentration of the input Tertiary brine. Magnesium and bromide are the least reactive of the major ions with CR values approximately 7 (Fig. 4). Thus, these values are nearly identical to the value independently calculated from the hydrologic data and are consistent with the hypothesis that most of the solutes are derived from the underlying continental brines.

The loss of halite from the solution is clearly shown by the low CR values and the concentration of sodium compared to that of chloride (Fig. 5). Similarly, the loss of sulfate and calcium from solution results from precipitation of gypsum and anhydrite. Mass-balance calculation (Wood et al. in press) shows that most of the missing sulfate and calcium are present in the unsaturated zone; halite, however, is not present in sufficient mass in the aquifer, in the unsaturated zone, or on the surface.

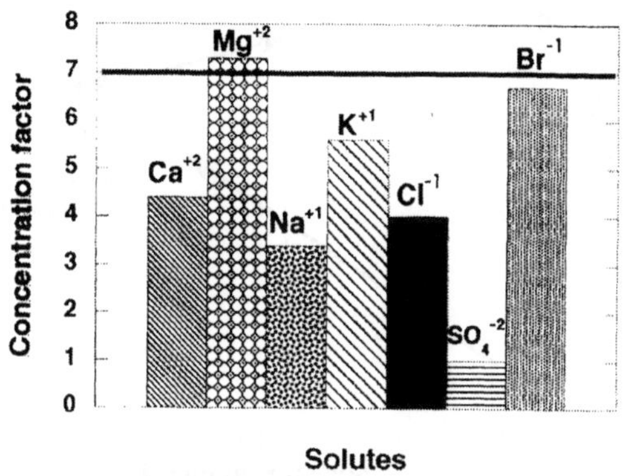

Figure 4. Concentration factor (ratio of existing concentration to input concentration) of Tertiary brines entering the coastal-sabkhat aquifer of the Emirates of Abu Dhabi. Line at a concentration factor of 7 is that calculated from physical hydrologic information.

The mechanism or combination of mechanisms by which this halite is lost from the system is not known. Halite may be removed from the surface by eolian processes or solute transport. The salts tend to be firmly cemented, however, and more "ice-like" rather than loose grains, so that our eolian samplers did not detect any significant mass transport in the two years of the study. Additionally, the concentration ratios for potassium and magnesium ions are different from those for sodium and chloride (Fig. 4), suggesting selective removal rather than loss of all surface salts by eolian erosion. Solute transport into the adjacent Arabian Gulf could occur by flooding of Gulf water whereby the flood water dissolves the salts and returns them to the Gulf as the flooding recedes. Because the halite is missing over the entire area, not just seaward of the trash line approximately 3 km inland, it appears that periodic flooding by the Gulf waters does not explain the missing salt. It is speculated that occasional major rainfall events dissolve the surface salts and transport them to the Gulf. It is known that approximately 200 mm of rain occurred in a 24-hour period at least once in the 30 years for which records exist (Bottomley 1996). In the rainfall scenario, recharge preferentially selects the most soluble salts and returns them to the aquifer before they are transported to the Gulf. This selective process is aided by the geometry of the polygonal surface features that tend to trap the first rainfall, forming millions of disconnected pools. This trapped water becomes thermodynamically saturated with the most soluble minerals. Transport of this water into the small volume of the unsaturated zone quickly fills the available space, preventing less soluble salts from being recharged. Only when the rain exceeds a certain flux do the pools coalesce and flow toward the Gulf. Thus, it is possible to selectively lose significant amounts of halite without the loss of associated, more soluble salts on the crusts.

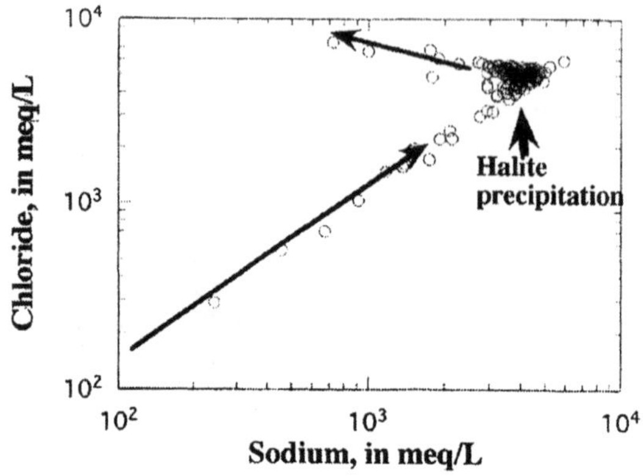

Figure 5. Concentration of sodium as function of chloride, in milliequivalents per litre. Arrows indicate concentration evolutionary path before and after halite precipitation.

The pH of the water in this sabkha-aquifer system appears to be controlled bythe formation of $Mg(OH)_2$ complex rather than carbonate chemistry. Because of the large concentration of calcium and magnesium, there are almost no free, un-complexed bicarbonate or carbonate ions in solution. The concentration of magnesium is so great that it almost certainly complexes a significant number of hydroxyl ions from water molecules. This complexing of the hydroxyl ions creates free hydrogen ions that, without the carbonate buffer, reduce the pH (Bodine 1976). Thus, increasing magnesium concentration results in a decreasing pH (Fig. 6).

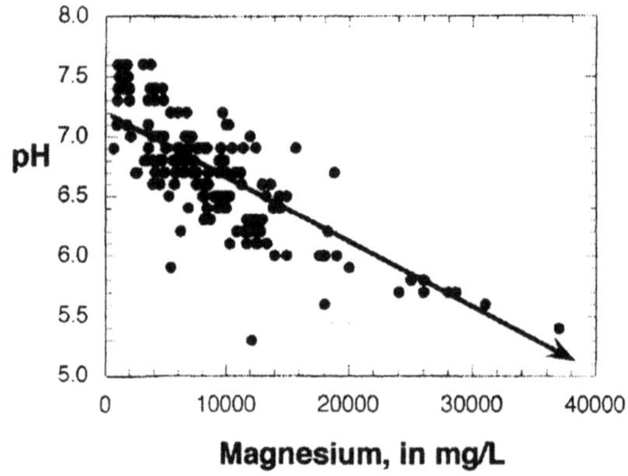

Figure 6. Decline in pH as a function of increasing magnesium concentration in the coastal-sabkha aquifer of the Emirates of Abu Dhabi.

The relative concentrations of calcium and sulfate also can be used to define the source of the solutes. It is known from the chemical divide concept (Hardie & Eugster 1970) that in a closed system, mineral precipitation will cause the concentration of one element of a precipitating mineral to approach zero, while the other element or elements will increase in concentration. The initial input concentration of the elements determines which element will be limiting. In the case of sea water, there is a greater equivalent of sulfate than calcium; thus, with precipitation of calcite and/or anhydrite and gypsum, the concentration of calcium will approach zero while sulfate concentration will increase. In the case of the ascending Tertiary brines, the average equivalents of calcium are greater than the average equivalents of carbonate and sulfate; thus, precipitation of anhydrite will result in sulfate approaching zero while calcium increases in concentration. In our analyses, calcium is increasing in most samples while sulfate is approaching zero (Fig. 7), which is consistent with a calcium and sulfate source from ascending Tertiary brines rather than from sea water. The few samples in figure 7 in which the sulfate is greater than calcium represent the supratidal zone near the distal end of the sabkha.

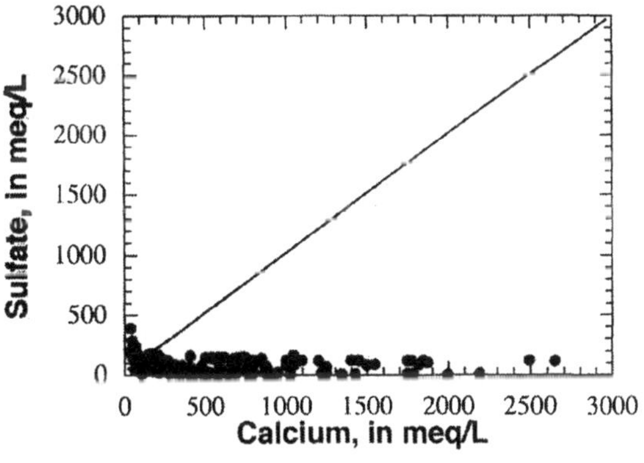

Figure 7. Concentration of sulfate as a function of calcium suggesting the decrease in sulfate with increasing calcium for most samples during evaporative concentration in the coastal-sabkhat aquifer of the Emirates of Abu Dhabi.

The density of ten samples of water ponded on the sabkha averaged 1.2675 g/cm^3 (Wood et al. in press), while the average density of 161 sabkha-aquifer samples averaged 1.1896 g/cm^3 (Tab. 1). The high density and solute concentration of this surface water are a result of dissolution of soluble chloride and nitrate minerals in the efflorescent surface crusts and concentration by evaporation. This difference in density suggests that when surface water is recharged, it will descend to the bottom of the sabkha-aquifer. Sanford & Wood (2001) show that the heads and Rayleigh numbers in 12 paired piezometers at the top and bottom of the aquifer are consistent with this instability of the two solutions. The observation that tritium exists throughout the vertical profile (Wood et al. in press and McKenzie et al. 1980) is consistent with vertical circulation resulting from this density instability.

RESULTS AND DISCUSSION

In the ascending-brine model of the coastal sabkha aquifer proposed here, most of the solutes are derived from continental brines, while the water is provided from local rainfall. When the sabkha-aquifer was formed between 7000 and 8000 years BP, it contained sea water in the interstitial pores. Brines from the underlying Tertiary formations then began discharging water and solutes to the newly-formed aquifer. Water and solutes were brought to the surface, precipitating anhydrite, calcite, and gypsum in the unsaturated zone, and soluble salts on the surface, forming the sabkha. Annual rainfall selectively dissolved the salts on the surface causing the water to become more dense and sink to the bottom of the aquifer. These recycled solutes were added to the input of solutes from the Tertiary brine, resulting in an increase of solutes with time and the accumulation of anhydrite, calcite, and gypsum in the unsaturated zone and halite on the surface. Over time, the solute ratios and concentrations of the original sea water were modified by the addition of brines from the underlying Tertiary formations and the precipitation of minerals. Because of progradation of the aquifer and the lack of time for recycling the surface salts, the sea water's "chemical signature" is greater at the distal end of the aquifer. This coastal-sabkha system has not yet reached chemical equilibrium and will not for many thousands of years under the present-day climatic, hydraulic, and sea level conditions. While the details of this model are specific to the sabkha of Abu Dhabi, the concepts, techniques, and methods of evaluating solute and water transport are universally applicable to current sabkhat in other areas.

ACKNOWLEDGEMENTS

We thank Mohamed B. Al-Qubaisi, Abu Dhabi National Oil Company, UAE, for his interest and support. We thank Craig Simmons and Michael Rosen for reviews that enhanced both the clarity of thought and the presentation. Publication is approved by the Director of the U. S. Geological Survey and the Abu Dhabi National Oil Company.

References

Alsharhan, A.S. & Nairn, A.E.M. 1997. Sedimentary basins and petroleum geology of the Middle East. Elsevier. Amsterdam.

Bottomley, N. 1996. Recent climate in Abu Dhabi. In: Osborne, P.E. (ed.), Desert ecology of Abu Dhabi. Pisces Publications, Newbury UK.

Bodine, M.W. Jr. 1976. Magnesium hydroxchloride: a possible pH buffer in marine evaporite brines? *Geology* 4: 76-80.

Butler, G.P. 1969. Modern evaporite deposition and geochemistry of coexisting brines, the sabkha, Trucial Coast, Arabian Gulf. *Journal of Sedimentary Petrology* 39: 70-81.

Butler, G.P., Krouse, R.H. & Mitchell, R. 1973. Sulphur-isotope geochemistry of an arid, supratidal evaporite environment, Trucial Coast. In: Purser, B.H. (ed.), The Persian Gulf. Springer-Verlag, New York.

Evans, G. 1995. The Arabian Gulf: a modern carbonate-evaporite factory; a review. In: Cuadernos de Geologia Iberica: Servicio de Publicaciones, Universidad Complutense, Madrid. 61-95.

Hadley, D.G., Brouwers, E.M. & Bown, T.M. 1998. Quaternary paleodunes, Arabian Gulf coast, Abu Dhabi Emirates: age and paleoenvironment evolution. In: Alsharhan, A.S., Glennie, K.W., Whittle, G.L. & Kendall, C.G. St. C. (eds.), Quaternary Deserts and Climate Change. A. A. Balkema, Rotterdam.

Hardie, L.A. & Eugster, H.P. 1970. The evolution of closed-basin brines. Mineral Society of America Special Publication No. 3: 273-290.

Hsü, K.J. & Siegenthaler, C. 1969. Preliminary experiments on hydrodynamic movement induced by evaporation and their bearing on the dolomite problem. *Sedimentology* 12:11-25.

Hsü, K.J. & Schneider, J. 1973. Progress report on dolomitization - hydrology of Abu Dhabi sabkhat, Arabian Gulf. In: Purser, B.H. (ed.), The Persian Gulf. Springer-Verlag, New York.

Kinsman, D.J.J. 1969. Modes of formation, sedimentary associations, and diagnostic features of shallow-water and supratidal evaporites. *American Association of Petroleum Geologists Bulletin* 53: 830-840.

McKenzie, J.A., Hsü, K.J. & Schneider, J.F. 1980. Movement of subsurface waters under the sabkha, Abu Dhabi, UAE, and its relation to evaporative dolomite genesis. SEPM Special Publication No. 28.

Müller, D.W., McKenzie, J.A. & Mueller, P.A. 1990. Abu Dhabi sabkha, Persian Gulf, revisited: application of strontium isotopes to test an early dolomitization Model. *Geology* 18: 618-621.

Patterson, R.J. & Kinsman, D.J.J. 1977. Marine and continental groundwater sources in a Persian Gulf coastal sabkha. *Studies in Geology* 4: 381-397.

Patterson, R.J. & Kinsman, D.J.J. 1981. Hydrologic framework of a sabkha along Arabian Gulf. *American Association of Petroleum Geologists Bulletin* 65: 1457-1475.

Patterson, R.J. & Kinsman, D.J.J. 1982. Formation of diagenetic dolomite in coastal sabkha along Arabian (Persian) Gulf. *American Association of Petroleum Geologists Bulletin* 66: 28-43.

Sanford, W.E. & Wood, W.W. 2001. Hydrology of the coastal sabkhat of Abu Dhabi, United Arab Emirates. *Hydrogeology Journal* 9(4): 358-366.

Stannard, D.I. 1988. Use of a hemispherical chamber for measurement of evapotranspiration. U.S. Geological Survey Open-File Report 88-452.

Stokes, S. 2000. Written communication. Department of Geography, Oxford University, Oxford, UK.

Whybrow, P.J. 1989. New stratotype; the Baynunah Formation (Late Miocene), United Arab Emirates: lithology and paleontology. *Newsletter on Stratigraphy* 21: 1-9.

Wood, W.W., Sanford, W.E. & Al Habshi, Abdul Rahman S. In press. The source of solutes in the coastal sabkhas of Abu Dhabi. Bulletin of the Geological Society of America.

HOLOCENE CARBONATE/EVAPORITES OF ABU DHABI, AND THEIR JURASSIC ANCIENT ANALOGS

A. S. ALSHARHAN[1] & C. G. St. C. KENDALL[2]

[1]*Faculty of Science, U.A.E. University, P.O. Box: 17551, Al-Ain, United Arab Emirates*
[2]*Department of Geological Sciences, University of South Carolina, Colombia, SC 29208, U.S.A.*

Abstract

The Holocene shallow-water carbonate and supratidal evaporite tract of Abu Dhabi consists of seaward reefs, barrier islands and tidal flats. As with similar sedimentary sequences in the subsurface of the Arabian Gulf, and elsewhere, the sediments of this coastal region pass landward into continental facies and seaward into basinal facies. In eastern Abu Dhabi oolites form on the inter-island tidal deltas and coral reefs are restricted to small patches along channels and just seaward of the center of the islands. To the west, coral reefs grow along the northern edges of most of the offshore banks north of the Khor al Bazam. Eastward, in the protected lagoons, carbonate muds and pellets are accumulating, whereas to the west of Al Dhabaiya Island, carbonate muds only accumulate in a narrow belt south of the offshore bank. Grapestones and skeletal debris are the dominant components. South of this bank, supratidal flats are encroaching on the lagoons through the development of beach ridges and cyano-bacterial flats. The similar relationship of these sedimentary facies to some of the Upper Jurrassic Period found in the subsurface of the Central Arabian Gulf, suggests that Abu Dhabi's coastal area can be used as a comparative model for understanding some ancient carbonate/evaporite depositional and diagenetic processes. In the Late Jurassic facies relationships at the pinch out between the Hith Anhydrite and the Asab Oolite in Central Abu Dhabi suggest that these sediments accumulated in a nearshore to sabkha setting. Eastward of this margin the Hith changes its character and appears to have accumulated in a standing body of water.

Barth & Böer (eds.), SABKHA ECOSYSTEMS, 187-202

INTRODUCTION

The Arabian Gulf covers approximately 226,000 km^2 with an average depth of 35 m and a maximum depth of 100 m at the Strait of Hormuz (Fig. 1). The coastal flats of Abu Dhabi are part of the southeastern Arabian Gulf carbonate province, and are characterised by a series of carbonate barrier islands and associated shallow lagoons. The southern coasts of these lagoons are flanked by long cyano-bacterial flats that pass landward into supratidal salt flats or sabkhas (Fig. 2). The almost completely landlocked Holocene sedimentary province of the Abu Dhabi coast provides a contrast to the isolated platform of the Bahamas and is the arid equivalent of the shallow shelf settings of the Bahamas, Yucatan, British Honduras and Florida.

Depositional settings include seaward reefs, islands and tidal deltas that protect sheltered saline lagoons and supratidal evaporite/carbonate flats. The sediments of the coastal region pass landward into a continental facies and seaward into a basinal facies. A similar relationship can be seen in the shallow-water carbonate/evaporite facies of the subsurface Jurassic of the Arabian Gulf.

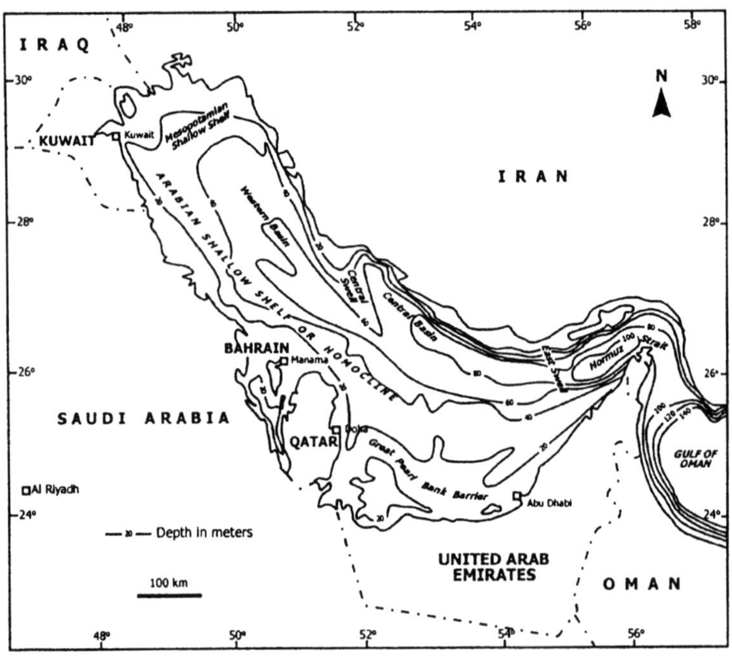

Figure 1. Arabian Gulf and its gross morphologic features (modified from Seibold & Vollbrecht 1969).

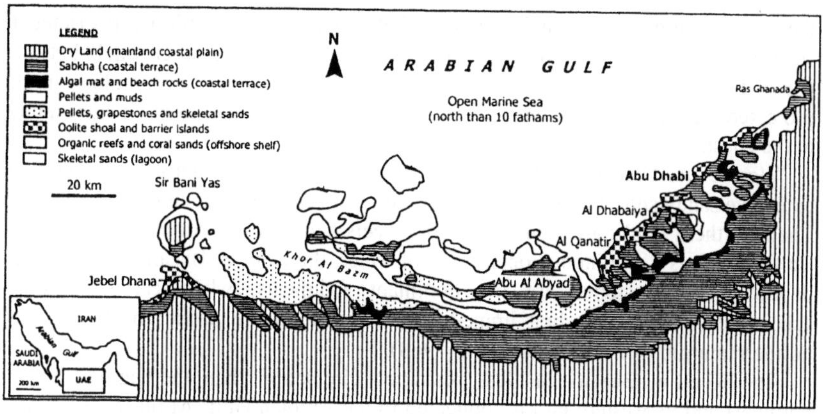

Figure 2. Map of the general marine facies of the coast of Abu Dhabi in the United Arab Emirates.

Lying close to the tropic of Cancer, the UAE has an arid, subtropical climate, in which the high temperatures of the summer months often reach 45-50°C, while temperatures in the winter can be as low as 3°C. The average rainfall of United Arab Emirates coastal region is less than 40 mm, and occurs in autumn, winter and spring, and although infrequent, may be torrential. The higher temperatures, coupled with the aridity, partially explain the widespread occurrence of carbonates and evaporites. The combined affect of strong winds, high temperature and low rainfall on the coast results in significant evaporation and the development of high salinity which explains the widespread occurrence of carbonates and the precipitation of the evaporite minerals beneath the sediment surface of the sabkha settings.

This paper is focused on the description of three Holocene carbonate/evaporite sediments localities of the Abu Dhabi area. They are representative of the diverse depositional settings that occur in close proximity to each other in the Abu Dhabi Emirate, and also represent potential depositional analogs for some of the Jurassic sediments found in the immediate subsurface.

HOLOCENE COASTAL FLAT SEDIMENTS

Traced from seaward to landward the coastline of Abu Dhabi consists of a complex of barrier islands, tidal deltas, lagoons, and tidal flats that all are constantly changing. Narrow tidal channels and deltas separate the barrier islands east of Al Dhabaiya Peninsula. The seaward margins of the "T" shaped islands parallel the coast while their stems lie at right angles to the dominant wind direction (Evans et al. 1964a, b) and separate the protected lagoons to the south. Oolites accumulate on the higher energy inter-island ebb tidal deltas, while coral reefs are restricted to small patches at the seaward centre of the islands and along tidal channels between them (Evans et al. 1964a, b). Carbonate muds and pellets accumulate in the protected lagoons. Tidal movement within these lagoons tends to be at right angles to the coast.

West of Al Dhabaiya Peninsula, the protecting barrier islands are situated on extensive sandy shoals more widely separated from each other than those to the east. Their northern flanks are rimmed by coral reefs and they are cut by tidal channels. To the south is a continuous open body of water, the Khor al Bazam. Here skeletal debris and grapestones are the dominant sediments, while carbonate muds accumulate in a narrow belt immediately adjacent to the barrier bank along its northern margin (Kendall & Skipwith 1968).

The Khor al Bazam has less restricted circulation than the lagoons to the northeast and its western end is connected to the Arabian Gulf. At the west end it is 20 meters deep. Eastward, the lagoon shallows, so that south of the island of Abu al Abyad, it is intertidal, and the circulation of the tidal waters becomes restricted. The dominant tidal movement in the Khor al Bazam is east-west. South of the bank and the Khor al Bazam, in the protected waters of the eastern lagoon, supratidal flats are encroaching northward into the lagoon through the development of beach ridges and cyano-bacterial flats. The situation is similar to south of the Abu Dhabi lagoons where maximum protection from the dominant north-northwest wind means that cyano-bacterial flats to form, accreting northward in the intertidal zone. Without any break in slope these flats are flanked by the supratidal sabkha (Evans et al. 1964a, b).

In contrast, along the more exposed coasts of the western Khor al Bazam, small beach ridges are common at the landward edge of the intertidal platform. Here the beaches on the islands are protected from waves by broad coral reefs and shoals that form wide frontal intertidal flats that are backed by small subaerial dunes 1 to 2 m in height. In contrast the barrier islands to the east around Abu Dhabi are positioned on a narrow shelf with little protection from heavy seas. Here, before the extensive industrial development of this coastline, these easternmost islands developed steep beach faces, landward of which large subaerial dunes with heights greater than 5 m formed.

CARBONATE/EVAPORITES FACIES OF AL QANATIR

Al Qanatir is a barrier island marking the eastern end of the Khor Al Bazam. Here, along a road just to the south of this island (Fig. 3), a traverse can be made from the earliest Holocene beach ridges to the present shoreline.

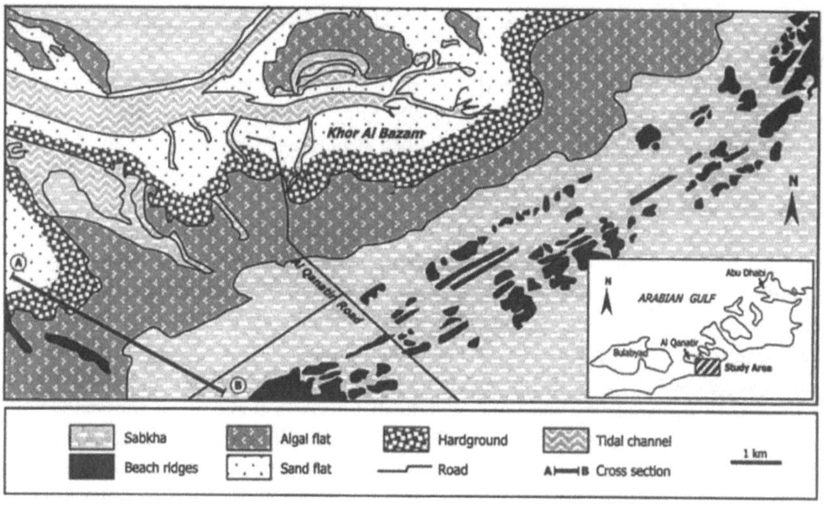

Figure 3. Coastal sabkha and tidal flats in the vicinity of Al Qanatir Island. Note the east-west road at the top of the tidal flat. This acts as a barrier of floodwaters.

Traced from landward to seaward the surface sediments on the traverse (Fig. 4) vary from: 1) a thin (3 mm) polygonally-cracked halite surface on the upper sabkha (upper supratidal); 2) a convoluted halite crust on the mid sabkha (mid supratidal); 3) a moist carbonate mud and gypsum mush on the lower sabkha (lower supratidal); and 4) a series of distinct cyano-bacterial mat surfaces (upper intertidal). In the vicinity of the city of Abu Dhabi this variation in surface sediment type and morphology was determined to reflect the local topography and a relationship to the tidal range (Butler 1969, Butler et al. 1982, Kendall et al. 1998a, and Kirkham 1998).

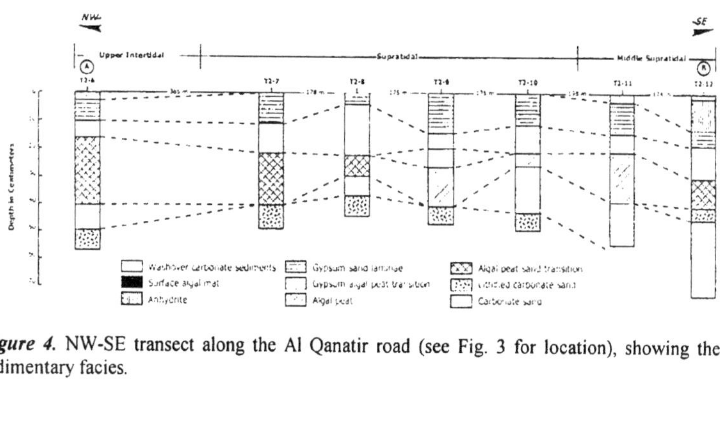

Figure 4. NW-SE transect along the Al Qanatir road (see Fig. 3 for location), showing the different sedimentary facies.

As in the vicinity of Abu Dhabi, where the upper supratidal zone has been trenched, a vertical sequence in these sediments can be seen that catches the character of their progradation at Al Qanatir. Notable exceptions however do occur where the sabkha slope adjacent to beach ridges is disrupted by tidal channel depressions and by elevated hard-ground tepee structures.

The upper sabkha facies landward of beach ridges

The upper sabkha facies south of the cerithid beach ridges is capped by a widespread polygonally-cracked surface crust of halite. Locally the halite may be crumpled into convoluted 5 cm high sand-rich ridges. The halite crust overlies a sand dominantly composed of carbonate with some quartz. Both the carbonate and the quartz were transported to their location by a combination of wind and storm wash-over from the tidal flats to the north, and from the Tertiary and Quaternary sediments to the south by the wind and flash floods. Caught up in the sand, and traced from the surface down, are dispersed anhydrite nodules (1 to 5 cm in diameter), which downward are locally replaced by diapiric anhydrite layers, 1 to 10 cm thick. Beneath this anhydrite zone, with it's predominantly carbonate sand matrix, is the water table. Close to the water table at about a meter in depth, gypsum rosettes become common.

The stranded beach ridge facies

Three to four kilometers from the intertidal cyano-bacterial flats to the north are a series of stranded beach ridges that form topographic highs about a meter above the adjacent sabkha (Fig. 2). These features are visible on aerial and remotely sensed images and are used by travelers to avoid the adjacent soft sabkha. Their most striking component is cerithid tests, but other gastropod and pelecypod tests are common too, along with encrusting bryozoan (that once coated long since decayed bladder-wrack), cuttle fish bones, and foraminifera. This beach ridge belt is some two to three kilometers wide. It is thought to represent a shoreline formed some 3000 years B.P. when the Khor al Bazam was more open and deeper (Kirkham 1998). No gypsum or anhydrite have been found within these ridges, and we speculate that this is probably because the groundwater trapped within them moves freely through them and is constantly replaced by marine waters. Carbonate cementation, however, is quite common.

The upper sabkha facies seaward of the beach ridges

The upper sabkha facies is defined primarily by a thick upper sequence of anhydrite (Fig. 4). This anhydrite may locally form the sediment surface or lie just beneath a polygonally-cracked surface crust of halite. Underlying the anhydrite or halite surface is, in descending order, a sequence of storm washover carbonates, anhydrite, minor gypsum, and cyano-bacterial peat. Lying beneath the peat is a 2-5 cm layer of unconsolidated carbonate sand. Locally, a burrowed carbonate mud may replace this. Beneath this layer is a cemented 2-5 cm thick layer of medium-to very coarse-grained carbonate sand (an intertidal hardground) whose composition is identical to the overlying unconsolidated sand. Below the hardground is carbonate sand that is over one meter thick. Locally, all but the carbonate sand facies may be absent. The interbedded nature of the carbonate mud and gypsum mush often creates a mottled texture, particularly where the gypsum alters to anhydrite through dewatering. Large (up to 6 cm) gypsum crystals are common within or immediately below the lowest cyano-bacterial peat layer, often preserving the original cyano-bacterial laminations.

The mid sabkha facies

The mid sabkha is covered with a convoluted crust of halite that overlies a thin layer of gypsum mush that is being replaced by anhydrite (Fig. 4). The gypsum may be interlayered with storm washover carbonate sediment, a thicker layer of gypsum, a well-developed cyano-bacterial peat, a carbonate hardground, or a thick sequence of carbonate sand. The cyano-bacterial peat is commonly laminated and displays desiccation polygons on the bedding plane. The cyano-bacterial peats overlie hardgrounds that are cemented by carbonates or gypsum. The hardgrounds are common at the top of the carbonate sand, but are not always present.

The gypsum mush facies

Gypsum actively precipitates at a depth of a centimeter and less beneath the flat surface of the lower supratidal zone, and beneath crenulated mats of cyano-bacteria in the upper intertidal zone (Fig. 4). The sediment is moist, preventing formation of anhydrite, except where dewatering of the sediment occurs, especially where footsteps and vehicle tracks have compressed the sediment. The surface carbonate washover sediment may contain halite, but it does not form a crust at the surface. The disc like crystals of the gypsum vary in size from a few millimeters to a centimeter in diameter. The crystals overlie a cyano-bacterial peat, commonly forming a transition zone of organic rich gypsum layers. Below the peat are the ubiquitous hardgrounds and unconsolidated carbonate sands.

The cyano-bacterial facies

The cyano-bacterial mats of Abu Dhabi have been described and subdivided into groups, based upon gross surface morphology of the mats (e.g. Kendall & Skipwith, 1968). The morphologic variations represent differences in microbial species that form the mats, or modifications in the environmental conditions, primarily the relative amount of exposure to the atmosphere between tides. Thus, mat types can be directly correlated to the topographic level of the flat.

Lower intertidal sand/mud flats and hardgrounds

In the more elevated portions of the tidal flat, closer to the cyano-bacterial flat, the sediment is composed of carbonate sands (Fig. 4). This is commonly cemented and forms extensive laterally continuous hardgrounds. Often these are buckled into megapolygons 2 to 3 m in diameter. The edges of these megapolygons protrude above the adjacent sands and are coated by a blackened-surface of cyano-bacteria. These crumpled margins are filled by several generations of sediment and cement. We believe the initial cracking is the product of thermal expansion, but subsequent fracturing and fill are related to generations of cementation and sedimentation within the cracks, forcing further expansion.

Local dissolution of the hardground surface appears to be driven by infestation by cyano-bacteria. The best examples occur where all the hardground but crumpled margins have dissolved. Commonly occurring in association with the hardgrounds are ubiquitous colonies of mitolids and cerithids. Additionally, occasional single mangroves colonise small areas of the sand and mud flats. Seaward of the cyano-bacterial flat and hardgrounds is an undulating surface of carbonate sands and muds. Locally, in the lower lying areas, tidal creeks dissect this surface and are incised into carbonate mud. Along the tidal channels mud-cracks are common.

BEACHROCK AND OOLITIC FACIES AT JEBEL DHANNA PENINSULA

Beach rock facies

Southwest of the Jebel Dhanna Peninsula (Figs. 2 and 5) the higher portions of the tidal zone are dominated by beachrock sediments. Here, aragonitic marine cement forms coatings on cerithid gastropod shells washed into recurved coastal spits by southward longshore transport and local westerly storms. Marine carbonate cement crusts (up to 3 cm in thickness) are present and best developed to the lee of these sand spits. Uncoated cerithid shells occur seaward of the easternmost spit. This is a low energy region where shells are cemented into beachrock (see also Purser & Loreau 1973 and Kendall et al. 1994).

Beachrock occurs as cemented slabs of bioclastic beach sand coated with a micritic marine aragonite. This cement resembles the flowstones of meteoric settings (Whittle et al., 1998). Pisolites (which are similarly coated) are present locally, usually nucleated about cerithid shells and other mollusc fragments. The beachrock crusts are buckled into megapolygons which can reach up 1 to 2 meters in diameter, most commonly on the berms of the beach ridges. The buckled margins of the megapolygons form the juvenile forms of tepees, similar to those described by Asseretto & Kendall (1977). These megapolygon slabs of beachrock have an incomplete marine travertine coating. The beachrock shows a zonation from the low intertidal, where dark gray uncoated to partially coated slabs occur in the lowest position, and cement coated beachrocks that occur in a higher position (some of which have a polished texture). Coated cerithids occur on the berm of the spits. Ridge and runnel bedform structures are preserved in cemented beachrock along the leeward edge of the beach ridges and are also coated with micritic aragonite measuring 5 to 30 mm in thickness. The thickness, microscopic characteristics and morphology of these cement coatings from Jebel Dhanna suggest incremental deposition of aragonite co-existing with traces of sulfate minerals. Most of these cement coatings are micritic, but the layers encrusting cyano-bacterial flat hardgrounds have a more radially fibrous microstructure and are composed solely of aragonite. The stable isotope composition of these coatings falls within the compositional range for modern nonskeletal aragonite. This suggests that the marine travertines precipitate from agitated, slightly hypersaline Arabian Gulf seawater through repeated immersion and evaporation.

The topographically higher areas tend to be colonised by the halophyte *Chenopodacea* that stabilises the local dunes. Among the coated cerithids, *Chenopodacea* also occurs. The topographically lower areas are sites of local mangrove growth. Sediment grain size ranges from fine in the lower intertidal to coarse in the upper intertidal. In the protected bay east of the coastal spits, the intertidal zone is characterised by cyano-bacterial flats. These flats consist of tufted cyano-bacterial mats.

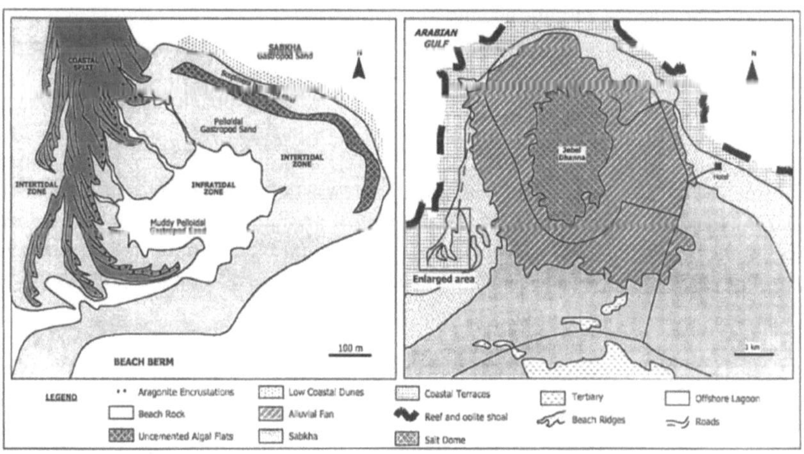

Figure 5. Facies distribution in the vicinity of Jabel Dhana Peninsula, western Abu Dhabi, UAE (modified from Purser 1973, Kendall et al. 1994).

Oolitic facies

Oolitic facies are accumulating on the northern shore of the Jebel Dhanna Peninsula (Fig. 5), while reefs form on seaward edges of the adjacent coastal terrace.

Here the intertidal sand flats are present on the eroded surface of ancient reef flats and wave-cut benches of pre-Holocene rocks. These sediments occur on shoals, the offshore bank and the coastal terrace, and have a gradient of approximately 1:5000 and vary in width from 0.5 km to 8 km. They are generally covered by varying thicknesses of unconsolidated bioclastic sand which may be underlain by alternating layers of poorly cemented limestone (1 to 3 cm thick) and unconsolidated carbonate sand (3 to 6 cm thick). The cemented layers have flat upper surfaces and irregular lower surfaces. Some layers are contemporaneous beachrock and are forming from the unconsolidated sediment of the sand flat (Whittle et al. 1998). They are probably cemented by calcium carbonate that is precipitated by cyano-bacteria (Nesteroff 1956) and from evaporating capillary water during low tide (Ginsburg 1953). This latter process is probably especially active during the height of summer. Before additional sediment accumulates, the upper surface of the beachrock is often truncated by storm action. Sugden (1963) also observed similar crusts in these western portions of Abu Dhabi.

Not all the layers are contemporaneous and some are of different composition than the loose sand above them. These particular horizons often extend from 2 m below the low water mark to a few meters above the high water mark (in the latter case they are often overlain by beach and cyano-bacterial sediments). These limestones probably represent sediments deposited as sea level transgressed across the shelf during the last rise in sea level. They were cemented to form beachrock that was contemporaneous with the change in sea level.

The surface of this intertidal sand flat is ripple marked or covered by growing seaweed, break-point bars, and runnels and bars, all of which are controlled in their distribution by the direction of wave approach. Intertidal spits and beaches on which oolites are forming delimit the landward edge. The surface may be divided on the basis of fauna into *Cerithium* flats on the lower area and crab flats on the upper (Kinsman 1964). *Scopimera* sp., the crab characteristic of the latter zone, produces radial aligned lines of feeding balls. Crabs, worms and molluscs, extensively burrow the sediments of the flats so that primary sedimentary structures are destroyed. Ginsburg (1957) reported the same phenomenon in carbonate sediments of Florida.

MANGROVE SWAMPS AND CARBONATE LAGOONS AT AL-DHABAIYA PENINSULA

The Abu Dhabi coast west of Al Dhabaiya consists of a barrier/lagoon complex in which widely spaced barrier islands protect the continuous Khor al Bazam lagoon (Fig. 6). Coral reefs and shoals, creating gentle beach faces and small dunes, protect the islands. The supratidal on either side of Al Dhabaiya displays different characteristics. In the areas protected by the islands like Abu Dhabi and Abu Al Abyad there are the extensive cyano-bacterial flats that pass landward into sabkha.

However, along the more exposed coast along the western portions of the Khor al Bazam there are small beach ridges flanking the intertidal zone that are backed by sabkha.

West of the Al Dhabaiya Peninsula there is a region of dominant mangrove vegitation. Here a series of Pleistocene outcrops protect the region from heavy seas and as a result stands of *Avicennia marina* mangrove grow in the form of small bushes and trees, either lining the edges of creeks draining cyano-bacterial flats, or in narrow strips parallel to the shore at the very top of the intertidal flats.

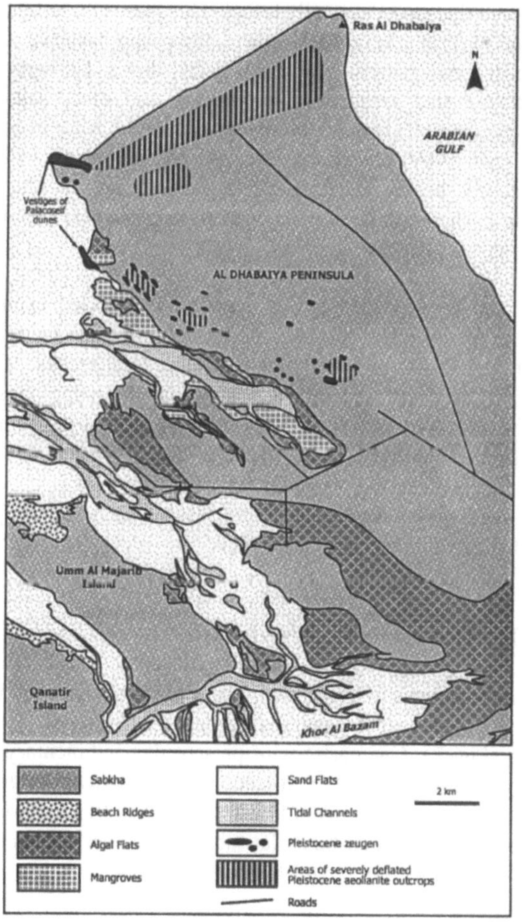

Figure 6. Al Dhabaiya Peninsula showing the mixed mangrove/cyano-bacterial tidal flat association..

This mangrove has a characteristic system of protruding roots or "pneumatophores". It grows in areas that experience frequent tidal interchange, so that the mangrove roots and lower trunk become covered at high tide. As Kendall & Skipwith (1969) observed the larger mangroves grow on the edges of channels where there is better tidal interchange and diminish in height and distribution inland as described by Allen (1965) on the Niger Delta. In the Khor al Bazam, when these channels become sediment-choked, the area is colonised by cyano-bacterial mat and the mangroves die. The cyano-bacterial mats tend to grow landward of the mangroves, and, where they grow on creek banks, they form a crinkled blue-green cyano-bacterial mat that is occasionally tufted. This mat is underlain by gypsum and is often cemented to form a beachrock. Inland, the cyano-bacterial mat is marked by a superficial polygonal pattern and backed by dunes covered by halophytes, including *Salicornia* sp., *Arthocnemum glaucum* and other salt tolerant plants. Carbonate mud accumulates around the mangroves of the Khor al Bazam, as it does in Florida (Vaughan 1909) and in the Bahamas (Newell et al. 1951). This muddy facies is commonly highly burrowed by crabs. Some of the mud probably precipitates *in situ*. There are three main characteristic successions of facies that can be seen in the area:

1. The sabkha can be traversed into a gypsiferous crinkled cyano-bacterial flat, which in turn, passes into a burrowed crab flat (with some halophytes colonising local highs), and finally to mangroves lining the tidal creeks. Cementation can be observed in the form of hardgrounds around the halophytes and extending down into the burrowed crab flats. Mud-cracked zones are common here.
2. Traverses from halophyte fixed gypsum dunes into cyano-bacterial flats, followed by burrowed crab flats and mangrove-lines creeks.
3. Halophyte fixed dunes lead into a bioclastic beach, abruptly separated by mangrove stands within a fine-grained mud.

JURASSIC DEPOSITIONAL ANALOGS

The Holocene sediments of the United Arab Emirates have been used to interpret portions of the sediments of the underlying upper portions of the Upper Jurassic Arab Formation (Fig. 7). Comparisons of present day sedimentary sequences with this part of the formation suggests that some sequences of litho-facies of the Arab Formation, particularly in central Abu Dhabi where the Hith Anhydrite pinches out, belong to an inner shelf-lagoon-tidal flat-sabkha sequence (Alsharhan & Kendall 1994, Alsharhan & Whittle 1995, Al-Silwadi et al. 1996).

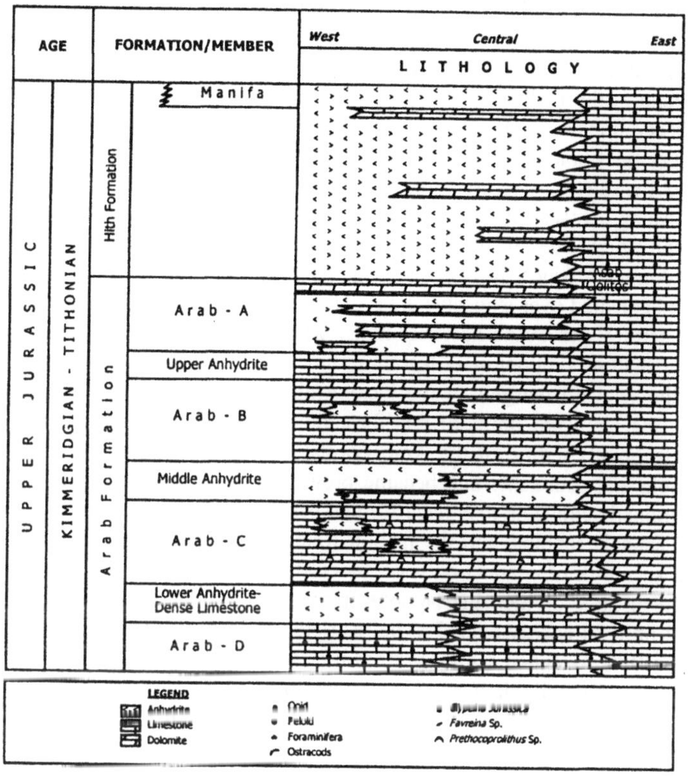

Figure 7. Lithostratigraphy of the Arab and Hith Formations in Abu Dhabi, UAE.

However it should be recognised that the Tithonian aged Hith Anhydrite, a major hydrocarbon seal in the Arabian Gulf region, most frequently contains less than 20% carbonate over much of the area it occupies, often in the form of thin laminations. This lack of carbonate, local layers of salt, and the predominance of anhydrite argue for a playa setting (Warren & Kendall 1985) for this extensive deposit, which covers a much greater geographical area than the current coastal area of the Southern Arabian Gulf, and has a much thicker sequence of sediments dominated by evaporites, with much less carbonate than in the Holocene sabkha sediments of Abu Dhabi. In contrast to central Arabia, the Hith beneath the southern Arabian Gulf appears to have accumulated in a sabkha setting matching the present-day supratidal evaporite depositional settings along the Arabian Gulf coastal area. Like these settings these deposits have a relatively limited areal extent. The interbedded relationship of the Hith anhydrites with tidal flat carbonates and the local predominance of horizontally flattened nodules and enterolithic layers of anhydrite support this interpretation. These latter features match some of the characteristic fabrics found in the Holocene coastal sabkhas of the United Arab Emirates. Based on our observations of the Recent sedimentary setting and cores from the subsurface, it is our personal contention that the setting of the Hith Anhydrite was apparently dominated by salt flat-encrusted plains and playas formed in an essentially arid climate. Along its seaward margins we believe storm activity accompanied by exceptionally high tides resulted in flooding of the sabkha surface and deposited "storm sediments". Since the sabkha had a very flat surface, such sediments would have been deposited over extensive areas. The formation mudflats characterised this intertidal setting and cyano-bacterial mats, while the periodic flooding produced influxes of shallow subtidal inner lagoon sediments. Beach bars probably developed close to the seaward margins of the local Arab Formation shelf, and shoals were formed at the landward limits of this formation by high storm tides. The shallow subtidal (lagoonal) setting of the Arab Formation in the UAE is characterised by marine biota and sediments that became progressively coarser grained landwards. Coarser grained sediments may have formed proximal to offshore bars in the outer lagoon. Channels running into the lagoon from the intertidal zone may also have produced an influx of channel outwash bioclastic coarse-grained sands. Shoals are interpreted to have divided the lagoons from the intertidal zones. Channel sands dependent on tidal energy tended to be restricted to lagoons and intertidal settings. The shelf setting extended seawards from the outer reaches of the lagoon, with the inner boundary generally delineated by offshore bars.

The Asab oolite (the lateral equivalent of the Hith Formation) in eastern Abu Dhabi is composed mainly of well-sorted oolitic grainstone. The ooids have cortices composed of thin laminae, and the nuclei composed of bioclastics and occasional foraminifera (Alsharhan 1989; De Matos 1994). There is local evidence of vadose meniscus cements and absence of marine cements, suggesting accumulation on high energy wave-dominated tidal barrier island complex. This is analogous to the present-day barrier island complex located along eastern coast of Abu Dhabi. The Manifa reservoir in central Abu Dhabi represents the back-barrier facies to the Asab oolite. It consists of dolomitic lime mudstone and oolitic limestone and has a restricted fauna. This is broadly analogous to the present-day Abu Dhabi lagoon east of Al Dhabaiya and the Khor al Bazam at Al Qanatir Islands where there is currently a low diversity hypersaline fauna.

CONCLUSIONS

Where the eastern Abu Dhabi Holocene coast trends northeast-southwest and the barrier/lagoon complex narrows, oolites form on inter-island tidal deltas adjacent to small patches of coral reef restricted to channels and just seaward of the centre of the islands. To the west, coral reefs grow along the northern edges of most of the offshore banks north of the Khor al Bazam.

Protected lagoons occur south of the barriers of eastern Abu Dhabi and carbonate muds and pellets accumulate, whereas to the west of Al Dhabaiya Island, carbonate muds only accumulate in a narrow belt south of the offshore bank. Grapestones and skeletal debris are the dominant components in the lagoon to the south. The offshore bank is accreting seaward through a combination of coral growth and tidal delta progradation. South of this bank, supratidal flats with characteristic suites of carbonate/evaporate facies are encroaching on the lagoons through the development of beach ridges and cyano-bacterial flats.

A comparison of the Holocene sedimentary facies of the surface to those of the subsurface Upper Jurrassic of the Central Arabian Gulf, suggests that Abu Dhabi coastal area can be used as a comparative model for understanding some ancient carbonate/evaporite depositional and diagenetic processes. Certainly at the pinch out between the Hith Anhydrite and the Asab Oolite there is reason to believe that these sediments accumulated in a nearshore to sabkha setting. Eastward of this margin the Hith changes its character and appears to have accumulated in a standing body of water as a playa.

References

Al-Silwadi, M.S., Kirkham, A., Simmons, M.D. & Twombley, B.N. 1996. New insight into regional correlation and sedimentology, Arab Formation (Upper Jurassic), Offshore Abu Dhabi. *Geo Arabia: Middle East Petroleum Geosciences* 1: 6-27.

Allen, J.R.L. 1965. Coastal geomorphology of Eastern Nigeria: beach ridges, barrier islands and vegetated tidal flats. Geologie en Mijnbouw 44 (1): 1-21.

Alsharhan, A.S. & Kendall, C.G.St.C. 1994. Depositional setting of the Upper Jurassic Hith anhydrite of the Arabian Gulf: An analogue to Holocene evaporites of the United Arab Emirates and Lake McLeod of Western Australia. *American Association of Petroleum Geologists Bulletin* 78: 1075-1096.

Alsharhan, A.S. & Whittle, G.L. 1995. Carbonate-evaporite sequences of the Late Jurassic, southern and southwestern Arabian Gulf. *American Association of Petroleum Geologists Bulletin* 79 (11): 1608-1630.

Alsharhan, A.S. 1989. Petroleum geology of the United Arab Emirates: *Journal of Petroleum Geology* 12: 253-288.

Asseretto, R.L.A.M. & Kendall, C.G.St.C. 1977. Nature, origin and classification of peritidal structures and related breccias. *Sedimentology* 24: 153-210.

Butler, G.P. 1969. Modern evaporite deposition and geometry of co-existing brines, the sabkha, Trucial Coast, Arabian Gulf. *Journal of Sedimentary Petrology* 39: 70-89.

Butler, G.P., Harris, P.M. & Kendall, C.G.St.C. 1982. Recent evaporites from Abu Dhabi coastal flats. In: Hanford, C.R. Robert, G.L. & Graham, R.D. (eds), Depositional and diagenetic spectra of evaporites, SEPM Core Workshop no. 3, Tulsa, Oklahoma, 33-64.

De Matos, J.E. 1994. Upper-Jurassic-Lower Cretaceous stratigraphy: The Arab, Hith and Rayda formations in Abu Dhabi. In: M.D. Simmons, (ed.), Micropalaeontology and hydrocarbon exploration in the Middle East: London, Chapman and Hall, 81-111.

Evans, G., Kendall, C.G.St.C. & Sir P. Skipwith, 1964. Origin of the coastal flats, the sabkha of the Trucial Coast, Persian Gulf. *Nature*: 759-761.

Evans, G., Kinsman, D.J.J. & Shearman, D.J. 1964. A reconnaissance survey of the environment of recent carbonate sedimentation along the Trucial Coast, Persian Gulf. In: van Straaten, L.M.J.U. (ed.), Deltaic and shallow marine deposits. Amsterdam: Elsevier, 129-135.

Ginsburg, R.N. 1953. Beachrock in South Florida. *Journal of Sedimentary Petrology* 23 (2): 85-92.

Ginsburg, R.N. 1957. Early diagenesis and lithification of shallow water carbonate sediments in south Florida. In: Le Blanc, R.J. & Breeding, J.G. (eds.), Regional aspects of carbonate deposition. *Society of Economic Paleontologists and Mineralogists Special Publication* 5: 80-100.

Kendall, C.G.St.C., Saad, J.L. & Alsharhan, A.S. 1994. Holocene marine cement coatings on beachrocks of the Abu Dhabi coastline (U.A.E.), an analog for cement fabrics in ancient limestone. *Carbonates and Evaporites* 9 (2): 119-131.

Kendall, C.G.St.C. & Skipwith, P.A. 1968. Recent algal mats of a Persian Gulf lagoon. *Journal of Sedimentary Petrology* 38: 1040-1058.

Kendall, C. G. St. C. & Skipwith, P. A. 1969. Geomorphology of a Recent shallow water carbonate province: Khor al Bazam, Trucial Coats, southwest Persian Gulf. *Geol. Soc. Amer. Bull.* 80: 865-891.

Kendall, C.G.St.C., Alsharhan, A.S. & Whittle, G.L. 1998b. The flood re-charge sabkha model supported by recent inversions of anhydrite to gypsum in the UAE Sabkhas. In: A.S. Alsharhan, K.W. Glennie, G.L. Whittle & C.G. Kendall (eds.), Quaternary Deserts and Climatic Change, Balkema, Rotterdam, The Netherlands, 29-42.

Kinsman, D.J.J. 1964. Recent carbonate sedimentation near Abu Dhabi, Trucial Coast, Persian Gulf. Unpublished Ph.D. Thesis, University of London.

Kirkham A. 1998. Pleistocene Crarbonate Seif Dunes and their Role in the Development of Complex Past and Present Coastlines of the U.A.E. *Geo Arabia* 3 (1): 19-32.

Nesteroff, W.D. 1956. La substratum organique dans les depots calcaries, sa signification. *Geological Society of France Bulletin* 6 (6): 381-389.

Newell, N.D., Rigby, J.K., Whiteman, A.J. & Bradley, J.S. 1951. Shoal water geology and environments Eastern Andros Island, Bahamas. *American Museum Natural History Bulletin* 97: 7-26.

Purser, B.H. & Loreau, J.P. 1973. Aragonitic supratidal encrustations on the Trucial Coast, Persian Gulf. In: Purser, B.H. (ed.), The Persian Gulf, Holocene Carbonate Sedimentation in a Shallow Epeiric Continental Sea. Springer-Verlag, New York, 343-376.

Purser, B.H. (ed.) 1973. The Persian Gulf, Holocene Carbonate Sedimentation in a Shallow Epeiric Continental Sea. Springer Verlag, New York.

Seibold, E. & Vollbrecht, K. 1969. Die Bodengestalt des Persischen Golfs. „METEOR" Forsch. Ergebnisse, Reihe C, No. 2: 29-56.

Sugden, W. 1963. Some aspects of sedimentation in the Persian Gulf. *Journal of Sedimentary Petrology* 33 (2): 355-364.

Vaughan, T.W. 1909. Geology of the Florida Keys and the marine bottom deposits and Holocene corals of southern Florida. Carnegie Institute of Washington Yearbook, no. 7: 131-136.

Warren, J.K., & C.G.St.C. Kendall 1985. Comparison of Marine Sabkha (subaerial) and Salina (subaqueous) settings: Modern and Ancient. *American Association of Petroleum Geologists Bulletin* 69: 1013-1023.

Whittle, G.L., Alsharhan, A.S. & Kendall, C.G.St.C. 1998. Petrography of Holocene beachrock and hardgrounds, Abu Dhabi, United Arab Emirates. In: A.S. Alsharhan, K.W. Glennie, G.L. Whittle & C.G. Kendall (eds.), Quaternary Deserts and Climatic Change, Balkema, Rotterdam, The Netherlands, 57-70.

SABKHAT IN COASTAL AREAS WITH SPECIAL EMPHASIS ON THE UAE

HELMUT LIETH[1] & UWE MENZEL[2]

[1] *University of Osnabrück, Institute for Environmental Systems Research, Artilleriestr. 34, D-49069 Osnabrück, Germany*
[2] *University of Osnabrück, Artilleriestr. 34, D-49069 Osnabrück, Germany*

Abstract
Large areas of the coastal plain in the United Arab Emirates and Saudi Arabia have a special sabkha type. This sabkha is formed by the interplay of seasonal inundation with Gulf water and rain water and by sandstorm deposits during the hot dry season. Under these environmental conditions in the UAE soil profiles develop in which horizontal deposits of different texture and colour appear. The horizons are quite different from place to place and may with additional analyses provide new information about the genesis of the landscape at any particular point. The chemistry of this sabkha type differs between coastal- and sabkhat further inland. Among the most conspicuous differences in the ocean near sabkhat are the large composite gypsum crystals. These are dark blue grey with lots of other inclusions in areas under continuous salt water inundation to transparent brown in places out of the continuous ocean influence. The impact of human activities is clearly visible everywhere along the coastal regions. Further investigations in the sabkhat should be initiated before the signals of past historical events are destroyed. The designation of suitable areas as sabkha research reserves is highly recommended.

INTRODUCTION

Sabkhat are a common phenomenon in deserts and semi deserts all over the world. Their genesis and appearance is largely dependent on the availability of sand, soil and water in the region. Orography, soil types and the balance between water input and evaporation create an amplitude of patterns and dictate the vegetation type

Barth & Böer (eds.), SABKHA ECOSYSTEMS, 203-214

and density. This can range from barren soil where only a few algae can exist to a sparse vegetation canopy of succulent chenopods to a close cover consisting of a multitude of halophytes.

The coastal plain in the UAE consists mainly of a sabkha type which is subject to regular flooding by winter rains and spring tide intrusions. In the area between Abu Dhabi and Tarif in the UAE this sabkha type is almost coherent behind the coastal dunes and mangroves with the exception of some rocky outcrops and a few small dune ridges cemented with salt deposits after evaporation. The sabkha type we treat in this article is based on this region. This sabkha type exists in similar forms in many other desert areas of the world but each region has specific forces which create a sabkha type different from our example and determine also the vegetation on it.

SURFACE STRUCTURE AND APPEARANCE

The coastal sabkha in Abu Dhabi Emirate appears like a system of shallow pans each of which extending for several square kilometres. After rain showers which occur mostly in February (see Fig. 1) these pans are filled with shallow water. This water evaporates within a few weeks and leaves behind a salt crust which stains the pan bottoms bright white. This colour turns into brown to yellow, after a few dust storms have blown fine material from the vast moving inland sand dunes.

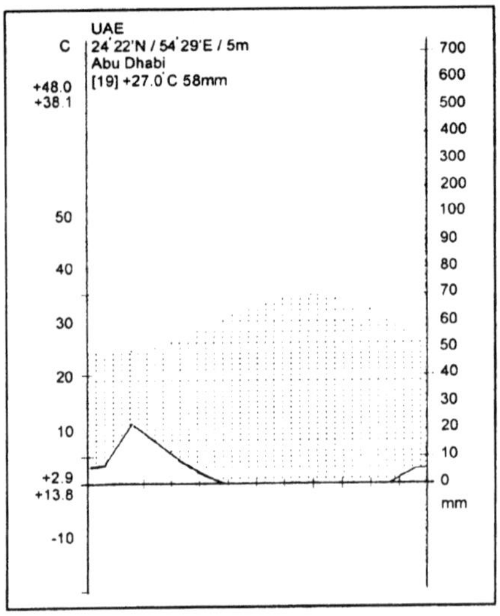

Figure 1. Climate diagram of Abu Dhabi according the conventions of Walter and Lieth now available in Lieth et al. (1999).

The absence of relief and the altitude of only a few meters above mean ocean level allows ocean water to intrude for several kilometres inland, washing during winter spring tides all sorts of debris into these shallow pans where they remain when the ocean water retreats. This part of the sabkha has usually a coarser surface than the pan bottoms further inland. In both sabkha types is the surface salinity almost alike, the surface salt crust is thicker, however, closer to the sea. The sea itself is separated from this sabkha region by either some heavily eroded outcrops or by the catena: sandy beach, sand dune or mangrove swamp and debris fringe. The surface is covered by marine shell debris which are also abundant in the profiles, as far as the ocean influences reach.

We have analysed the structure of two profiles for this paper. For the comparison of ion concentrations in soil water we took samples from 2 additional locations. Figure 2 shows the geographical position of the two profiles and the locations of additional soil water sampling. The ocean influence described for profile 1 and 2 is presently restricted to the ocean side area of the highway Abu Dhabi / Tarif. The sabkhat extend, however, far further inland, for approx. 5 km.

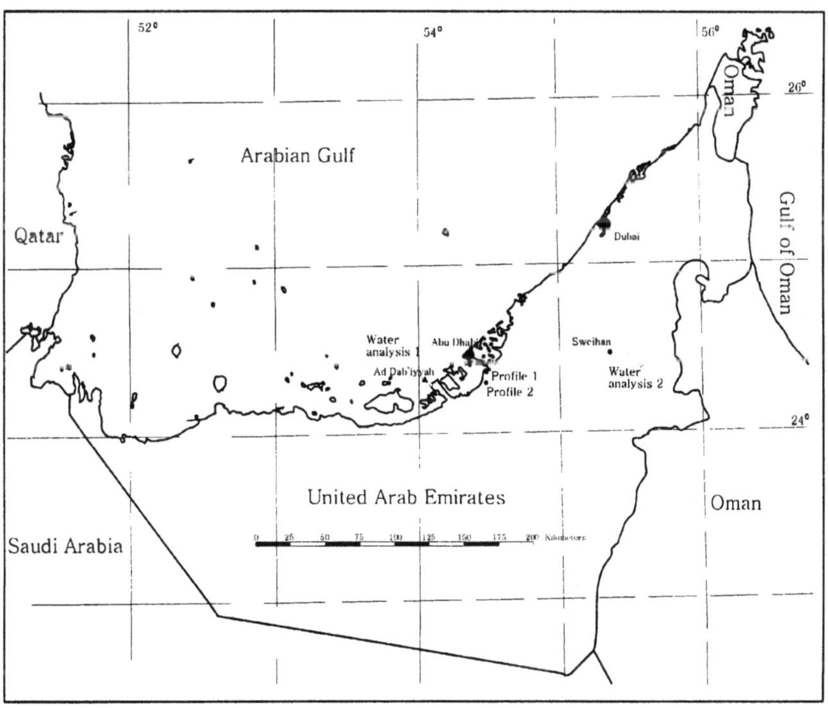

Figure 2. Map of the United Arab Emirates showing the location of all points from where data were taken for this study.

SUBSTRATE STRUCTURE

The structure of the "soil" in this substrate depends largely on the level of the land surface relative to the ocean. Depending on the location of a profile one can distinguish horizons of finer or coarser material, which reflect the alternating forces controlling the sedimentation process. Shells are found throughout the substrate. Chemical composition in the soil water (saturated soil extract) of the two profiles is compared in tables 1 and 2. Additional analyses were made for the water salinity at two places near shore and further inland to support our explorations of the two profiles.

Several layers of iron deposits (red ribbons in Tab. 2, profile 2, and Fig. 3) indicate the influence of ground water for longer periods. Gypsum crystals of considerable sizes occur at places in large quantities. These vary in colour from grey black to light brown, the latter being quite often transparent. Soil pH and nitrogen content in the profile is shown in table 2.

Table 1. The quantity of major ions dissolved in the soil water of the two profiles.

Profile 1

Horizon	[cm]						
Soil depth	0-45	45-75	75-83	83-95	95-113	113-120	120-175
Ions in the saturated soil extract	Quantity in [mg/l]						
Cl^-	112318	127078	104323	66335	61182	84838	7016
Na^+	58101	57626	49888	23979	30447	44905	4280
SO_4^{2-}	3130	3197	3066	2667	3132	3035	3331
Mg^+	4725	4088	4068	4296	1901	2395	178
Ca^+	5775	4266	4666	4726	2479	2979	957
K^+	1017	1044	1052	1491	583	910	138
HCO_3^-	400	450	370	140	160	200	20
Br-	554	508	518	655	492	625	17
F^-	622	207	201	107	114	155	19
PO_4^{2-}	<1	<1	<1	<1	<1	<1	1
total ions	186641	198464	168152	104397	100491	140043	15956
Error ion balance [%]	-1,5	-17,5	-9,1	-15,0	-10,8	-6,4	-6,4
Ion / l [%]	18,7	19,8	16,8	10,4	10,0	14,0	1,6
Electrical conductivity (E.C.) [mS]	278	273	273	173	166	224	32
E.C. x 0,07 [=% salt]	19,5	19,1	19,1	12,1	11,6	15,7	2,3
Refractometer: [% salt]	20	19	19	12	11	15	2
Water–soluble salt [% of soil dry weight]	5,2	5,2	5,1	3,5	3,5	7,0	0,4

See legend under tab. 1 / profile 2

Profile 2 (Tab. 1 continued)

Horizon	[cm]				
Soil depth	0-20	20-80	red ribbons between 20-80	80-95	95-120
Ions in the saturated soil extract	Quantity in [mg/l]				
Cl^-	147615	113872	93388	148991	143203
Na^+	58610	48265	41559	60734	47952
SO_4^{2-}	1707	2145	2627	2356	1622
Mg^+	14713	9870	6024	13573	18760
Ca^+	10082	8076	4729	9454	14054
K^+	3421	2186	1286	2868	3480
HCO_3^-	430	480	290	500	550
Br-	58	38	805	11	0
F^-	196	235	171	186	221
PO_4^{2-}	29	12	13	19	0
total ions	236861	185180	150891	238692	229841
Error ion balance [%]	3,1	2,8	-5,3	0,8	7,6
Ion / l [%]	23,7	18,5	15,1	23,9	23,0
Electrical conductivity (E.C.) [mS]	340	287	232	339	328
E.C. x 0,07 [=% salt]	23,8	20,1	16,3	23,7	23,0
Refractometer: [% salt]	25	20	16	25	25
Water–soluble salt [% of soil dry weight]	6,5	5,8	5,0	6,5	6,9

All results refer to the saturation extract. Soluble salts are quoted as electrical conductivity (E.C.) in mS at 25^{0}C of a saturation extract. NO_2, NO_3; NH_4 are provable for < 1mg/l. The soil salt content is also given as per cent of dry weight soil as here defined. The soil salt content was calculated from the average E.C. value, from the sum of ions and refractometer values based on the quantities of water and soil used for the saturation extracts. The salt content in % of soil dry weight refers to soil without stones (particle size below 2 mm).
All values refer to soil fractions lower then 2mm (stones were not included)

Table 2. pH values and total nitrogen in different horizons of the soil profiles.

Profile 1

Horizon	[cm]						
Soil depth	0-45	45-75	75-83	83-95	95-113	113-120	120-175
pH [KCl]	8,38	7,91	7,86	8,13	7,98	7,88	8,15
pH [soil paste]	6,97	7,08	7,30	7,49	7,45	7,44	8,10
H_2O/soil [ml/100g soil]	26,81	27,09	27,91	31,08	31,69	47,05	23,78
mg/kg N	223	183	103	248	408	1298	125

Profile 2 (Tab. 2 continued)

Horizon	[cm]					
Soil depth	0-3	0-20	20-80	red ribbons between 20-80	80-95	95-120
pH [KCl]	7,82	7,98	8,31	8,23	8,01	7,98
pH [soil paste]		6,78	7,18	7,32	6,64	6,55
H_2O/soil [ml/100g soil]		26,70	29,92	31,98	26,87	29,33
mg/kg N	301	254	316	114	342	467

pH is measured on saturated soil paste and also in 1:2.5 soil/KCl ratio. Total nitrogen is measured by Kjeldahl (soil dry weight). H_2O/soil gives the ratio between dry soil and water quantity for the saturated soil paste.

Figure 3. Photographic documentation of the two coastal sabkhat analysed in greater details in this study. The scale of both profiles is different. The bars along side are divided in 10 cm fractions.

Figure 3 shows two profiles in which different horizons, animal deposits and gypsum crystals were found. The absence of a low positioned hard pan from iron compounds, so conspicuous in inland sabkhat of Abu Dhabi Emirate is clearly documented in this profile. At high tide are some ocean side sabkhat so soft that it is difficult to even walk on it. The particle size distribution throughout the profile shown in table 3 explains the softness of the substrate especially when moist.

Table 3. Percent soil particle size in different horizons of the two soil profiles.

Profile 1

	Horizon				[cm]			
	Soil depth	0-45	45-75	75-83	83-95	95-113	113-120	120-175
Particle size [µm]					%			
> 4000	Stones	13,6	1,2	1,3	4,7	12,2	2,0	0,0
2000-4000	Stones	6,6	2,0	2,9	5,9	5,2	1,9	0,0
630-2000	Coarse sand	19,3	11,7	1,1	17,0	9,6	10,5	0,6
200-630	Medium sand	25,9	21,1	8,2	25,9	20,7	23,4	28,7
63-200	Fine sand	27,9	17,7	60,9	31,8	37,8	31,2	59,9
20-63	Coarse silt	6,8	17,5	8,2	14,2	12,2	12,8	0,5
6-20	Medium silt	14,3	21,7	11,3	10,7	9,7	13,4	0,7
<6	Fine silt a. clay	5,8	10,3	10,2	0,4	10,1	8,8	9,6

Profile 2

	Horizon				[cm]		
	Soil depth	0-3	0-20	20-80	red ribbons between 20-80	80-95	95-120
Particle size [µm]					%		
> 4000	Stones	1,0	1,8	3,6	6,8	0,0	6,2
2000-4000	Stones	4,1	0,7	0,6	2,7	0,0	0,0
630-2000	Coarse sand	14,6	1,2	1,1	10,9	1,7	0,6
200-630	Medium sand	28,4	5,0	8,8	20,3	8,6	7,4
63-200	Fine sand	30,4	61,1	64,2	35,6	65,1	70,9
20-63	Coarse silt	6,0	7,3	3,2	15,7	3,1	4,9
6-20	Medium silt	10,4	17,6	13,4	11,8	12,8	9,4
<6	Fine silt a. clay	10,2	7,9	9,3	5,7	8,7	8,6

The particle size values for all horizons are in percent of the total substrate mass in horizon. Sand, silt and clay (particle size < 2000 µm) together are defined as 100% soil after the salt was washed out. The stone fraction (size > 2000 µm) has been calculated on this base.

Little is known about the stratigraphy of the coastal sabkhat in the UAE. It differs widely in colour, texture and chemistry from place to place. The only common feature is the appearance of sedimentation layers which may tell us something about the history of the soil formation at any particular place. For this purpose we compare two profiles of contrasting places. The positions of the profiles are shown in figure 2 and descriptions of both profiles are compared in tables 1-5. From these two profiles we analysed salinities, ion concentration, particle sizes, pH and nitrogen contents, soil water salinities and from two additional places we analysed the ion composition of the soil water to demonstrate differences between seaside and inland sabkhat.

The particle sizes of each profile are compared in figure 4. Profile 1 contains coarser material than profile 2 which indicates that profile 1 is closer to an underground river bed than profile 2 where most of the substrate is obviously provided by the tidal action of the Gulf. This observation is supported by the higher salinity levels, in the upper part of profile 1 starting to decrease below 80 cm depth.

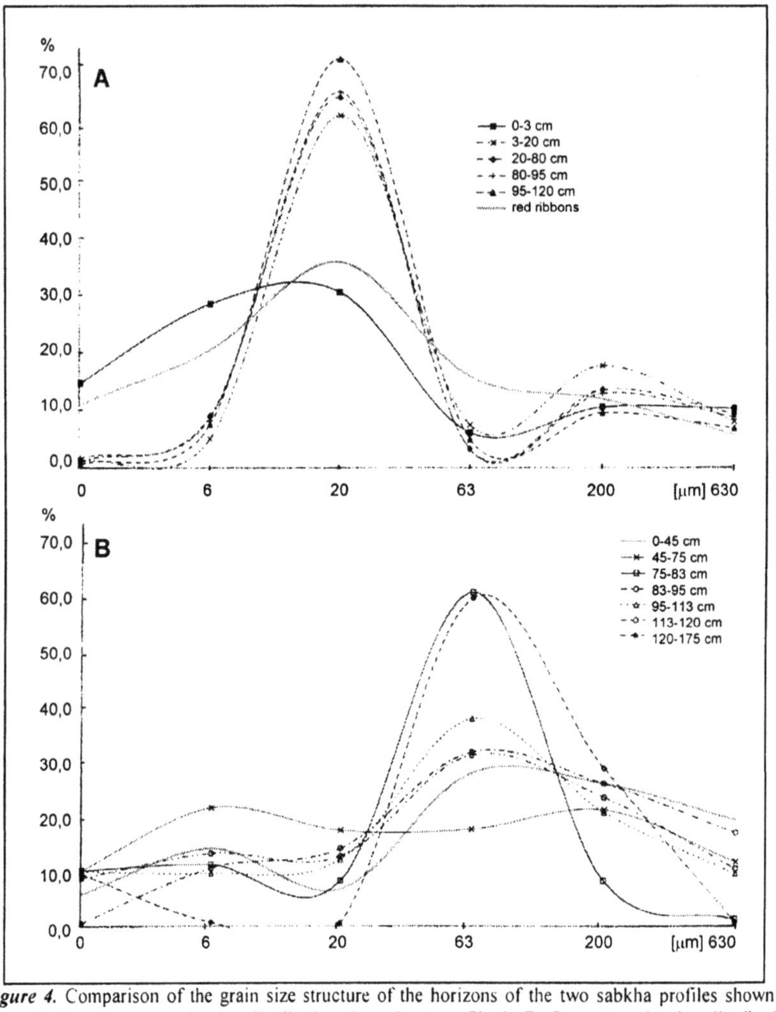

Figure 4. Comparison of the grain size structure of the horizons of the two sabkha profiles shown in figure 3. **A:** Percent grain size distribution throughout profile 1. **B:** Percent grain size distribution throughout profile 2. The European standard sieve sets for particle size were used. The comparison shows clearly the coarser substrate in profile 1.

Table 4. Comparison of the ion content in water from two additional sabkhat in Abu Dhabi Emirate. An ocean side (1) and an inland location (2). See figure 2 for location.

Location	**1**	**2**
Ion	Quantity in[mg/l]	
Cl^-	23990	53627
Na^+	14980	38227
SO_4^{2-}	3659	11886
Mg^+	1227	1592
Ca^+	460	987
K^+	341	581
HCO_3^-	177	396
Br-	61	31
? ions	44894	107326
Error ion balance [%]	3,7	5,0
Ion /l [%]	4,5	10,7
Refractometer: [% salt]	4,6	11,0

1. Peninsula Ad Dab'iyyah east side.
2. Tawi near Suweihan (inland sabkha)

Coastal sabkhat differ from inland sabkhat among other parameters in the ion content of the soil water as demonstrated in this table from 2 places in Abu Dhabi Emirate. The ion composition in the inland sabkha is richer in sulfate and lower in Bromium.

The surface salinities of sabkhat vary with rainfall. During the dry season salt crusts appear on the coastal sabkhat. In inland sabkhat where soil water may drain sideways over the hard pan these salt crusts are not as distinct. The soil water salinity throughout the profile ranges in the ocean near sabkhat close to sea level salinity. Further inland the soil water salinity is higher throughout the profile (see values in table 1 and 2). The ground water salinity level varies with the import of brackish water arriving at the coastal area from inland ground waters. The salinity in inland sabkhat shows strong differences between the water above and below a 1 to 2 m deep lying hard pan. The water above the hard pan may reach levels reported for sabkha 2 while the water below that hard pan my have 1 - 2 % salinity only, depending on the input of rain water from the Oman mountains.

Table 5. Geographical parameters for the coastal sabkha soil profiles analysed for this study near Abu Dhabi (city).

(Saline gypsiferous coastal desert soil, saline water)

Profile 1	
Location:	24^0 21'N, 54^0 27'E
Altitude:	< 2 m
Physiography:	plain plateau
Drainage:	gently undrained
Parent material:	sand with small gypsum crystals between 3 and 45 cm, with sea shells and sea snails layers
Vegetation:	occasional *Arthrocnemum macrostachyum*
Profile 2	
Location:	24^0 14'N, 54^0 25'E
Altitude:	< 2 m
Physiography:	plain plateau
Drainage:	gently undrained
Parent material:	sand with gypsum crystals
Vegetation:	vegetation free

Water analysis

Point 1	
Location:	24^0 19'N, 54^0 6'E
Altitude:	< 2 m
Physiography:	plain plateau behind coastal dunes
Drainage:	rapid on top, slow close to the saline ground water
Parent material:	mostly grain size crushed sea animal shells
Vegetation:	occasional chenopods
Point 2	
Location:	24^0 26'N, 55^0 34'E
Altitude:	proxy 25 m
Physiography:	plain sabkha plateau between sand dunes
Drainage:	water logged after rains
Parent material:	silicatious and calcareous sand
Vegetation:	sparse cover of different halophytes

GENESIS OF THE PROFILE

As we can see from figure 4 the profile structure tells us much about some generalities of the dynamics of the landscape development. The usual forces of water seepage down through a substrate once it is deposited is almost completely missing. In this sense the ocean side sabkha could be classified as regosol or paternia. Only at places where fine deposits prevail, they might be classified as solonchak related soil. The components of the substrate are either coarse gravel which comes from the inland mountains or coarse marine debris. The only sign of longer lasting chemical transfers are the iron bands (see Fig. 3) and gypsum crystals. The latter are formed when the sulfur containing decaying bodies of ocean animals produce sulfuric acid which dissolves the calcium carbonate of the substrate and converts it into calcium sulfate. This crystallises in situ anywhere in the profile. The size and form of these crystals are increasing with depth and can get great ornamental objects. The amount of animal debris and tenure of the flooding appears to control the size of the gypsum crystals which we found at some places in sizes up to 40 cm high and 60 cm wide.

We try to show the different genesis of profiles with the two examples in figure 4. The only common features are at both places the thick top salt crust (see Fig. 3), the coarse gravel and sand mixture, the oxidising and reducing horizons throughout the profile, especially prominent in profile 1 of figure 3, and the presence of animal shells and gypsum crystals. All of these features vary widely in quality and quantity as the material was imported or exported here by the geologic and environmental forces existing through history as indicated in the legend to figure 4.

Water content, salinity and vegetation

The water content in this soil type fluctuates greatly throughout the year. During floods the entire pore volume is filled with water, because drainage can be extremely low, due to the fine particle size. Most of the water evaporates from the surface leaving a surface salt crust for the summer months. This consists mostly of sodium chloride mixed with other salts dissolved in ocean water. Below this salt crust the salinities are close to seawater levels, which is on the Arabian Gulf shores between 4 and 5%. Depending on the distance from the shoreline and the position within of the shallow water bodies the salt content varies from below to above 4 %. This makes it very difficult to predict the salinity pattern in these sabkha systems. Slight elevations of coarse sandy material can cause localised desalinisation which then allows xerophytic semihalophytes such as *Arthrocnemum macrostachyum*, *Halocnemum strobilaceum* and *Halopeplis perfoliata* to grow on such places. The ion composition of the soil water varies with the distance from the ocean and the influence of rain water. Table 4 adds values from two places in Abu Dhabi Emirates. Together with the values presented in table 1 it becomes clear that the soil water ion content in the UAE sabkhat varies widely depending on the balance between the various inputs of water.

The vegetation on these salt flats is ephemeral as far as the algal crusts is concerned. During floods and drying periods the surface can have a green shine from algal mats the composition of which is described elsewhere in this volume. This turns black as soon as the surface dries. Higher plants can only exist on elevated ripples as they develop when wind and shallow water form the surface of the otherwise flat landscape. Closer to the ocean edge of the sabkhat can the activity of crabs create surface patterns which may allow the settlement of a few chenopods such as *Cornulaca sp.*

The construction of the highway across this landscape has brought an additional aspect into the vegetation cover. Along the highway Abu Dhabi / Tarif grow now salinity tolerant trees and shrubs with irrigation. The water provided is of different quality since much of it comes from the sewage treatment plants of the nearby city complexes around Abu Dhabi.

Human impact

In the course of the development of Abu Dhabi Emirate this sabkha area is affected in many ways by human activities. Close to Mussafah covers a mountain of solid waste a previously ocean side sabkha. The weight of this material has pressed the partly soft sabkha substrate further down into the groundwater level. The annual rains wash the multitude of chemicals contained in the waste into the adjacent flat land and into the ocean. Birds and rodents are now regular occupants of such places adding new elements to the otherwise trist appearance of this flat salt desert.

In spite of the presently high salt content the area may become utilised by man in the near future. Research carried out by various groups in the UAE have shown that seawater irrigation can remove the salt crust and reduce substrate salinity to ocean levels. This allows under continued ocean water irrigation the plantation of the local mangrove species *Avicennia marina* as well as other halophytes. This plant can provide leaves for animal forage, edible fruits as vegetable and honey from its flowers and many more uses. Landscape improvement studies carried out by several new institutions in Abu Dhabi and Dubai let us expect a sharp change of this barren landscape to a green forested area.

ACKNOWLEDGEMENTS

We thank M. Moschenko for typing and editing the manuscript. We thank Dr. B. Böer for the encouragement and the help rendered to the junior author during the field work. The work was partly supported by the CEC under the contract IC18CT96-0055 "Sustainable halophyte utilisation in the Mediterranean and subtropical dry regions".

References

Barth, H.-J. 1998. Sebkhas als Ausdruck von Landschaftsdegration im zentralen Küstenland der Ostprovinz Saudi-Arabiens. *Regensburger Geographische Schriften*. 29. Selbstverlag, Libri Druck, Regensburg, 279 pp.

Böer, B. 1997. An introduction to the climate of the United Arab Emirates. *Journal of Arid Environments* 35: 3-16.

FAO – Unesco 1977. FAO-Unesco Soil map of the world, Vol. VII South Asia, Prepared by the Food and Agriculture Organization of the United Nations. Unesco Paris. 117 pp.

Lieth, H., Berlekamp, J. Fuest, S. & Riediger, S. 1999. Climate Diagram World Atlas on CD. SPB Academic Publishers e. V. Leiden, The Netherlands.

Menzel. U., Lieth, H. 1999. Tabulation of Halophytes reported as utilised in different publications and handbooks. In: Hamdy, A., Lieth, H., Todorovia; M. & Moschenko, M. (eds.), Halophyte use in different climates II. *Progress in Biometeorology*: 14: 127-133.

DYNAMICS OF SABKHA EDGES - EXAMPLES FROM THE JUBAIL AREA, SAUDI ARABIA

HANS-JÖRG BARTH

University of Regensburg, Department of Physical Geography, Universtiätsstr. 31, D-93040 Regensburg, Germany

Abstract

Within areas of low vegetation densities at several sabkha edges deflation processes were observed. In other places sabkha surface disappeared due to sand accumulation. Sedimentation and erosion processes at sabkha edges demonstrate that sabkhat can change within the short time of years or decades. These processes were studied by satellite image analysis and field measurements. In regions where vegetation cover was below 3% sabkha surface areas expanded up to 6 meters in one year. Remains of older dead vegetation in sabkhat indicate the former extent of sand sheets. The inland sabkhat in the study area show a significant orientation within the main wind direction which is from north to south. Their average length-width ratio is 2.8. These observations lead to the conclusion that the origin of inland sabkhat has to be related to the recent geomorphological history and seems to be a result of a negative sand budget in a supraregional wind action system much intensified by heavy grazing.

INTRODUCTION

Observations in the field revealed that some of the inland sabkhat are no long lasting geomorphological features. Signs of sedimentation and erosion processes at sabkha edges demonstrate that sabkhat can change within years or decades. This study was designed to find out which processes change sabkha edges and how fast these changes take place.

Barth & Böer (eds.), SABKHA ECOSYSTEMS, 215-226

GENERAL GEOGRAPHIC DESCRIPTION OF THE AREA

The study area is located at the Arabian Gulf coast, north of Jubail Industrial City and covers about 900 km² (Fig. 1). It is part of the central coastal lowlands of the Eastern Province of Saudi Arabia. The relief is weak, although some limestone exposures in form of minor escarpments and little domes are quite common. A general feature in the topography of the area are flat and very extensive sand sheets and widely rolling inactive dune systems which are stabilised by vegetation and are therefore referred to as "fossil dunes" (Barth & Niestlé 1992). The longitudinal sand ridges, reaching heights of 5 to 10 m and lengths of more than 100 m are oriented in a general north south direction. The flat and low lying sand sheets are undisturbed layers of wind blown sand also covered by protective vegetation. The Eastern Province belongs to the arid part of the subtropical desert belt. Precipitation is confined mainly to the winter season and varies widely in amount and spatial occurrence. The long term average annual precipitation in the area is around 80 mm. The average annual temperature is about 26°C with extreme values of 51°C in summer and 3°C in winter.

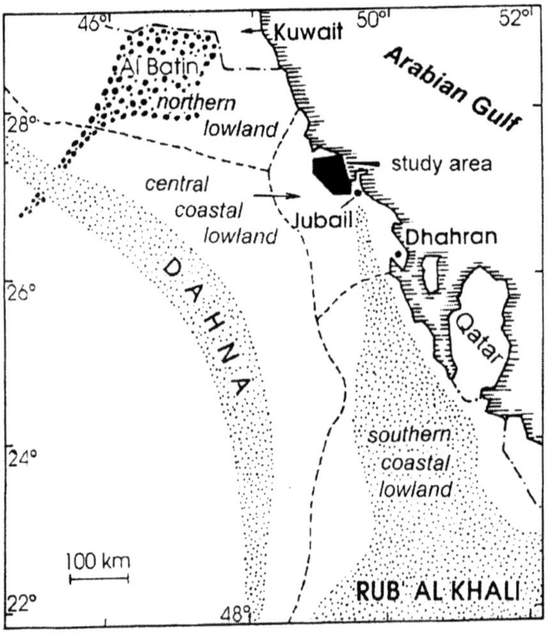

Figure 1. Location of the study area

DETECTION OF SABKHA CHANGES WITH SATELLITE DATA

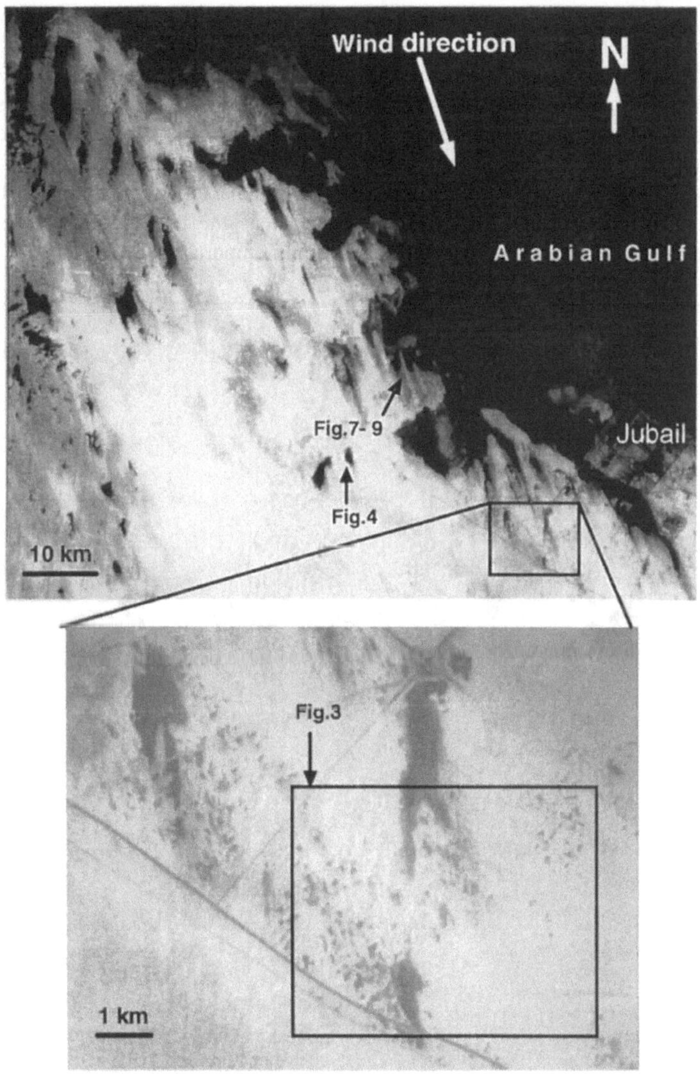

Figure 2. Landsat Thematic Mapper image (4-5-7) shows sabkha surface in black between the bright sand sheets. The longitudinal axis of most sabkhat correspond with the main wind direction. Note the wind streaks south of the sabkhat. The lower section shows an area of active sand movement.

To gain an overall impression of distribution and shape of the sabkhat in the area of interest Thematic Mapper satellite images (164/041) were visually interpreted (Fig. 2). The satellite image shows sabkhat in black. Wind streaks in downwind direction of sabkhat indicate strong unimodal winds coming from north-north-west (340°). The wind streaks are detectable because the sand has a higher salt content than the surrounding sand sheets. This is due to the location downwind of sabkhat where significant amounts of salt are removed by deflation. It is obvious that the orientation of most sabkhat corresponds to the main wind direction. This and the average length-width-ratio of 2.8 indicate that sabkha formation is related to wind action. To locate changes of sabkha surface areas the above mentioned satellite images from 1984 and 1991 were used for a multitemporal image analysis.

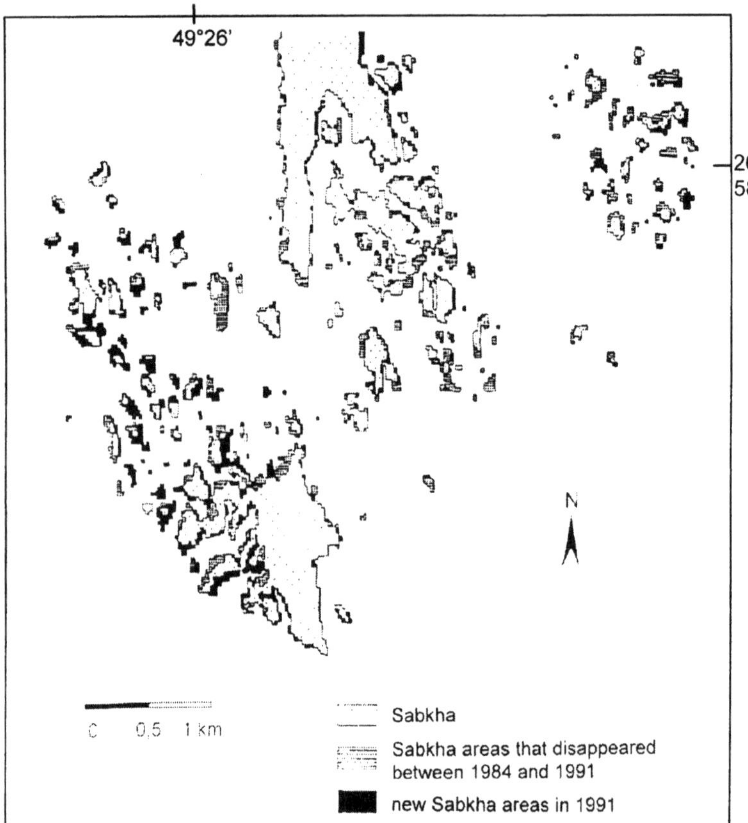

Figure 3. The multitemporal image analysis shows changes between 1984 and 1991.

Image processing was carried out on Sun workstations and ERDAS Imagine 8.2 software. Both images (4-5-7, RGB) were enhanced to distinguish sabkha surfaces from the surrounding sand sheets. The fact that sabkha surfaces hold moisture facilitated their separation from the dry sand sheets. Road intersections (which are abundant in the area of interest) and landmarks like power plants etc. were chosen as ground control points for image rectification. The total RMS error was 0.52. Classification and two arithmetical operations (subtraction) lead to a 4 - signature image (Fig. 3). It shows sabkha areas and sand sheets which have not changed within the seven years, new sabkha surface areas and sabkha areas that were covered by sand sheets (Fig. 3). Many small sabkhat especially in the southern part of the study area disappeared after seven years due to an increase in sand supply probably caused by the reduction of vegetation cover (Fig. 3, upper right side). Other places show strong deflation at the southern parts and only little deposition at the northern fringes of the sabkhat so that the overall sabkha surface increased (Fig. 3, left side).

FIELD MEASUREMENTS OF SABKHA CHANGES

Changes of the extension of inland sabkhat were measured in the field along permanent transect lines. Several iron poles located at the narrow transition zone between the sabkha surface and the sand sheet, were used as permanent markers. They allowed to measure the sabkha edges on future field trips. The vegetation of the whole area is degraded due to heavy grazing. Average vegetation densities were below 3%. Figure 4 shows the sabkha border (line where surface salt crust changes into dry sand) on 2.4.1995, 11.5.1995 and 25.11.1995. Deflation leads to a retreat of the sand sheet which was between 2.5 and 6 meters. Also some of the Aeluropus-hummocks disappeared.

VEGETATION INDICATING FORMER SAND SHEET LOCATIONS

When deflation processes remove the dry sand and therefore sand sheets retreat in downwind direction, remains of deep routing vegetation in the sabkha are the only signs of the former sand sheet locations. Because even wood decays fast in the damp and salty environment these remains only last for decades. But what we find helps to estimate the intensity of deflation. In the study area several stages of sand sheet deflation around the southern and south-eastern sabkha edges could be observed.

Stage 1: deflation

At this initial stage the top sand layer is being removed by the force of wind. Exposed roots of different species help to quantify the erosion. On exposed sites even annuals like *Plantago boissieri* Hausskn. et Bornm. were found with roots exposed to a length of 5 cm (Fig. 5). Because the plant is not older than 2 months this implies an erosion rate of 80 kg m^{-2} in two months (1 cm^3 sand is about 1.6 g; if 5 cm sand are eroded this makes 50,000 cm^3 per m^2 which makes a total weight of 80 kg).

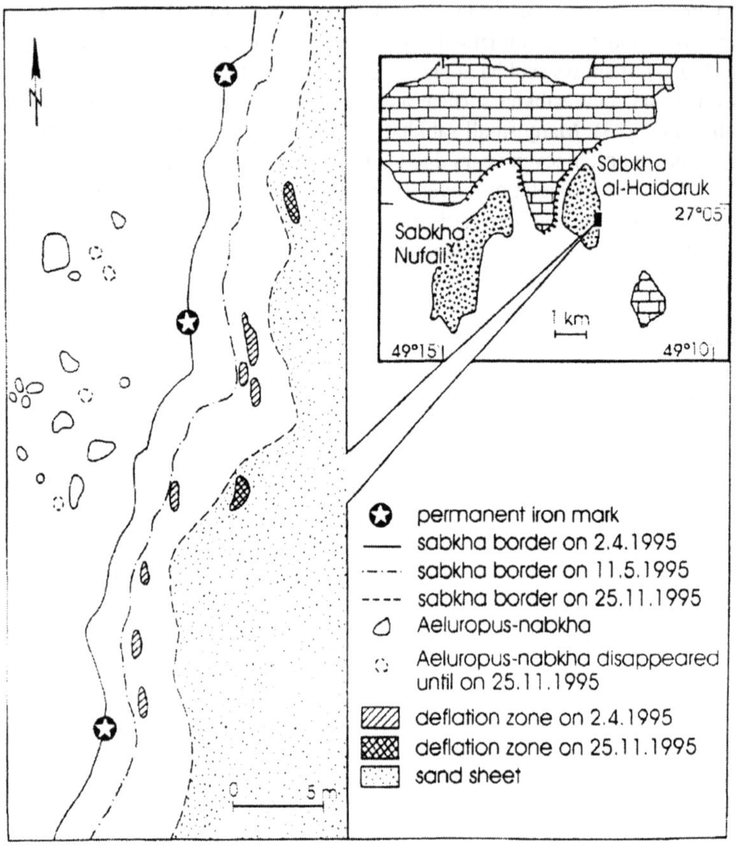

Figure 4. Sabkha border between April and December 1995.

Stage 2: isolation of sand patches

At this stage the dry sand is removed almost completely. Small patches displaying wind ripples are the only remains together with exposed root systems of plant individuals like *Calligonum comosum* L'Hér. (Fig. 6).

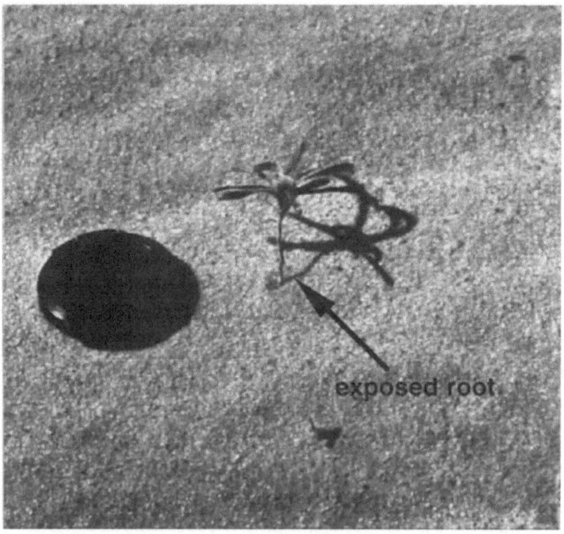

Figure 5. *Plantago boissieri* with roots exposed to a length of 5 cm.

Figure 6. Last remains of a former sand sheet. In the foreground *Calligonum comosum* with exposed roots surrounded by salt crusts.

Stage 3: total deflation of dry sand

In this stage dry sand is completely removed. The only remains indicating the former presence of a sand sheet are dead plant individuals who could never have grown in a sabkha because of the unfavourable soil and groundwater chemistry. Figures 7-9 show the southern edge of an inland sabkha with remains of *Phoenix dactylifera* L. as far as 100 m in the present sabkha (Fig. 7). Groundwater salinity at these sites was from 79 to 88 mS/cm, by far too high to allow for the trees to grow there (at other Phoenix-sites a salinity between 10 and 20 mS/cm was observed). The present sabkha - sand sheet transition zone is patchy and resembles stages 1 and 2 (see aerial view, Fig. 8). Further downwind about 100 m from the sabkha edge large nabkhas around *Phoenix dactylifera, Tamarix arabica* Bge. and *Calligonum comosum* demonstrate a high mobility of the sand. Even the nabkhas in the bordering sand sheets are subjected to erosion at some places (Fig. 9).

Figure 7. Remains of Phoenix dactylifera 100 m in the sabkha.

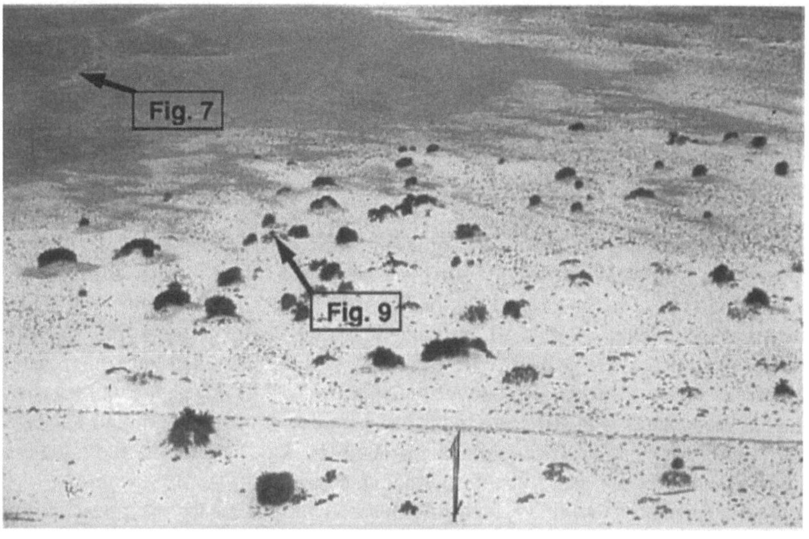

Figure 8. Aerial view shows the patchy pattern in the sabkha - sand sheet transition.

Figure 9. Heavy erosion at nabkha exposes roots of Tamarix sp. (see arrow).

GYPSUM LAYERS

One of the characteristic features regarding inland sabkha sediments of this region is the presence of various horizontal layers rich in gypsum crystals up to cm size. They usually occur in depths between the upper capillary fringe and the saturated zone. These gypsiferous sediments most probably originate in the highly mineralised ground water in sabkhat. That would explain why we never found gypsum in the sediments below the adjacent sand sheets. Interesting is, that in sabkhat with significant deflation processes at their southern edge, the gypsum concentration in the sediment declines rapidly towards the southern end even 15-30 m before the sabkha - sand sheet transition (Fig. 10). Usually the gypsum concentration decreases only 3-5 m from the sabkha edge. On the other hand gypsiferous sediments were abundant at the northern sabkha edges, sometimes even below the sand sheet (Fig. 10). Groundwater chemistry at the southern and northern sabkha edge was similar. The explanation could be sand erosion at the southern sabkha edges and deposition at the northern counterpart.

Where deposition dominates, gypsiferous sabkha sediments disappear below aeolian sand. That is why gypsum was found even at the sabkha - sand sheet transition, usually at the northern sabkha edges. At the southern sabkha edges deflation dominates. The sand sheets retreat and new sabkha surface develops due to high evaporation rates. Within these areas gypsum was rare because there was not time enough for it to develop.

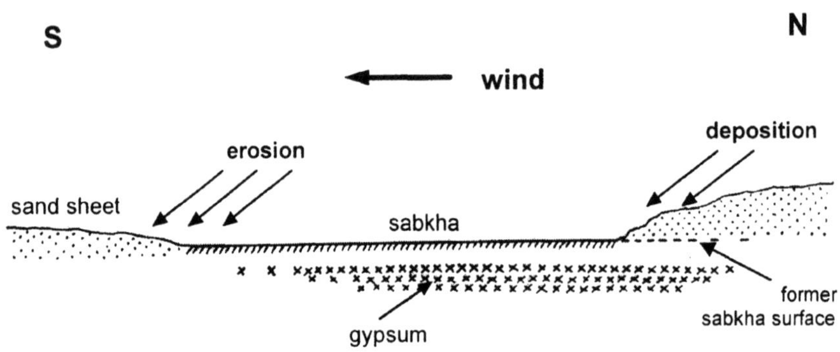

Figure 10. Erosion at the southern edge creates new sabkha surface. Gypsum crystallisation is at the initial stage there, whereas there is plenty of gypsum in the sabkha sediment at the northern edge.

DISCUSSION

These observations covering a time period from months until years demonstrate that inland sabkhat in the Eastern Province of Saudi Arabia are subjected to rapid changes. These changes were only observed, where the vegetation cover was reduced due to heavy grazing. The origin of the inland sabkhat obviously lies in the recent geomorphological history and seems to be the result of a negative sand budget in a supra-regional wind system (Barth 2001). The dynamic of a sabkha is highly influenced by the geo-ecological condition of the surrounding area (sand supply, hydrology, vegetation and climate). Because this condition can change within years it is unlikely that the inland sabkhat in the Eastern Province of Saudi Arabia were formerly coastal sabkhat which originated in the Pleistocene (Holm 1960). Instead, they are rapidly changing geomorphological features and therefore further indicators representing the ecological situation of areas in the Eastern Province of Saudi Arabia.

The following results support this conclusion:

1.) The spatial extension of gypsum layers which are typical sabkha sediments is often underrepresented in the southern and southeastern margins of the sabkhat (downwind side, where deflation dominates).

2.) Exposed roots, vegetational debris and nabkhas are evidence of the former presence of sand sheets in an area that is today covered by sabkhat

3.) Erosional surfaces often occur in sabkhat which are aeolian in nature.

4.) Measurements proved a maximum shift of the southeastern sabkha edges of up to six meters in nine months.

5.) A correlation between vegetation cover of the surrounding sand sheets and the expansion of sabkhat in southern and southeastern direction is evident.

6.) The length-width-ratio of all the inland sabkhat in the study area is 2.8 and their orientation from north to south and north-northwest to south-southeast respectively corresponds with the main wind direction in the area.

7.) Satellite images prove the strong influence of the aeolian component in the morphology of the study area. Significant wind streaks downwind of almost all the sabkhat indicate erosion of salt and dust within the sabkhat and downwind deposition.

8.) The northern central coastal lowlands are located within a negative sand budget area.

ACKNOWLEDGEMENTS

I am grateful to NCWCD, Riyadh for the help they provided during several visits. Especially I would like to thank H. E. Prof. Dr. Abdulaziz H. Abuzinada for his support and encouragement. Assistance, continuous help and cooperation both in the field and the laboratory in Jubail were provided by numerous individuals.

References

Barth, H. K. & Niestlé, A. 1992. Physical and ecological mapping. In: Establishment of a Marine Habitat and Wildlife Sanctuary for the Gulf Region. Final Report for Phase I, pp. 28-51, Frankfurt, Jubail.

Barth, H.-J. 2001. Characteristik of the wind regime north of Jubail, Saudi Arabia, based on high resolution data. *Journal of Arid Environments* 47: 387-402.

Holm, D. A. 1960. Desert geomorphology in the Arabian Peninsula. *Science* 132: 1369-1379.

BIOLOGICAL SOIL CRUSTS OF ARABIAN SABKHAT

JAYNE BELNAP

US Geological Survey, Forest and Range Ecosystem Science Center, 2290 S. West Resource Blvd., Moab, UT 84532, USA

Abstract
Biological soil crusts are a critical component of many arid regions. Where vegetation is extremely limited, such as Arabian sabkhat, these soil crusts may play many important ecological roles. These roles include nitrogen and carbon inputs, maintaining plant-essential nutrients in a bio-available form, influencing local hydrology, affecting establishment of vascular plants, and decreasing soil erosion from either wind or water. Surface disturbance can heavily impact biological soil crusts, and often leads to reduced nutrient input and increased soil erosion. Little work has been done on Arabian sabkhat, although they may be an essential part of these ecosystems.

INTRODUCTION

Most arid and semi-arid landscapes have a limited cover of vascular vegetation. In these ecosystems, the soil spaces between the vascular plants generally contain a community of cyanobacteria, green algae, microfungi, and other bacteria. The cyanobacterial and microfungal filaments weave among soil grains in the top few millimetres of soil, gluing them together and forming a biological crust which stabilises and protects soil surfaces from erosive forces (Belnap & Gardner 1993). In regions with sufficient precipitation and soil stability, lichens and mosses colonise these spaces as well. These biological soil crusts occur in all arid and semi-arid regions and may constitute up to 70% or more of the living cover of a region. They are also referred to as cryptogamic, microbiotic, or microphytic crusts (Harper & Marble 1988).

Worldwide, biological soil crusts have many similarities in species composition, structure, and function. They occur in all types of desert and semi-desert plant

Barth & Böer (eds.), SABKHA ECOSYSTEMS, 227-237

communities. They also occur in steppes and tundra, and are frequently present in xerothermic formations in temperate regions. Despite huge differences in climate and soil in these different habitats, soil crust species within each of the above communities are often of the same genera, if not the same species. The structure of soil crusts is also similar among deserts of the same type: hot deserts of the world (e.g., Atacama, Chihuahuan, Sahara) have flat, smooth crusts, while in cool and cold deserts (e.g., Colorado Plateau, Great Basin, Arctic, Patagonia), frost-heaving gives the soil crusts a bumpy morphology.

SPECIES COMPOSITION AND COVER

Species composition of soil crusts is dependent on soil characteristics, local topography, vegetation, climate, disturbance regimes (including wind and water erosion, water pooling, as well as soil surface disturbance) and other ecological factors (e.g., hydrology). Soil characteristics that influence species composition encompass chemistry (including salinity), texture and stability. Cyanobacteria are halotolerant, with some species able to grow in NaCl-saturated solutions. However, different species have varying tolerances to salinity and the species composition of soil crusts in sabkhat is often different and more diverse at the sabkha edge than in the center (Ullmann & Büdel 2001). A sabkha in Tunisia had no cyanobacteria when Na levels were 15-19 g kg dw^{-1} soil. Once Na dropped to 7-10 g kg dw^{-1} soil, *Chroococcidiopsis* sp., *Xenococcus* sp., *Microcoleus chthonoplastes* and *M. paludosus* appeared, but were infrequent. When Na dropped further, these species became abundant, and six other species appeared. In samples from sabkhat of the United Arab Emirates, *Microcoleus* and *Nostoc* were commonly found (Belnap, pers. obs.) where salinity was low. In sabkhat with higher salinity, several genera that likely belong to the *Halothece* phylogenetic clusters were found (Garcia-Pichel, pers. comm.). These extremely halotolerant cyanobacteria have been reported from hypersaline marine mats, lagoons and inland evaporitic lakes such as the Great Salt Lake in the United States (Garcia-Pichel et al. 1998).

Crust organisms also avoid high shrink-swell clays, while preferring >10% silt (Verrecchia et al. 1995, Belnap, unpubl. data). Soil stability is also critical, as moving sediment will quickly bury crustal organisms. Where wind carries sharp sand that scours the soil surface, organisms only survive in locally-protected areas. In places of more stable sand, or with windbreaks (plant material, outcrops, local macro- or microtopography), crusts are better developed. Pole-facing slopes have greater cover compared to equator-facing slopes (Alexander & Calvo 1990). However, as sabkhat are flat, the local topography's influence on soil crusts is more through larger climate patterns than local solar radiation or evapo-transpiration rates.

Generally, crust cover is correlated with the amount of soil surface not occupied by plants, rocks or actively moving soils (Harper & Marble 1988, Johansen 1993).

This means most soil surfaces are available for crust colonisation in sabkhat. Crusts are often better developed around plants in sabkhat, as they mitigate the extreme environment by protecting crusts from wind scouring, increasing rainfall capture, providing shade, and capturing wind-blown silt.

Climate also influences crust development. Because crust organisms are only metabolically active when wet, the amount and timing of rainfall is critical in determining species cover and composition. When the potential evaporation rate (PET) is high, crust cover and species diversity are low. As the PET declines, metabolic activity times increase, with a concomitant increase in cover and diversity (Alexander & Calvo 1990, Jeffries et al. 1993, Rychert et al. 1978). As sabkhat are found in areas with high PET, crust cover is generally well below 50% and diversity limited to a few species of cyanobacteria. Timing of precipitation is also important. When rain occurs at high air temperatures, it may be available to the soil surface organisms only for a short time, which can result in net carbon losses for species that take longer to reach their net compensation point (Jeffries et al. 1993).

Disturbance regimes, including intensity, type and time since disturbance also influence crust composition. Nothing is known about the successional patterns of soil crusts on sabkhat; however, it is likely they are similar to other crusts. Newly or severely disturbed surfaces are generally bare. Colonisation by different crust species occurs in a manner analogous to terrestrial plant succession (Johansen 1993), with nitrogen-fixing species or species best able to stabilise soils appearing first. In crusts, the large filamentous cyanobacteria are first to stabilise soils, followed by smaller cyanobacteria and green algae. Further soil stabilisation (assuming salinity levels are low enough) allows the colonisation of the lichens *Collema* sp. and *Catapyrenium* sp., followed by other lichens and mosses where soil moisture or precipitation is adequate (Johansen 1993, Harper & Marble 1988).

While studies on what controls species distributions in sabkhat have been very limited, low water availability, wind-scouring and salinity appear important in determining crust composition and cover. In areas with nearby mobile sand sources and frequent winds, soil crusts are generally absent or limited to protected areas, such as rock outcrops, microtopography, or plants, and salinity controls local occurrence patterns (Ullmann & Büdel 2001, Belnap unpubl. data). In sabkhat throughout the Arabian Peninsula, as well as other hyperarid regions, cyanobacteria are generally absent from very hypersaline soils. As salinity drops, *Chroococcidiopsis* sp., *Microcoleus chtonoplastes* and *M. paludosus* are most abundant. With further declines in salinity, *Calothrix parietina*, *Nostoc* sp., and *Gloeocapsa* sp. are found. Further decreases in salinity often allow colonisation of lichens that include *Collema* sp., *Heppia* sp., and *Catapyrenium lacinulatum*. Less frequent, but sometimes seen around sabkhat are *Squamarina lentigera*, *Psora decipiens*, *Diploschistes diacapsis*, and *Fulgensia fulgens*. Worldwide, it has been observed that lichens increase with the level of carbonate, gypsum and silt in soils, and this has also been observed for sabkhat in Tunisia, Australia, and the Arabian Peninsula (Ullmann & Büdel 2001, Belnap unpubl. data).

MICROSTRUCTURE

In arid lands, most relatively undisturbed soils contain both non-biotic (physical) and biological crusts. The presence of any crust alters soil surface characteristics, and thus can play a defining role in many ecosystem functions. Physical crusts, created by raindrop impact, trampling, and/or accumulation of evaporated salts, are found in most arid soils, and their effects are often confused with those of biological crusts. They are transient soil-surface layers ranging from <1 mm to a few centimetres in thickness. They consist mostly of clays, silts, and evaporites, and thus are often structurally different from soils immediately beneath them. The presence of a physical crust provides stability to the soil surface, and the clay and silt enhance soil surface moisture retention. Consequently, the presence of a physical crust hastens colonisation of soil crust organisms. The presence of a physical crust can also seal and smooth surfaces, thus increasing the volume and velocity of water runoff, as well as often inhibiting establishment of vascular vegetation.

Worldwide, the microstructure of biological crusts is influenced by climate (frequency and depth of soil freezing and PET), species composition, soil texture, chemistry and physical crusts, disturbance history, and aeolian deposition. Unlike crusts in regions where freezing is common or lichen cover high, crusts in sabkhat are generally smooth. Smooth crusts occur in hyperarid and arid regions where rainfall is low, temperatures are high, and soils never freeze (e.g., Arabian, central Sahara, Negev, Atacama deserts). Smooth crusts consist mostly of endedaphic (living within the soil) cyanobacteria, algae and microfungi that bind together mostly mineral particles, with limited lichen cover. High carbonate or gypsiferous soils often have lichens and mosses, thus creating a rugose surface (Belnap & Lange 2001). Soils with a high content of shrink-swell clays (e.g. bentonite) are extremely unstable, and often support only large filamentous cyanobacteria (e.g., *Microcoleus*). In such soils, physical/chemical soil characteristics control crust morphology, with little contribution from biological components.

Soils crusts are dominated by photosynthetic organisms that require sunlight. Most cyanobacteria are found at 0.2-0.5 mm, where sufficient light for net carbon gain is available, but UV exposure is reduced. Well-developed crusts show a vertical layering of cyanobacterial species (Davey & Clarke 1992). Smaller, less mobile organisms (e.g., *Chroococcidiopsis, Nostoc, Scytonema*) are most often found at the soil surface, and contain specialised pigments for UV-protection. The larger cyanobacteria (e.g., *Microcoleus, Phoridium*), lacking UV pigments, are found below the soil surface. When soils are wetted, these large cyanobacteria glide up to the soil surface. Cyanobacteria are found closer to the surface in finer soils than coarse soils, as light is quickly extinguished by soil particles (Garcia-Pichel & Belnap 1996). Intensity, type, and time since soil surface disturbance often control the external morphology of soil crusts, as surface roughness is lost. Airborne silt and clay are trapped by sticky cyanobacterial sheaths and protruding moss stems and lichen thalli. This results in a thin layer of silt and clay on the crust surface. Silts increase soil fertility and water-holding capacity (Davey & Clarke 1992).

ECOLOGICAL ROLES

While no studies have directly addressed the ecological role of soil crusts on sabkha surfaces, there are many studies from hot desert ecosystems that are applicable to these systems.

Carbon fixation

Crust organisms respond quickly to water, with respiration beginning less than 3 minutes after wetting (Garcia-Pichel & Belnap 1996). Photosynthesis begins shortly afterwards, reaching full activity within 10-30 minutes. When soils are moist, steep oxygen gradients result in anoxic zones 1-4 mm from the surface. Recently studies of soil crust lichens (Lange 2001) showed that lichens have several different photosynthetic strategies, and that carbon (C) gains are heavily dependent on thallus hydration levels and temperature. Photosynthesis rates in most species increase with temperature up to about 26-28°C. Biological soil crusts are an important source of fixed C on sparsely vegetated areas common throughout the western United States (Beymer & Klopatek 1991). In this region, C contributed by soil crust organisms helps keep plant interspaces fertile and supports other microbial populations.

Nitrogen fixation

Soil nitrogen (N) concentrations are known to be low in deserts relative to other ecosystems, and N is known to limit net primary productivity in deserts (Romney et al. 1978). Cyanobacteria and cyanolichens can be the dominant source of N in many desert ecosystems (Evans & Ehleringer 1993). Nitrogen fixation is highly dependent on past and present water and light regimes, as well as crust species composition (Belnap & Lange 2001). Maximum N fixation occurs at soil surface temperatures of 20-30°C, with 26-28°C optimal for most non-polar crusts. Timing, extent and type of past disturbance can also be critical factors in determining fixation rates (see disturbance section). Much of the fixed N leaks into the surrounding substrate, and is available to nearby vascular plants and microbes (Silvester et al. 1996). Vascular plants growing in crusted areas show higher tissue concentrations of N when compared to plants in uncrusted soils (Mayland & McIntosh 1966; Harper & Belnap 2001). Since rainfall events in desert areas are often too small to promote plant growth, but stimulate microbial community activity, soil crusts may contribute greater N than N-fixing plant species. In addition, crusts help maintain fertility of plant interspaces, counteracting nutrients concentrating around perennial plants. Although no N fixation studies have been done in sabkhat, N-fixing cyanobacterial species are present, and are expected to increase soil N in these ecosystems.

Vascular plants

No work has directly addressed the influence of soil crusts on vascular plants in sabkhat. There have been general studies on smooth crusts in hot deserts, but results are mixed. The probability of seed entrapment on smooth cyanobacterial crusts is low compared to frost-heaved crusts of cool deserts (Belnap & Lange 2001). Smooth crusts enhance, inhibit, or don't affect seed germination, depending on the species. Plant survival and fecundity on crusted surfaces in hot deserts are either enhanced or not affected, while biomass is increased.

Vascular plants growing on crusted surfaces generally have higher tissue concentrations of N, K, Ca, Mg, P, Fe, Mn, Cl and S when compared to plants on non-crusted soils (Harper & Belnap 2001). There are several mechanisms to explain this, reviewed in Belnap & Lange (2001). Clay particles stick to the mucilaginous sheath material, which in turn bind positively-charged plant macronutrients. Crusts increase N in the upper soil layers (0-20 cm) by up to 200%. Most crust organisms bind metals and secrete powerful metal chelators; cyanobacteria also secrete peptide nitrogen and riboflavin. These substances keep P, Cu, Zn, Ni, and Fe plant-available. Cyanobacteria secrete glycollate, which stimulates uptake of phosphate, vitamins, and auxin-like substances. Crusts also secrete fixed carbon, increasing soil carbon in surface layers (0-10 cm) by up to 300%, benefitting heterotrophic microbes which are often carbon limited. C additions lower the soil C:N ratio, increasing decomposition rates. Increases in soil organic matter ameliorate compaction, reduce inorganic soil crusting, reduce nutrient leaching losses, and increase soil moisture retention. There is also a strong correlation between mycorrhizal infections of seed plants and crusted surfaces (Harper & Pendleton 1993, Belnap & Lange 2001).

Water Relations

The effect of biological soil crusts on soil water relations is heavily influenced by soil texture and structure, as well as the growth form of the crusts. In hot deserts where frost-heaving is not present and soil crusts are relatively flat, infiltration rates are heavily dependent on soil type and local features. In general, smooth crusts reduce infiltration on sandy soils and increase infiltration on finer-textured soils (Belnap & Lange 2001). On sabkha surfaces, cracking of evaporite surfaces can influence local infiltration patterns. In hyperarid regions, local reduction of infiltration can be critical in supporting plants. In the Negev, studies showed that shrubs were dependent on water run-off generated from upslope, crusted soils (Yair 2001).

Soil stabilisation

The presence of soil crusts reduces both wind and water erosion in all environments that have been studied. Polysaccharides extruded by the cyanobacteria, green algae and fungi (Fig. 1), along with the rhizins and rhizomorphs of lichens and protonemata of mosses, entrap and bind soil particles together, aggregating them into larger particles. Larger aggregates are more difficult to move, thus reducing soil loss via wind or water erosion. Soil aggregation by crusts also enables otherwise loose sandy soils to stay in place on steep slopes and in areas of shallow bedrock or caliche (Belnap & Gardner 1993).

Figure 1. *Microcoleus vaginatus* in sandy soils. Sheath material binds sand grains together (note the sheaths are wound around and among sand grains [bar = 100 μm]).

EFFECTS OF DISTURBANCE AND RATE OF RECOVERY

Soils inarid regions are often highly erodible, and soil formation extremely slow, taking 5000 to 10,000 years (Dregne 1983). Animal, vehicle, and human foot traffic generate compressional and shear forces that can severely impact biological crusts. The effects of disturbance are similar worldwide, and include a reduction in species diversity, total crust cover, nitrogen and carbon fixation,as well as a reduction in soil stability. Well-developed crusts around sabkhat can contain up to 10 soil lichen, 3 moss and 10 cyanobacterial species, while disturbed areas often have only one cyanobacterial species (generally *Microcoleus*). Surface disturbance negatively affects the cohesion and coverage of cyanobacterial crusts, as crusts are brittle when dry and easily crushed (Belnap & Gardner 1993). Any compressional disturbances to these crusts leaves surfaces vulnerable to wind and water erosion (Belnap & Gillette 1998). Impacts can also be indirect: sediments from nearby disturbed areas can bury living crusts, often resulting in the death of the photosynthetic organisms (Belnap 1995). Because over 75% of the photosynthetic biomass, and almost all photosynthetic productivity, is from organisms in the top 3 mm of these soils (Garcia-Pichel & Belnap 1996), very small soil losses can dramatically reduce site fertility and further reduce soil surface stability. Consequently, trampling can greatly accelerate degradation processes through increased soil and water loss (Dregne 1983). C and N fixation is dramatically reduced in soil crusts when disturbed (Belnap & Eldridge 2001), and gaseous N loss increased (Peterjohn & Schlesinger 1990). This has large implications for deserts, where the crusts are the dominant source of C and N in plant interspaces (Evans & Belnap 1999). Decreased soil C and N alters soil food webs and thereby affects nutrient availability in these systems (Ingham et al. 1989).

Soil surface albedo is also affected by surface disturbance. Trampled surfaces have a 50% increased reflectance of wavelengths 0.25- 2.5 μm compared to crusted surfaces (Belnap 1995). Trampled surfaces can also have significantly lower temperatures than crusted surfaces, which lowers N and C fixation rates, microbial activity, plant nutrient uptake, seed germination time and seedling growth rates (Belnap & Lange 2001). Timing is often critical in deserts, and relatively small delays in germination can reduce plant fitness and seedling establishment, which may eventually affect community structure. Food and other resources are often partitioned among ants, arthropods and small mammals on the basis of surface temperature-controlled foraging times (Crawford 1991). Many small desert animals are weak burrowers, and soil surface microclimates are of great importance to their survival. Consequently, altering surface temperatures can affect nutrient availability and community structure for many desert organisms.

Because crustal organisms are only metabolically active when wet, re-establishment time is slow in arid systems. Cyanobacteria are mobile and can often quickly recolonise a site. Lichens and mosses, however, are much slower, and thus underlying soils are left vulnerable to wind and water erosion for at least 20 years (Belnap & Gillette 1998). Moving sediments further destabilise adjoining areas by burying adjacent crusts, or by "sandblasting" nearby surfaces, thus increasing wind erosion rates (Belnap 1995, McKenna Neuman et al. 1996).

Recovery rates of soil crusts depend on the type and extent of disturbance, the availability of nearby inoculation material, as well as on the temperature and moisture regimes that follow disturbance events. Estimates of time for visually-assessed recovery have varied from 5 to 100 years (Harper & Marble 1988).

However, Belnap (1993) showed that many components of recovery can not be assessed visually, and recovery may be much slower. Since recovery time is dependent on presence of nearby material to colonise the disturbed area, larger disturbed areas will take longer to recover.

Recovery of nitrogen fixation appears to be quite slow, and studies suggest the negative effects on N dynamics may persist for extended periods of time after disturbance ceases (Evans & Belnap 1999). Restoration of normal surface albedos and temperatures will depend on the restoration of cyanobacteria, lichens and mosses. The use of inoculants to speed up recovery of these crusts has been used repeatedly (St. Clair et al. 1984, Belnap 1993) and is very effective in hastening recovery of all soil crust components.

CONCLUSION

Biological soil crusts are critical components of the ecosystems in which they occur. Many activities of man are incompatible with the presence and well-being of biological soil crusts. The cyanobacterial fibers that confer such tensile strength to these crusts are no match for the compressional stresses placed on them by anthropogenic activities. Crushed crusts contribute less N and C to the ecosystem. Impacted soils are left highly susceptible to both wind and water erosion. Unlike vascular plant cover, crustal cover is not reduced in drought, and unlike rain and salt crusts, these organic crusts are present during and after rain events. Consequently, they offer stability over time and in adverse conditions, factors that are often lacking in other soil surface protectors, and should be considered in any environmental management activity. Little is known about biological soil crusts of sabkhat, and it is hoped further studies will determine more specifically how soil crusts affect their ecology.

References

Alexander, R.W. & Calvo, A. 1990. The influence of lichens on slope processes in some Spanish badlands. In: Thornes, J.B. (ed), Vegetation and erosion: processes and environments. John Wiley & Sons, Ltd., Chichester, England, pp. 385-398.

Belnap, J. 1993. Recovery rates of cryptobiotic soil crusts: inoculant use and assessment methods. *Great Basin Naturalist* 53: 89-95.

Belnap, J. 1995. Surface disturbances: their role in accelerating desertification. *Environmental Monitoring and Assessment* 37: 39-57.

Belnap, J. & Eldridge, D.J. 2001. Disturbance and recovery of biological soil crusts. In: Belnap, J. & Lange, O.L. (eds.), Biological soil crusts: structure, function, and management. *Ecological Studies Series* 150. Springer-Verlag, Berlin, pp. 363-383.

Belnap, J. & Gardner, J.S. 1993. Soil microstructure of the Colorado Plateau: the role of the cyanobacterium *Microcoleus vaginatus*. *Great Basin Naturalist* 53: 40-47.

Belnap, J. & Gillette, D.A. 1998. Vulnerability of desert biological soil crusts to wind erosion: the influences of crust development, soil texture, and disturbance. *Journal of Arid Environments* 39: 133-142.

Belnap, J. & O.L. Lange (eds.) 2001. Biological soil crusts: structure, function, and management. *Ecological Studies Series* 150. Springer-Verlag, Berlin, 503 pp.

Beymer, R.J. & Klopatek, J.M. 1991. Potential contribution of carbon by microphytic crusts in pinyon-juniper woodlands. *Arid Soil Research and Rehabilitation* 5: 187-198.

Crawford, C.S. 1991. The community ecology of macroarthropod detritivores. In: Polis, G. (ed.), Ecology of desert communities. University of Arizona Press, Tucson.

Davey, M.C. & Clarke, K.J. 1992. Fine structure of a terrestrial cyanobacterial mat from Antarctica. *Journal of Phycology* 28: 199-202.

Dregne, H.E. 1983. Desertification of arid lands. Harwood Academic Publishers, New York.

Evans, R.D. & Belnap, J. 1999. Long-term consequences of disturbance on nitrogen dynamics in an arid ecosystem. *Ecology* 80: 150-160.

Evans, R.D. & Ehleringer, J.R. 1993. A break in the nitrogen cycle in arid lands? Evidence from N^{15} of soils. *Oecologia* 94: 314-317.

Garcia-Pichel, F. & Belnap, J. 1996. Microenvironments and microscale productivity of cyanobacterial desert crusts. *Journal of Phycology* 32: 774-782.

Garcia-Pichel, F. Nubel, U. & Muyzer, G. 1998. The phylogeny of unicellular, extremely halotolerant cyanobacteria. *Archiv Microbiology* 169: 469-482.

Harper, K.T. & Belnap, J. 2001. The influence of biological soil crusts on mineral uptake by associated vascular plants. *Journal of Arid Environments* 47 (3): 347-357.

Harper, K.T. & Marble, J.R. 1988. A role for nonvascular plants in management of arid and semiarid rangelands. In: Tueller, P.T. (ed.), Vegetation science applications for rangeland analysis and management. Kluwer Academic Publishers, Dordrecht. pp. 136-169.

Harper, K.T. & Pendleton, R.L. 1993. Cyanobacteria and cyanolichens: can they enhance availability of essential minerals for higher plants? *Great Basin Naturalist* 53: 59-72.

Ingham, E.R., Coleman, D.C. & Moore, J.C. 1989. An analysis of food-web structure and function in a shortgrass prairie, a mountain meadow, and a lodgepole pine forest. *Biology and Fertility of Soils* 8: 29-37.

Jeffries, D.L., Link, S.O. & Klopatek, J.M. 1993. CO_2 fluxes of cryptogamic crusts. I. Response to resaturation. *New Phytologist* 125: 165-173.

Johansen, J.R. 1993. Cryptogamic crusts of semiarid and arid lands of North America. *Journal of Phycology* 29: 140-147.

Lange, O.L. 2001. Photosynthesis of soil-crust biota as dependent on environmental factors. In: Belnap, J. & Lange, O.L. (eds.), Biological soil crusts: structure, function, and management. *Ecological Studies Series* 150. Springer-Verlag, Berlin, pp. 217-240.

Mayland, H.F. & McIntosh, T.H. 1966. Availability of biologically fixed atmospheric nitrogen-15 to higher plants. *Nature* 209: 421-422.

McKenna Neuman, C., Maxwell, C.D. & Boulton, J.W. 1996. Wind transport of sand surfaces crusted with photoautotrophic microorganisms. *Catena* 27: 229-247.

Peterjohn, W.T. & Schlesinger, W.H. 1990. Nitrogen loss from deserts in the southwestern United States. *Biogeochemistry* 10: 67-79.

Romney, E.M., Wallace, A. & Hunter, R.B. 1978. Plant response to nitrogen fertilization in the northern Mohave Desert and its relationship to water manipulation. In: West, N.E. & Skujinš, J.J. (eds.), Nitrogen in desert ecosystems. US/IBP Synthesis Series 9. Dowden, Hutchinson and Ross, Stroudsberg, PA, pp. 232-243.

Rychert, R.C., Skujinš, J.J., Sorensen, D. & Porcella, D. 1978. Nitrogen fixation by lichens and free-living microorganisms in deserts. In: West, N.E. & Skujinš, J.J. (eds.), Nitrogen in desert ecosystems. US/IBP Synthesis Series 9. Dowden, Hutchinson and Ross, Stroudsburg, PA, pp. 20-30.

Silvester, W.B., Parsons, R. & Watt, P.W. 1996. Direct measurement of release and assimilation of ammonia in the *Gunnera-Nostoc* symbiosis. *New Phytologist* 132: 617-625.

St. Clair, L.L., Webb, B.L., Johansen, J.R. & Nebeker, G.T. 1984. Cryptogamic soil crusts: enhancement of seedling establishment in disturbed and undisturbed areas. *Reclamation and Revegetation Research* 3: 129-136.

Ullmann, I. & Büdel, B. 2001. Ecological determinants of species composition of biological soil crusts on a landscape scale. In: Belnap, J. & Lange, O.L. (eds.), Biological soil crusts: structure, function, and management. *Ecological Studies Series* 150. Springer-Verlag, Berlin, pp. 203-213.

Verrecchia, E., Yair, A., Kidron, G. & Verrecchia, K. 1995. Physical properties of the psammophile cryptogamic crust and their consequences to the water regime of sandy soils, north-western Negev Desert, Israel. *Journal of Arid Environments* 29: 427-437.

Yair, A. 2001. Effects of biological soil crusts on water redistribution in the Negev Desert, Israel: a case study in longitudinal dunes. In: Belnap, J. & O.L. Lange (eds), Biological soil crusts: structure, function, and management. *Ecological Studies Series* 150. Springer-Verlag, Berlin, pp. 303-314.

MACROALGAE ASSOCIATED WITH SABKHA

DAVID M. JOHN

The Natural History Museum, Department of Botany, Cromwell Road, London SW7 5BD, UK

Abstract

Sabkha represents an extremely inhospitable environment for macroscopic eukaryotic algae with the commonly referred to 'algal mats' usually consisting of prokaryotic cyanobacteria. Only three genera of green algae actively grow in shallow hypersaline pools, namely *Enteromorpha*, *Chaetomorpha* and *Rhizoclonium*. Other macroalgae, such as *Hormophysa cuneiformis* and *Sargassum decurrens*, are usually blown into sabkha areas or become carried into them along with other marine plant debris when flooded by seawater during high water spring tides and storm surges. These macroalgae are common in adjacent ecosystems including seagrass or mangrove-dominated lagoons. The macroalga *Enteromorpha* can be used as a supplement to the feed of domestic animals and has the potential for cultivation in seawater-irrigation systems, possibly as part of a sustainable agro-ecosystem based on halophyte crops or an aquaculture scheme.

INTRODUCTION

Many areas of sabkha are devoid of macroscopic algae and any mention of 'algae' is normally reference to the commonly extensive, thick mats of cyanobacteria. These photosynthetic, oxygen-producing organisms have been traditionally referred to as 'blue-green algae' or cyanophytes although they are prokaryotic and thus more closely related to bacteria. Only rarely recorded in sabkha areas are eukaryotic macroalgae, commonly known as seaweeds, with these frequently divided into 3 or 4 major taxonomic groups depending in part on pigmentation. Often these macroalgae are mentioned incidentally in studies whose primary focus is other biological groups (sometimes the cyanobacteria) or on physico-chemical features of

Barth & Böer (eds.), SABKHA ECOSYSTEMS, 239-245

the environment. Some of the most extensive and in-depth studies of sabkha ecosystems have been carried out in shallow hypersaline ponds on the coast of the Sinai peninsula in which *Enteromorpha* (Green Alga) is the most frequently recorded eukaryotic alga (see Friedman & Krubein 1985). All receive seawater through subterranean seepage except for the occasional flood. The sabkha shows its optimal development in the Arabian Gulf along the coast of Abu Dhabi in the United Arab Emirates where it has been described by Evans (Evans et al. 1964, 1969, 1995), although with no mention of 'algae' other than of extensive cyanobacterial mats.

DISTRIBUTION AND ECOLOGY

The greatest diversity of eukaryotic macroalgae or seaweeds is to be found where conditions are close to fully marine. Increasing or decreasing salinity and high seawater temperature cause stress to many seaweed with the result that many surviving in warm hypersaline habitats are adapted to withstand these conditions. Low algal diversity in the south-eastern part of the Arabian Gulf compared to the adjacent Indian Ocean has been attributed in part to high summer seawater temperature (>30°C) and to high salinities (>45 ppt). In the Arabian Gulf and elsewhere in Arabia macroalgae are significantly more abundant and diverse on open rocky shores compared to lagoon areas and creeks. Sometimes immediately adjacent to such habitats there are relatively flat, episodically or periodically flooded areas of salt marsh or sabkha characterised by hypersalinity and the deposition of carbonates, gypsum and salts.

Comparatively few macroalgae are tolerant of hypersaline environments including those commonly encountered in the more enclosed inner parts of lagoons. Absent in the Arabian area is the characteristic assemblage of macroalgae normally associated with mangroves elsewhere in the world. The few macroalgae in the inner reaches of tidal lagoons are often associated with sponges, other sessile invertebrate animals, seagrasses or on isolated, often partly sand-buried, bivalve molluscs or rock and shell fragments. One of the most conspicuous macroalgae to survive directly on relatively stable sediment is the siphonaceous *Avrainvillea amadelpha* (Montagne) Gepp & Gepp. The few macroalgae, known as 'Red Algae', that are present in quiet lagoon systems are usually bleached yellow, orange or straw-coloured. Green macroalgae are often the most abundant and seemingly most well-adapted to survive in seagrass-dominated areas within lagoon systems. Sometimes present are free-floating bright green filaments of *Chaetomorpha linum* (O.F. Müller) Kützing although these are more commonly entangled with sea grasses and the breathing roots of the black mangrove *Avicennia marina* (Forsskål) Vierhert. If these filaments and those of the finely divided green macroalga *Cladophora nitellopsis* Børgesen are very abundant then at low tide they form curtain-like shrouds over the breathing roots (personal observations). The green macroalgae *Enteromorpha intestinalis* and *Ulva* spp. have been recorded by Khoja (1998) from mangrove areas on the Gulf coast of Saudi Arabia (see Khoja 1998). In very sheltered areas in the Arabian Gulf there sometimes develops on thick organic mud a felty covering of the finely divided filaments of a species of *Vaucheria* (Yellow-Green Alga). Otherwise the only other macroalgae present are drift specimens that have probably arisen from populations growing under more fully marine conditions. Drift macroalgae are most common in the spring and early summer along the coasts of Abu Dhabi in the UAE and Saudi Arabia after which time many die and decay as the seawater temperature rapidly rises.

Low lying areas of sabkha, often separated from the sea by a low sand bar or periodically connected to it by channels and creeks, are sometimes flooded by seawater which might carry with it a considerable quatity of floating macroalgae along with seagrass. Such plant debris can be carried many km inland and includes robust brown macroalgae (e.g., *Sargassum decurrens* (Brown ex Turner) C. Agardh, *Hormophysa cuneiformis* (J.F. Gmelin) P.C. Silva), seagrasses (principally *Halodule uninervis* Forsskål), mangrove seeds along with various macroalgae that normally grow attached in adjacent lagoons, shallow bays, and creeks. Such flooding must bring into sabkha areas significant amounts of organic matter that must significantly contribute to soil formation where it accumulates in any quantity.

Most of the macroalgae carried into sabkha areas are unable to survive for any length of time in hypersaline pools with the exception of the green algal genera *Rhizoclonium, Chaetomorpha* and *Enteromorpha. Enteromorpha* is not common in the Arabian Gulf, unlike in the Red Sea from where Evenari et al. (1985) report it covering areas of a hypersaline pool to the north of Ras Muhammad (southernmost point of the Sinai) influenced by seawater seepage. In a pond (Gavish Sabkha) on the coastal plain between the shore of the Gulf of Aqaba and the Sinai mountains (few km north of the Strait of Tiran) there is a clear series of horizontal zones behind the sandbar. The first two zones, recognised by having a carbonate cementation plate and hyalite-gypsum crusts, are followed by a series of biologically defined zones: *Bledius*-zone (staphhilinid beetle), *Pirenella-Enteromorpha* zone, *Pirenella* zone without *Enteromorpha,* nodule zone (mostly cyanobacteria), seasonally air-exposed cyanobacterial mats and permanently water-covered blue-green algal mats. The only zone in which a eukaryotic macroalga is conspicuous is the first, characterised by the gastropod *Pirenella conica.* In this zone the green alga *Enteromorpha* forms thick scummy mats along with other unnamed green algae. This zone not only receives water from seawater springs on the eastern slope of Gavish Sabkha, but from small water outlets forming within gullies during the winter (see Gerdes et al. 1985). Such green algal mats are covered with epiphytic diatoms and cyanobacteria that are grazed by the *Pirenella.* As mentioned, green algal mats survive only within standing water and little is known of what changes occur when these hypersaline desert depressions dry out in summer, or the areas of standing water become greatly reduced in extent. According to Sheppard et al. (1992), *Enteromorpha* and *Chaetomorpha* cannot survive salinities above 65 ppt although *Rhizoclonium* has been recorded growing in salinities of 87-132 ppt. Other algae that occur in lagoon environments adjacent to sabkha, such as *Avrainvillea amadelpha* and *Sargassum decurrens,* are known to tolerate salinities of 56-58 ppt in Saudi Arabia (Basson et al. 1977). This contrasts to most salt-resistant diatoms and cyanophytes that can tolerate salinities of up to 190 ppt and 400 ppt respectively (see Ehrlich & Dor 1985).

The absence of suitable hard surfaces, high seawater temperature and hypersalinity make sabkha areas extraordinary inhospitable environments for benthic macroalgae. If macroalgae survive at all in such a harsh environment then they are confined to pools receiving seawater by subterranean seepage and/or during extreme high spring tides, periods of strong onshore winds or storm surges. In some places where the sabkha forms a low coastal plain just above sea level there are well-defined channels or creeks transporting seawater along with drift macroalgae and other plant material into these areas. If the sabkha is never flooded by the sea then macroalgal material can only be carried into it by birds or blown in from the strandline. As mentioned earlier, macroalgae often develop in adjacent ecosystems, especially from those close to the seaward margin of the sabkha. These adjacent ecosystems include seagrass-dominated or mangrove-fringed lagoons as well as shallow patch or fringing coral reefs. Often the most common macroalgae in sabkha areas are loose lying having been washed in on a flood tide and thus remain alive for only a very short time. No macroalgae live for long in such areas other than those green algae mentioned above. These usually form mats overlying a thin crust of silt and precipitated gypsum and/or salt. Occasionally artificial ponds supplied by freshwater become discoloured by large populations of microalgae and floating surface flocs might develop. On the landward side of cyanobacteria mats in the Abu Dhabi sabkha, algae have been observed (unpublished observations) developing in small pools formed by the overspill of water used for irrigating roadside trees.

SABKHA MACROALGAE

The few records of green macroalgae in sabkha pools are often referred to in studies concerned principally with microbial ecology. The macroalgal material is not described in detail and identification is given only to genus. The two most commonly recorded *Enteromorpha* species, *E. flexuosa* and *E. clathrata,* are also recorded growing in fully marine habitats adjacent to sabkha. It remains uncertain as to whether one or both of these species are those that are commonly recorded in sabkha pools simply as '*Enteromorpha*'. The '*Rhizoclonium*' species is probably *R. tortuosum* (Dillwyn) Kützing (see De Clerk & Coppejans 1996), the only species recognised in the Arabian Gulf (probably also in the Red Sea) if it is accepted that

R. riparium (Roth) Harvey and *R. implexum* (Dillwyn) Kützing are synonyms of it (see Burrows 1991). More uncertainty surrounds the identity of *Chaetomorpha* of which several species are reported from the Arabian region, including adjacent parts of the Indian Ocean (Silva et al. 1996). The unattached, free-floating mats in hypersaline environments in Abu Dhabi usually closely correspond to *C. linum* whilst attached forms are considered by some authors to be a separate species (*C. crassa* (Dillwyn) Kützing). *Rhizoclonium* might be sometimes mistaken for *Chaetomorpha* if its characteristic rhizoidal branches are sparsely developed and therefore some records might be based on misidentification.

POTENTIAL FOR MACROALGAL CULTIVATION

Seawater irrigation systems are a means of converting hypersaline land into halophyte production systems. If there is sufficient through-flow in seawater irrigation channels and in aquaculture ponds (especially those used for shrimp cultivation) then green macroalgae might grow rapidly and develop a significant biomass. Sabkha soils are known to be generally rich in nutrients known to enhance plant growth. It is possible that the algae may be harvested for use as a food supplement for domesticated animals. Moll & Deikman (1995) have obtained high yields of *Enteromorpha clathrata* in seawater-irrigated ponds in which the water was exchanged once in several days to maintain salinity within tolerable limits. They believe that *Enteromorpha* pond culture will enable 'highly productive agriculture in areas that are now desert' and 'will provide a new resource for feed, food and biomass'

CONCLUSION

The sabkha is an extraordinarily inhospitable environment for eukaryotic macroalgae with little known of the taxonomic status of the macroalgae that survive in these areas. The only areas where these green algae are present are those usually influenced by seawater seepage and where salinity normally remains below 70 ppt. Quantitative information is lacking on the quantity of macroalgae and other plant debris carried into sabkha areas and what contribution such accumulated organic material makes to any soil formation. The development of aquaculture ponds and the establishment of halophyte production systems using seawater irrigation opens up the possibility of growing in quantity and harvesting sabkha-living macroalgae.

RECOMMENDATIONS FOR FUTURE WORK

- Research is required to determine the type and quantity of macroalgae and other organic material of marine origin that is transported into sabkha.
- Needing to be established is the taxonomic identity of the macroalgae living within sabkha and whether there exist different physiological races.
- Needing to be determined is the fate and the role played by macroalgal and other material transported into sabkha compared to the algae that live in situ.
- Serious consideration needs to be given to the possibility of cultivating macroalgae in seawater irrigation channels and ponds, possibly linked to halophyte production systems and aquaculture schemes.
- Further research is needed on the food values of sabkha-grown macroalgae as a source of feed for domesticated animals.

ACKNOWLEDGEMENTS

Thanks go to Graham Evans for providing literature and to David George and Veryan Pappin who accompanied me on my many visits to marine habitats (including sabkha areas) in Abu Dhabi. The Abu Dhabi Company for Onshore Oil Operations (ADCO) supported The Natural Hisory Museum's research programme in the United Arab Emirates and British Airways assisted with an additional baggage allowance on their flights.

References

Basson, P.W., Burchard, J.E., Hardy, J.T. & Price, A.R.G. 1977. Biotopes of the Western Arabian Gulf: marine life and environments of Saudi Arabia. ARAMCO, Dhahran.

Burrows, E. 1991. Seaweeds of the British Isles. Volume 2. Chlorophyta. Natural History Museum, London.

De Clerk, O. & Coppejans, E. 1996. Marine algae of the Jubail Marine Wildlife Sanctuary, Saudi Arabia. In: Krupp, F., Abuzinada, A.H. & Nader, I.A. (eds.), A marine wildlife sanctuary for the Arabian Gulf. Environmental research and conservation following the 1991 Gulf War oil spill. NCWCD, Riyadh and Senkenberg Research Institute, Frankfurt, Germany, pp. 199-289.

Ehrlich, A. & Dor, I. 1985. Photosynthetic microorganisms of the Gavish Sabkha. In: Friedman, G.M. & Krumbein, W.E. (eds.), Hyperhaline ecosystems - The Gavish Sabkha. Ecological Studies 53. Springer, Berlin, Heidelberg, New York & Tokyo, pp. 296-321.

Evans, G. 1995. The Arabian Gulf: a modern carbonate-evaporite factory; a review. *Cuadernos de Geologia Ibérica* 19: 61-96.

Evans, G., Kendall, C.G.ST.C. & Skipworth, P.A. d'E. 1964. Origin of the coastal flats – the sabkha – of the Trucial coast, Persian Gulf. *Nature* 202: 759-761.

Evans, G., Schmidt, V., Bush, P. & Nelson, H. 1969. Stratigraphy and geologic history of the sabkha Abu Dhabi, Persian Gulf. *Sedimentology* 12: 145-159.

Evenari, M., Gutterman, Y. & Gavish, E. 1985. Botanical studies on coastal salinas and sabkhas of the

Sinai. In: Friedman, G.M. & Krumbein, W.E. (eds.), Hyperhaline ecosystems - The Gavish Sabkha. Ecological Studies 53. Springer, Berlin, Heidelberg, New York & Tokyo, pp. 145- 182.

Friedman, G.M. & Krumbein, W.E. (eds.) 1985. Hyperhaline ecosystems - The Gavish Sabkha. Ecological Studies 53. Springer, Berlin, Heidelberg, New York & Tokyo.

Geres, G., Krumbein, W.E. & Holtkamp, E. 1985. Salinity and water activity related zonation of microbial communities and potential stromatolites of the Gavish Sabkha. In: Friedman, G.M. & Krumbein, W.E. (eds.): Hyperhaline ecosystems - The Gavish Sabkha. Ecological Studies 53. Springer, Berlin, Heidelberg, New York & Tokyo, pp. 239-266.

Khoja, T.M. 1998. New records of open coast and mangrove algae on the Saudi Coast of the Arabian Gulf. *Nova Hedwigia* 67: 153-168.

Moll, B. & Deikman, J. 1995. *Enteromorpha clathrata* – a potential seawater-irrigated crop. *Bioresource Technology* 52: 255-260.

Silva, P. C., Basson, P.W. & .Moe, R.L. 1996 Catalogue of the benthic marine algae of the Indian Ocean. University of California Publications in Botany 79: 1-259.

Sheppard, C.R.C., Price, A. & Roberts, C. 1992. Marine ecology of the Arabian Region. Academic Press, London.

DIATOMS OF THE SABKHA

FRANK E. ROUND

School of Biological Sciences, The University of Bristol,
Woodland Road, Bristol BS8 1UG, U.K.

Abstract
The sabkha affected by seawater at Abu Dhabi (UAE) supported a rich diatom flora whereas inland sabkhat proved too extreme a habitat for this group of microscopic plants. However, the outflow pond at an inland sewage treatment works was found to support a non-marine flora. The results suggest that sites along the coastal sabkha are suitable for the rich growth of photosynthetic micro-organisms which could support the growth of more complex organisms and be exploited economically, e.g. for crustacean and fish culture. The probable biodiversity of this habitat is discussed in relation to the occurrence of *Mastogloia* species.

INTRODUCTION

The sabkha does not come to mind when considering the habitats occupied by diatoms. High temperature, drying and high salinity make for a harsh environment. However, sabkha is associated with sub-tropical coastlines and where these are exposed to flooding by seawater they can be colonised by microscopic algae. The flat topography of sabkhat means that colonisation can extend some distance inland. However only two groups of algae are at all prolific - blue-green algae (cyano-bacteria) and diatoms. In Abu Dhabi the sabkha extends far inland and where water stands (usually where hollows occur along roads) samples were taken but here no algae occurred, probably owing to high salt content. The areas where thick laminated cyanobacterial mats occur also proved very unfavourable for diatom growth though in places extensive patches of purple sulphur bacteria occurred. The sheaths of the cyanobacteria in some habitats are richly coated with diatoms but surprisingly this was not so in the regions visited in Abu Dhabi.

Barth & Böer (eds.), SABKHA ECOSYSTEMS, 247-253

Three separate microhabitats (epipelon, epipsammon and epiphyton) exist on the coastal sabkha and each was sampled. Only the community living at the sand surface (epipelon) yielded appreciable quantities of diatom cells, all of which were motile and therefore capable of moving up and down in the sediment according to tidal conditions. In the extensive fine sand these epipelic diatoms are the dominant photosynthetic organisms and must feature prominently at the base of the food chain. In spite of this, the community is almost unexplored and its contribution to carbon fixation unknown. The second community is non-motile and lives on the surface of sand grains – surprisingly, very few diatom species were found in this habitat. The epiphyton living attached to macroalgae (*Enteromorpha* etc.) was also sparse. In all, some 72 species were recorded from a relatively small number of samples. A few samples were also taken from the exposed sand on the edge of a pond (lake) used to receive the waste water from a sewage treatment plant – as expected, a rich diatom flora occurred at this site which differed in composition from that at the true sabkha sites.

METHODS

During a brief visit, merely to establish the presence or absence of a diatom flora, sand and *Enteromorpha* (Chlorophyta) were collected from the sabkha at sites along the coast around Abu Dhabi city. The diatoms were washed out of the sand by shaking in distilled water to separate the free-living diatoms from sand. The water was centrifuged to concentrate the diatoms which were then treated with concentrated potassium permanganate solution and hydrochloric acid, the mixture being heated until the solution became colourless. The samples were then washed in distilled water by centrifuging until the acid was removed. Finally, small samples were dried on cover glasses (and mounted in Naphrax) for light microscopy and on aluminium stubs (and gold coated) for scanning electron microscopy.

RESULTS

In spite of the small number of samples containing a diatom flora, the number of species recorded was high though they belonged to few genera. Only the genera *Amphora, Mastogloia* and *Nitzschia* are at all common, with a few species of *Navicula, Pleurosigma* and *Entomoneis* occurring sporadically. The true sabkha flora influenced by the ocean is dominated by *Mastogloia* – at least 12 species were seen (Figs. 1 & 2). This is a common genus with at least 123 known species all occurring along sub-tropical coasts and illustrated in Hustedt (1959) though no modern monographs exist.

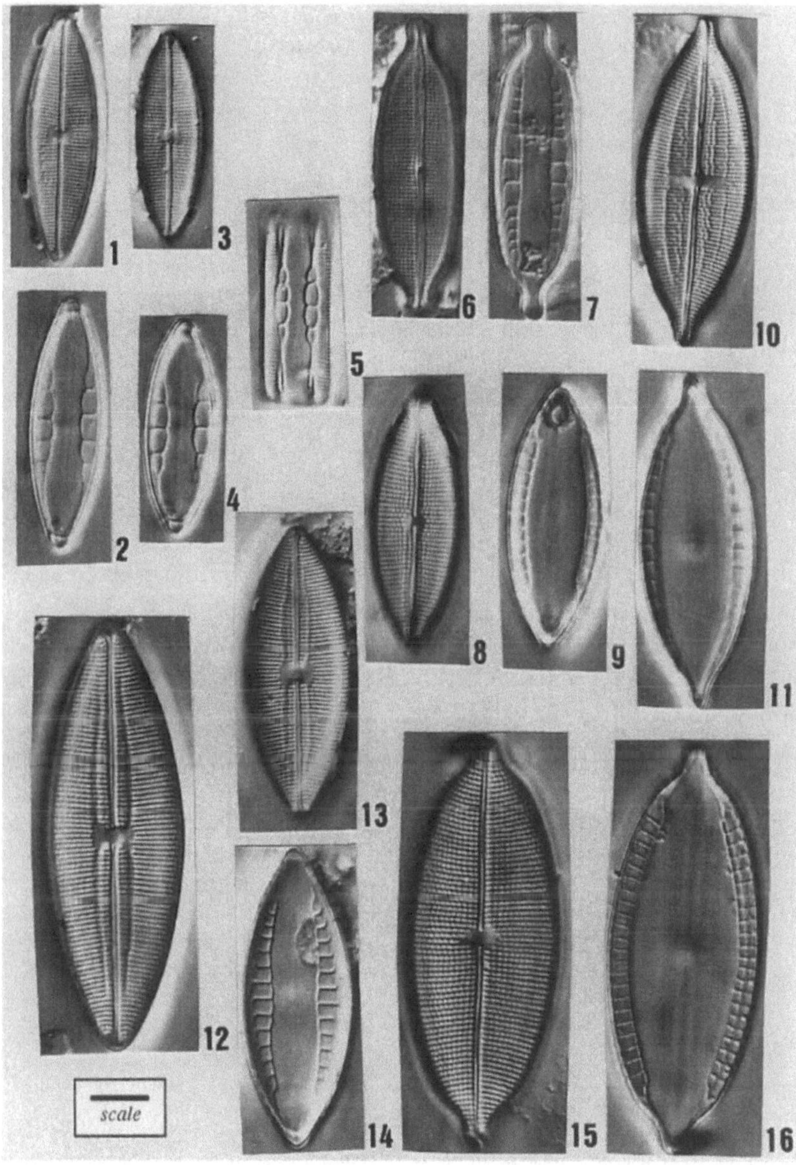

Figure 1. Species of *Mastogloia* living on the sabkha. 1-5: *M. exigua* Lewis. 1-4: Images of the upper valve surface and of the partectate girdle bands. 5: Side view to show the partecta. 6, 7: Valve and partecta of *M. paradoxa* Grun. in Cl. 8, 9: *M. recta* Hust. 10, 11: *M. macdonaldi* Grev. 12–14: *M. baldjickii* Grun. in Schmidt. 15, 16: *M. decipiens* Hust. Scale bar = 10 μm.

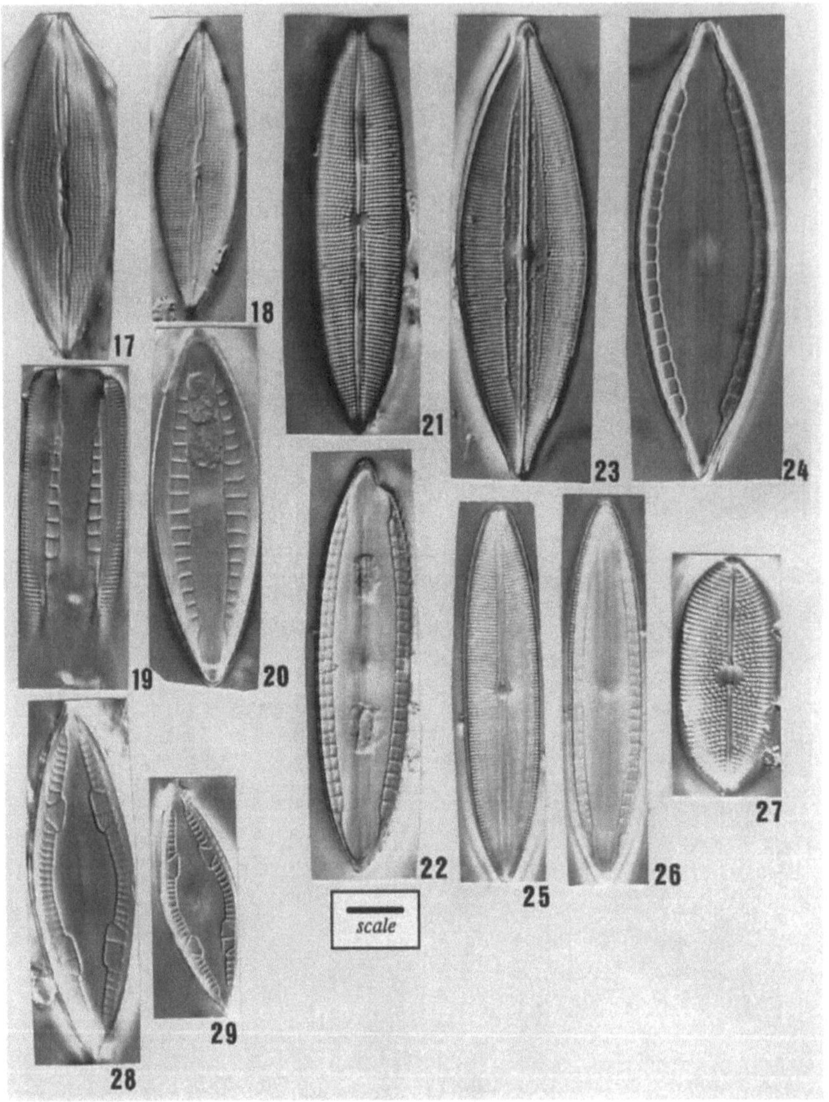

Figure 2. 17, 18: *M. punctifera* Brun. sp. 19, 20: *M. seychellensis* Grun. 21, 22: *M. acutiuscula* Grun. in Cleve. 23, 24: *M. pisiculus* Cl. 25, 26: *M. adriatica* v. linea Voigt. 27: *Mastogloia sp.* 28, 29: *M. erythraea* Grun. Scale bar = 10 μm.

The above figures are merely a fraction of the now recorded species - K. Kemp (personal unpublished computer database) has informed me that he has 539 taxa on his list - almost all are sub-tropical marine. The figures have been selected to show the range of species found in a single sample at a single site and it is worth noting that even in this small sample one form could not be identified suggesting that intensive collecting of sub-tropical material will reveal a great biodiversity in such habitats.

Mastogloia grows in a range of sub-tropical habitats and the epipelic species are presumably free-living but in any future study of the sabkha it will be necessary to study the live flora (this is rarely done by diatomists who rely on identifications from the cleaned valves). This observation of live material is necessary since empty cells from other communities (especially from the richer subtidal habitats) can be washed into the intertidal/supratidal sabkha. The paper by Paddock & Kemp (1988) gives excellent details of valve structure of this genus revealing the much greater morphological complexity than normally found in pennate genera.

Amphora. This also is a very common diatom genus in coastal regions and like *Mastogloia* is desperately in need of monographic treatment. Some species such as *Seminavis* (= *Amphora*) *gracilenta* (Grun. ex A. Schmidt) Mann which occur here has already been removed (Round et al. 1990) and many other modifications are needed.

Navicula. A wide range of species grow on the intertidal sabhka but few are identifiable except questionably to northern species.

Nitzschia. The only distinctive species was *N. sigma* (Kütz.) W. Sm. which was dominant in some samples.

A complete surprise in the flora amongst mangrove leaves was the occurrence of a small, live *Cosmarium* sp. I have never previously found any desmid species in such an environment (but again, surprisingly in some mountain ponds in the Oman a desmid species was recorded).

The flora of the sand around the pond receiving water from the Al Wathba sewage treatment plant outflow had, as would be expected, a completely different diatom flora to that of the marine influenced sabkha. Here forms of the ubiquitous *Achnanthidium minutissima* Kütz., *Cymbella affinis* Kütz, *Nitzschia* aff. *lanceolata* W. Sm. and a very long *Nitzschia* species were found growing in almost pure single-diatom stands. At one site a range of *Amphora* species was found. This artificial habitat is clearly a fruitful habitat for future applied research.

DISCUSSION

Previous work on the diatom flora of sabkhat is sparse. Erhlich & Dor (1985) report briefly on the flora around the non-tidal Gavish Sabkha in the Sinai but it is not clear exactly how the habitat compares with the intertidal zone of Abu Dhabi though some of the species are common to both sites. It is however interesting to compare the live floras e.g. only 3 *Mastogloia* species occur in the Gavish Sabkha and 12 at least are present in the sabkha of Abu Dhabi. *Navicula* species occur in both regions but Ehrlich & Dor wisely indicate only possible identifications. I prefer to wait until a revision of the marine *Navicula* is available since many of the species observed at the scanning electron microscope level have to be assigned to new genera. The *Amphora* and *Nitzschia* species are identified by Ehrlich & Dor but again there are too many taxonomic problems to be solved before these complex genera can be studied in detail. The overall impression is that the Gavish Sabkha is sufficiently isolated from the sea to exclude the wider range of intertidal species which require full seawater conditions to develop. The sites receiving the processed water from the sewage works support species which are more common in salt enriched freshwater e.g. *Nitzschia sigma* (Kütz.) W. Sm. The occurrence of the essentially freshwater *Cocconeis placentula* Ehr. in the Gavish Sabkha and its replacement by *C. scutellum* Ehr. in Abu Dhabi also confirms the physiological control over the two habitats. It is noteworthy that Ehrlich & Dor distinguish a group of four diatoms, *Rhopalodia gibberula* (Ehr.) O. Müll, *Mastogloia braunii* Grun., *M. pusilla* Grun. and *Achnanthes brevipes* Ag. which only occur at their site where there is seepage of seawater; these are usually more common in inland saline sites. Clearly statements in floras commenting on the occurrences of species in highly saline habitats should distinguish between heightened salinity in fresh or marine systems. In each, the flora is modified by species dropping out and not by replacement between the communities.

POSSIBLE FUTURE UTILISATION OF THE SABKHA

The coastal sabkha is a valuable habitat allowing the rich growth of photosynthetic micro-organisms (diatoms, cyanobacteria and other bacteria) and further investigation will certainly reveal other groups (e.g. flagellates). Macroscopic algae also grow in the region. Since many marine animals rely on the microscopic flora as a prime food source, it should be possible to develop shallow lagoons for crustacea/fish farming. To date there is no information on chemical products from the numerous diatom species though other algae are known to produce valuable products (e.g. carotene from the green alga *Dunaliella* which is present in pools on the sabkha). Several potent poisons are produced by a small number of diatoms and these may have industrial (pharmaceutical uses). These have only been investigated where they have caused medical problems and thus the bulk of diatoms are untested for chemical products which every species must elaborate. Organic matter (composts) are sparse in desert environments and large quantities of mixed organic material could be produced from the complete range of organisms living on and in water overlying sabkhat.

The rich flora found in the discharge water from sewage plants illustrates the possibility of using this for fish farming or for forms of aquaculture of higher plant species.

A large amount of experimental work is required to utilise the extensive areas of sabkhat. Research facilities and training of research staff is essential to utilise this unexplored habitat. Taxonomic studies of microalgae along tropical coasts are sparse or non-existent and this makes ecological and applied studies extremely difficult. Clear recognition of microhabitats is also necessary to facilitate detailed studies.

ACKNOWLEDGEMENTS

I wish to thank Dr. B. Boer for inviting me to contribute to this book and also for his hospitality during a stay in Abu Dhabi. I also thank Klaus Kemp for allowing me to use his database on *Mastogloia* and for his advice on the species. Mrs Gillian Lockett greatly assisted in the production of this paper.

References

Ehrlich, A. & Dor, I. 1985. Photosynthetic Microorganisms of the Gavish Sabka. In: Friedman, G.M. & Krumbein, W.E. (eds.), Hypersaline Ecosystems. The Gavish Sabka. Springer Verlag, Berlin, pp. 296–321.

Hustedt, F. 1959. Die Kieselalgen Deutschland, Österreichs und der Schweiz. Bd. 7. In: Dr. L. Rabenhorst's Kryptogamen Flora von Deutschland, Österreichs und der Schweiz. Akademische Verlagsgesellschaft, Leipzig. 845 pp.

Paddock, T.B.B. & Kemp, K.-D. 1988. A description of some new species of the genus *Mastogloia* with further observations on *M. elegans* and *M. goessii*. *Diatom Research* 3: 109–121.

Round, F.E., Crawford, R.M. & Mann, D.G. 1990. The Diatoms. Biology and morphology of the genera. Cambridge University Press, Cambridge. 737 pp.

VERTEBRATE FAUNA OF SABKHAT FROM THE ARABIAN PENINSULA: A REVIEW OF MAMMALIA, REPTILIA AND AMPHIBIA

RONALD A. LOUGHLAND[1] & PETER L. CUNNINGHAM[2]

[1]Commission of Environmental Research, Emirates Heritage Club, P.O. Box 41464 Abu Dhabi, UAE
[2]Polytechnic of Namibia, School of Natural Resources and Tourism, Department Nature Conservation, Private Bag 13388, Windhoek 9000, Namibia

Abstract
A review of mammals, reptiles and amphibians associated with sabkhat is limited in results with no vertebrate species specifically known to occur on sabkhat from the Arabian Peninsula. Ten mammal and twenty-one reptile species are associated with sabkhat from the Arabian Peninsula while seventeen mammals and nine reptile species are associated with similar landforms from Central Asia. *Vulpes vulpes* is often associated with sabkhat while *Phrynocephalus maculatus* and *Stenodactylus khobarensis* are classified as sabkha dwellers in the United Arab Emirates. *Bunopus tuberculatus*, a species not previously recorded on sabkhat, was observed hiding under rubbish on inland sabkha during daylight hours in the UAE. No amphibians are known to occur on sabkha areas from the Arabian Peninsula while *Bufo viridus* is associated with similar landforms from Central Asia.

INTRODUCTION

The Arabian Peninsula is characterised by having a hyper arid climate, and large areas of desert landscape. High temperatures, low precipitation and saline substrate are some limiting factors that contribute to the paucity of vegetation, and therefore the low density of vertebrate fauna in the region. Nowhere in Arabia are the limiting factors more apparent than within the sabkhat that occur along the flat coastal areas and throughout the inland deserts of the Peninsula.

Barth & Böer (eds.), SABKHA ECOSYSTEMS, 255-266

The sabkhat of the southern Arabian Gulf coast are the most extensive found any where on earth (Evans et al. 1964, 1969), with contiguous sabkha areas such as Sabkha Matti in the UAE and Saudi Arabia covering over 10,000 km^2. To the casual observer, sabkhat may appear as lifeless expanses of wasteland, but despite their barren appearance these areas do provide limited refuge and foraging habitat to a variety of vertebrate species.

This review will focus on mammals and reptiles associated with the sabkha landforms, especially in the UAE, and will highlight the general paucity of information available on habitat utilisation by fauna throughout the region.

METHODS

Extensive literature surveys were undertaken to locate records of mammals, reptiles and amphibians associated with sabkha landforms on the Arabian Peninsula. Anecdotal information was also sought from professional and amateur naturalists, and local people living near or adjacent to sabkha landforms.

Information concerning fauna associated with sabkha landforms in adjacent countries to the Arabian Peninsula (Central Asia) was also sought in order to examine data from areas where there has been greater historical access. The Central Asian data was compared with data collected from the Arabian Peninsula. Field data collection involving mammal and reptile survey was undertaken throughout sabkha and associated habitat in the UAE from 1996 - 2000.

Field survey was conducted on 5 coastal, and 4 inland sabkha areas. These sabkha areas were systematically searched for fauna (mammals and reptiles) during diurnal and/or nocturnal observations. Small and medium sized mammal traps were baited and set at 0.5 km intervals along a 4.5 km transect at 6 sabkha sites throughout Abu Dhabi Emirate from October until March each year. Nightly spotlight surveys were conducted along a 10 km transect at 6 sabkha sites. A gas lantern was used at 3 different sabkha sites in the warmer months to specifically observe nocturnal reptiles. It was noticed that reptile behaviour seemed less affected by this light than that of a bright spotlight.

Rubbish, which occured in 11 - 29% frequencies in 2.5 km radius circles in Abu Dhabi Emirate (Böer 1999), was also investigated as possible hiding places. Species encountered were physically caught to facilitate identification. Identifications were also made from tracks left by fauna, especially on sand covered sabkha.

RESULTS

The literature review confirmed the lack of ecological work previously undertaken, especially on the herpetofauna of the area.

The field surveys revealed that in the UAE, only a few species utilise sabkha landforms, with the majority of trappings and observations being made on the perimeter of sabkhat.

Ten mammal species are associated with sabkhat from the UAE. Four species are often associated with sabkhat while only one is classified as a sabkha dweller. 17 species are associated with sabkha landforms in Central Asia. 21 reptile species are associated with sabkha areas in the UAE. 12 species are occasionally associated with sabkhat while seven species are often associated with this landform. Only two species are classified as sabkha dwellers. Nine species are associated with sabkhat in Central Asia.

No amphibian species are associated with sabkha areas in the UAE while one is associated with sabkha landforms in Central Asia.

The mammal, reptile and amphibian species associated with sabkha areas are presented in Tables 1 and 2. Appendix 1 indicates reptile species associated with sabkhat as presented by different authors.

DISCUSSION

The taxonomy and distribution of the mammalian fauna of the region is generally well documented by Büttiker (1980), Gasperetti et al. (1985), Harrison (1983), Harrison & Bates (1991) and Nader (1990). However detailed data on habitat utilisation is lacking for most Arabian species. The literature contained few references to mammals utilising sabkha exclusively, with most species recorded on sabkha, also recorded in adjacent desert habitats. The only record of sabkha habitation is by Drew (2000), who reports that within Sabkha Matti (UAE) Cape Hare (*Lepus capensis*) forage on vegetation growing on sand accumulations, and utilise raised salt crust for shelter (Fig. 1).

Small and medium sized mammals were often observed on the edge of, or near sabkhat, and this is most probably related to foraging strategy. During the field survey a total of ten mammal species were observed on sabkha (Tab. 1). The most commonly observed species was the Cape Hare (*Lepus capensis*), occurring at six sabkha sites. This species was observed both in the day and at night foraging at the edge of sabkha. The next most commonly encountered species was Cheeseman's Gerbil (*Gerbillus cheesmani*) observed foraging at the edge of sabkha during the night at five sites. One other foraging species observed on the edge of sabkhat was the Lesser Jerboa (*Jaculus jaculus*). Baluchistan Gerbil (*Gerbillus nanus*) has also previously been recorded on sabkhat (Harrison & Bates 1991, Zahavi & Wahrman 1957).

Figure 1. Sabkha salt crust "tepee" formation used as refuge by mammals and reptiles.

Predators such as Red Fox (*Vulpes vulpes*), Sand Fox (*Vulpes rueppellii*), Sand Cat (*Felis margarita*) and Feral Dog (*Canis familiaris*) were also observed foraging on the edges of sabkhat at seven different locations. *Vulpes* species were the most common predators encountered, occurring at six sites with *V. vulpes* more often observed than *V. rueppellii*, especially at coastal sites.

Larger mammals such as Oryx (*Oryx leucoryx*) and Sand/Goitred gazelle (*Gazella subgutturosa*) have also been observed to occasionally utilise sabkha areas (Thesiger pers. com.), with sightings having been made of these animals and/or their tracks crossing Sabkha Matti in the UAE. Thesiger's data are from the 1950's, however, before a rapid land-development process began which affected the landscapes, habitat and wildlife.

Scientific studies regarding the herpetofauna of the UAE are limited. Arnold (1984), Baha El Din (1996), Cunningham (2000), Jongbloed (2000) and Leviton & Anderson (1967) are some of the few people who have conducted formal studies on reptiles within the UAE. Other publications on the reptiles of the region include work by Anderson (1999), Arnold (1986) and Leviton et al. (1992). Arnold (1987), Joger (1987) and Werner (1987) published information concerning the zoogeographical aspects of the Arabian herpetofauna. Other publications from the region are limited mostly to taxonomical work and/or relevant to specific species and/or records and sighting.

Table 1. Fauna associated with sabkha landforms in the UAE.

	Reptiles		**Mammals**
B	*Acanthodactylus gongrorhynchatus*	B	*Canis familiaris*
C	*Acanthodactylus opheodurus*	C	*Felis margarita harrisoni*
C	*Acanthodactylus schmidti*	B	*Gazella subgutturosa*
B	*Bunopus tuberculatus*	C	*Gerbillus cheesmani arduus*
C	*Cerastus cerastus gasperettii*	C	*Gerbillus nanus arabium*
C	*Eryx jayakari*	C	*Jaculus jaculus vocator*
C	*Lytorhynchus diadema*	B	*Lepus capensis omanensis*
B	*Malpolon moilensis*	C	*Oryx leucoryx*
C	*Meselina adramitana*	B	*Vulpes rueppellii sabaea*
B	*Meselina brevirostris*	A	*Vulpes vulpes arabica*
A	*Phyrynocephalus maculatus*		
C	*Phyrynocephalus arabicus*		
C	*Scincus mitranus*		
C	*Stenodactylus arabicus*		
C	*Stenodactylus doriae*		
A	*Stenodactylus khobarensis*		
C	*Stenodactylus leptocosymbotes*		
C	*Stenodactylus slevini*		
B	*Trapelus flavimaculata*		
B	*Uromastyx aegyptius microlepis*		
B	*Varanus griseus*		

Key: (A) sabkha dwellers, (B) often associated with sabkha, (C) occassionaly associated with sabkha.

52 species of terrestrial reptiles are known from the UAE, of which 23 species are classified as rare and 18 species as endemic to Arabia (Hornby 1996). 35 species occur within the Abu Dhabi Emirate (Baha El Din). Reptiles that are specifically associated with sabkha as a landform are limited, with a total of 21 species being reported to occur on, or be associated with sabkha areas throughout the UAE (Tab. 1). Of these species, only the banded toad-head agamid (*Phyrynocephalus maculatus*) and the Khobar Gecko (*Stenodactylus khobarensis*) are known to be specifically associated (sabkha dwellers) with sabkha areas (Anderson pers. com., Anderson 1999, Arnold pers. com., Arnold 1977, Arnold 1980, Arnold 1984, Baha El Din pers. com., Baha El Din 1996, Brown 1992, Feulner pers. com., Gallagher pers. com., Gallagher 1971, Gallagher & Arnold 1988, Jongbloed pers. com., Jongbloed 2000, Osborne 1994, Papenfuss pers. com. and Ross 1989). *P. maculatus* is however also known to occur on hard substrates other than sabkha while *S. khobarensis* mostly utilises the edges of sabkhat. The Stone Gecko (*Bunopus tuberculatus)* was observed on sabkhat during this study hiding under human refuse

during daylight hours. This species has not previously been documented from sabkha areas throughout its range.

Meselina guttulata has also been observed on sabkha near Siwa Oasis (Western Egypt) and *Acanthodactylus boskianus* on the edge of sabkhat in Egypt (Baha El Din pers. com.). V*aranus griseus* has also been observed crossing coastal sabkha in pursuit of shore birds while a unique sighting was that of a *Chalcides ocellatus* specimen, a skink not normally associated with sabkha areas found as a salt impregnated specimen by the same author. It could however have been transported there by a raptor.

Most of the reptile species reported as being associated with sabkhat in the UAE are however, only occasionally associated with sabkhat (Baha El Din 1996) and are usually found on other substrate adjacent to sabkha areas (Tab. 1 and Appendix 1). Hence there are only nine species (*P. maculatus, T. flavimaculata, U. aegyptius microlepis, B. tuberculatus, S. khobarensis, M. brevirostris, V.griseus, M. moilensis* and *A. gongrorhynchatus*) that are often associated (two species classified as sabkha dwellers) with sabkha areas in the UAE (Tab. 1).

Amphibian fauna is sparse in Arabia with the natural distribution limited to the periphery of the peninsula. Nine amphibian species occur in Arabia, which include six Bufonidae (*Bufo arabicus*, *Bufo dhufarensis*, *Bufo hadramutinus*, *Bufo scorteccii*, *Bufo tihamicus* and *Bufo viridus*), one Hylidae (*Hyla savigni*) and two Ranidae (*Rana ehrenbergii* and *Rana ridibunda*). *Bufo viridus*, *Hyla savigni* and *Rana ridibunda* have their main populations to the north of Arabia while *Bufo tihamicus* and *Rana ehrenbergii* have other subspecies in North Africa and south-west Asia, respectively (Arnold 1980). Six of the above-mentioned species (*Bufo arabicus*, *Bufo dhufarensis*, *Bufo hadramutinus*, *Bufo scorteccii*, *Bufo tihamicus* and *Rana ehrenbergii*) are endemic to Arabia (Balletto, Cherchi & Gasperetti 1985). Amphibians are apparently absent from much of the dry centre of the Arabian Peninsula (Arnold 1980) with the exception of *Rana ehrenbergii* and *Bufo dhufarensis* occurring in the Riyadh area in Saudi Arabia. These species are however thought to be introduced to the area as escaped and/or released individuals (Balletto et al. 1985). A single locality of *Bufo arabicus* in central Saudi Arabia is also known (Leviton et al. 1992). All nine species known from the peninsula occur in the comparatively moist western drainage area (Arnold 1980). Amphibians occurring in the United Arab Emirates include *Bufo arabicus* and *Bufo dhufarensis* (Hornby 1996).

No reference could be found indicating that amphibians inhabit or even temporarily occur on sabkha areas from the Arabian Peninsula. This is possibly due to amphibians not typically being suited morphologically to inhabit saline areas. Bufonidae are however hardy with *B. arabicus* known to be able to aestivate for periods of up to three years or more (Schätti & Gasperetti 1994) while *B. dhufarensis* often occur on gravel plains far from water (Arnold 1980, Arnold & Gallagher 1975, Hornby 1996). It is thus possible that one or either of the species may occur in sabkha areas from time to time. However, it is not expected that amphibians would normally utilise sabkha areas due to the extreme saline conditions usually associated with this landform. The amphibian record from Central Asia (*Bufo viridus)* occurred on raised substrate adjacent to sabkha (Gubin pers. comm.).

The majority of the data published on Central Asian zoology is in Cyrillic language. Fortunately, The Institute of Zoology in Khazakhstan has translated species lists of fauna associated with large sabkha landforms throughout the country into English (Tab. 2).

Table 2. Fauna associated with sabkha landform in Khazakhstan. Source: Gubin (pers. com.), Institute of Zoology, Almatty, Khazakhstan.

Reptiles	Mammals	Amphibians
Agama sanguinolenta	*Allactaga elater*	*Bufo viridus*
Agrionemys horfieldi	*Canus lupus**	
Agkistrodon halys	*Cricetalus migratorius*	
Elaphe dione	*Diplomesodon pulchellum*	
Elaphe quatuorlineata	*Ellobius talpinus*	
Eremias arguta	*Gazella subgutturosa**	
Eremias intermeida	*Hemiechinus auritus*	
Eremias persica	*Lepus tolai*	
Phyrynocephalus heliosopus	*Meriones meridianus*	
	*Meriones lybicus**	
	Microtus arvalis	
	Mustela eversmanni	
	Rhombomys opimus	
	Saiga tatarica	
	Spermatophilus fulvus	
	Vormela peregusna	
	*Vulpes vulpes**	

* Also occur in the UAE.

Like the Arabian data, the Central Asian data do not indicate exclusive use of sabkha by any particular species, but suggests that the perimeter of sabkha is utilised by an array of species for foraging (Berezovikov et al. 1999). There is also anecdotal information from Central Asia that sabkha is used by mammals and birds as nocturnal roosting/resting sites in an apparent attempt to avoid and/or detect mammalian predators (Gubin pers. com.).

CONCLUSION

The sabkha landform provides limited habitat with few opportunities for refuge, and virtually no foraging resources throughout the year. The edge of the sabkha, where the saline substrate gives way to either sandy or gravel sediment is where increased opportunities for foraging and refuge are available. Along the edges of sabkhat there is usually halophytic vegetation (Deil 1998 & Roshier et al. 1996) that has germinated in accumulated wind blown sand. This vegetation is usually dominated by one or more species such as *Zygophyllum mandavillei* and *Haloxylon salicornicum*, which are able to utilise the shallow saline groundwater often associated with sabkhat. The sabkha edges, and raised vegetated islands within sabkhat in Khazakhstan also contain a high diversity of fauna. The sabkha landscape however is reported as being less diverse. However, migrating ungulates such as *Gazella subgutturosa* and *Saiga tatarica*, which are followed by *Canis lupus*, have been recorded (Berezovikov et al. 1999).

The empty nature of sabkhat may actually assist to increase the foraging resources on the edge. Wind blown seeds and insects often accumulate in higher densities on the edge of the sabkha where the vegetation and change in substrate reduces the wind speed, resulting in the deposition of detritus. These detritus windrows are often easy to locate. Hence it would be efficient for small granivorous mammals and insectivorous reptiles to forage on the edge of sabkha. Likewise it would also be efficient for predators such as *Vulpes* sp. to hunt on the sabkha edge where the likelihood of encountering prey would be increased. This was evident during the surveys were there was a multitude of tracks (invertebrates and vertebrate) on the periphery of sabkhat.

Small mammals may also benefit from the accumulated sand at the edge of sabkhat, which could be used, for burrowing. In an area with predominately hard substrate such as gravel plains, the sabkha edge would provide a unique opportunity for burrowing. It could also be expected that other species, not normally associated with sabkhat, could invade sabkha areas if dry and covered by wind-blown sand. The depth and extent of this sand would then determine which species could inhabit the sabkha areas. Sabkha edges are expected to harbour a greater diversity of reptile species that could occasionally utilise sabkhat for foraging, especially when wind-blown sand has accumulated on the sabkha. Comparative bio-geographical studies are urgently required, in order to determine, whether or not those mammal and reptile species, that occur in the edge-habitat, require or benefit from the presence of sabkha.

The increased vegetation often occurring on the sabkha edge would also benefit larger herbivores providing foraging and limited shelter. Despite sabkhat low inherent value as habitat, it is an integral part of the arid ecosystem of the region, and together with adjacent habitats, provides unique opportunities for foraging, refuge, display, mating and other forms of animal behaviour.

There is limited information on the ecological role sabkhat play in the arid ecosystem, and less information concerning the utilisation of these landforms by mammals and reptiles. More work is required in order to determine the extent of sabkha utilisation throughout the seasons, especially after rain events that may result in inundation of the sabkha. Sabkha-edges also need to be carefully assessed and studied for habitat, plant and animal life and their associations.

ACKNOWLEDGEMENTS

The authors would like to thank the following people and organisations for their valued assistance: H.H. Sheikh Sultan Bin Zayed Al Nahyan, Deputy Prime Minister of the UAE, and Chairman of the Emirates Heritage Club for his ongoing support of environmental research in the UAE. The Environmental Research and Wildlife Development Agency (ERWDA), Abu Dhabi, for support during this work. Steve Anderson, Nick Arnold, Sherif Baha El Din, Gary Feulner, Mike Gallagher, Boris Gubin, Marijcke Jongbloed, Ted Papenfuss and Wilfred Thesiger for their valuable contributions.

References

Anderson, S.C. 1999. The Lizards of Iran. Society for the study of Amphibians and Reptiles. 442 pp.

Arnold, E.N. 1977. Little-known geckoes (Reptilia: Gekkonidae) from Arabia with descriptions of two new species from the Sultanate of Oman. Journal of Oman Studies Special Report 1: 59-80.

Arnold, E.N. 1980. The Reptiles and Amphibians of Dhofar, Southern Arabia. Journal of Oman Studies Special Report 2: 273-332.

Arnold, E.N. 1984. Ecology of lowland lizards in the eastern United Arab Emirates. *The Zoological Society of London* 204: 329-354.

Arnold, E.N. 1986. A Key and Annotated Check List to the Lizards and Amphisbaenians of Arabia. *Fauna of Saudi Arabia* 8: 385-435.

Arnold, E.N. 1987. Zoogeography of the Reptiles and Amphibians of Arabia. Proceedings of the Symposium on the Fauna and Zoogeography of the Middle East (eds. F. Krupp, W. Schneider & R. Kinzelbach): 245-256. Beihefe zum TAVO A 28.

Arnold, E.N. & Gallagher, M.D. 1975. Reptiles and Amphibians from the Mountains of Northern Oman. The Scientific Results of the Oman Flora and Fauna Survey 1975. A Journal of Oman Studies Special Report, Ministry of Information and Culture, Sultanate of Oman.

Baha El Din, S. 1996. Terrestrial Reptiles of Abu Dhabi. In: Osborne, P.E. (ed.), Desert Ecology of Abu Dhabi - a review and recent studies. Pisces Publications, Newbury, UK. pp. 124-147.

Balletto, E.M., Cherchi, M.A. & Gasperetti, J. 1985. Amphibians of the Arabian Peninsula. *Fauna of Saudi Arabia* 7: 318-392.

Berezovikov, N.N., Gubin, B.M., Gul', I.R., Erokhov, S.N., Karpov, F.F. & Kovalenko, A.W. 1999. The Birds of Taukum Desert. Ukrainian Academy of Medical and Biological Science, Kiev, 116 pp.

Böer, B. 1999. Ecosystems, anthropogenic imacts and habitat management techniques in Abu Dhabi. *Paderborner Geographische Studien* 12, 141 pp.

Brown, B. 1992. Reptiles and amphibians. *Tribulus* 2.2: 44-45.

Büttiker, W. 1980. Zoological collections in Saudia Arabia. *Fauna of Saudi Arabia* 1: 1-22.

Cunningham, P.L. 2000. Daily activity pattern and diet of a population of the Spiny-tailed Lizard, *Uromastyx aegyptius microlepis*, during summer in the United Arab Emirates. *Zoology in the Middle East* 21: 37-46.

Deil, U. 1998. Coastal and sabkha vegetation. In: Ghazanfar, S. & M.Fisher (eds), Vegetation of the Arabian Peninsula. *Geobotany* 25, Kluwer Academic Publishers, Dordrecht/Boston/London, pp. 209-228.

Drew, C. 2000. The distribution of the Cape Hare, *Lepus capensis*, in Abu Dhabi Emirate, United Arab Emirates, *Zoology in the Middle East* 20: 15-20.

Evans, G., Kendal, C.G.St.C & Skipwith, P.A.d'E. 1964. Origin of the coastal flats, the sabkha, of the Trucial Coast, Persian Gulf. *Nature* 202: 759-761.

Evans, G., Schmidt, V., Bush, P. & Nelson, H. 1969. Strategy and geologic history of the sabkha, Abu Dhabi, Persian Gulf. *Sedimentology* 12: 145-159.

Gallagher, M.D. 1971. The Amphibians and Reptiles of Bahrain. Bahrain, 37 pp.

Gallagher, M.D. & Arnold, E.N. 1988. Reptiles and amphibians from the Wahiba Sands, Oman. *Journal of Oman Studies Special Report* 3: 405-413.

Gasperetti, J., Harrison, D.L. & Büttiker, W. 1985. The carnivora of Arabia. *Fauna of Saudi Arabia* 7: 397-461.

Harrison, D.L. 1983. The mammal fauna of Oman, with spcial reference to conservation and the Oman flora and fauna surveys. *The Journal of Oman Studies* 6(2): 329-339.

Harrison, D.L. & Bates, P.J.J. 1991. The mammals of Arabia. Harrison Zoological Museum, Kent, 354 pp.

Hornby, D. 1996. A checklist of amphibians and reptiles of the UAE. *Tribulus* 6.1: 9-13.

Joger, U. 1987. An Interpretation of Reptile Zoogeography in Arabia, with Special Reference to Arabian Herpetofaunal Relations with Arabia. Proceedings of the Symposium on the Fauna and Zoogeography of the Middle East (eds. F. Krupp, W. Schneider & R. Kinzelbach): 245-256. Beihefe zum TAVO A 28.

Jongbloed, M. 2000. Wild about Reptiles. Barkers Trident Communications, London.

Leviton, A.E. & Anderson, S.C. 1967. Survey of the reptiles of the Sheikhdom of Abu Dhabi, Arabian Peninsula. Part II. Systematic account of the collection of reptiles made in the Sheikhdom of Abu Dhabi by John Gasperetti. Proceedings of the California Academy of Sciences, 4th Series, Vol. XXXV, No. 9: 157-192.

Leviton, A.E., Anderson, S.C., Adler, K. & Minton, S.A. 1992. Handbook to Middle East Amphibians and Reptiles. Society for the study of Amphibians and Reptiles. 252 pp.

Nader, I.A. 1990. Checklist of the mammals of Arabia. *Fauna of Saudi Arabia* 11: 329-381.

Osborne, P. 1994. Reptiles. *Tribulus* 4.1: 29.

Roshier, D. A., Böer, B.B. & Osborne, P.E. Vegetation of Abu Dhabi and a preliminary classification of its plant associations. In: P.E. Osborne (ed.), 1996. Desert Ecology of Abu Dhabi - a review and recent studies. Pisces Publications, Newbury, UK. p.50-65.

Ross, W. 1989. Notes on the Ecology and Behaviour with Special Reference to Tail Signalling in *Phyrynocephalus maculatus* (Reptilia: Agamidae). *Fauna of Saudi Arabia* 10: 417-422.

Schätti, B. & Gasperetti, J. 1994. A Contribution to the Herpetofauna of Southwest Arabia. *Fauna of Saudi Arabia* 14: 348-423.

Werner, Y.L. 1987. Ecological Zoogeography of the Saharo-Arabian, Saharan and Arabian Reptiles in the Sand Deserts of Southern Israel. Proceedings of the Symposium on the Fauna and Zoogeography of the Middle East (eds. F. Krupp, W. Schneider & R. Kinzelbach): 245-256. Beihefe zum TAVO A 28.

Zahavi, A. & Wahrman, J. 1957. The cytotaxonomy, ecology and evolution of the gerbils and jirds of Israel (Rodentia: Gerbillinae) *Mammalia* 21: 341-380.

Appendix

A list of reptiles occurring on or associated with sabkha areas in Arabia and adjacent areas.

This Study	Anderson (*pers.com.*)	Anderson (1999) 1	Arnold (*pers.com.*)	Arnold (1977) 2	Arnold (1984)	Baha El Din (*pers.com.*) 3	Baha El Din (1996)	Brown (1992)	Feulner (*pers.com.*)	Gallagher (*pers.com.*)
A	**A**	**A**	**A**	**A**	**A**	**A**	**A**	**A**	**A**	**A**
P. maculatus *S. khobarensis* *B. tuberculatus**	*P. maculatus* *S. khobarensis*	*P. maculatus*	*P. maculatus* *S. khobarensis*	*S. khobarensis*	*P. maculatus* *S. khobarensis*	*P. maculatus* *S. khobarensis*	*P. maculatus* *S. khobarensis*	*P. maculatus*	*P. maculatus*	*P. maculatus* *S. khobarensis*
B	**B**					**B**	**B**			**B**
U. microlepis *M. brevirostris* *V. griseus*	*L. diadema* *C. cerastes* *E. jayakari*					*A. boskianus* *M. guttulata* *V. griseus*	*P. maculatus* *T. flavimaculata* *U. microlepis* *B. tuberculatus* *S. khobarensis* *S. slevini* *S. leptocosymbotes* *A. opheodurus* *A. gongrorhynchatus* *M. adramitana* *M. brevirostris* *V. griseus* *M. moilensis* *L. diadema* *C. gasperettii*			*Colubrid* Snakes
C						**C**				
Sandy areas: -*P. arabicus* -*A. schmidti* -*S. doriae* -*S. arabicus* -*S. mitranus* -*V. griseus* -*E. jayakari* -*C. gasperettii* *(tracks)* **Gravel areas:** -*U. microlepis* -*M. adramitana* -*V. griseus*						-*C. ocellatus* (transported by raptor?)				

A - Sabkha dwellers **B** - Marginal sabkha dwellers **C** - Occasionally associated with sabkhat

1 – Iran (Anderson 1999), **2** – Sultanate of Oman (Arnold 1977), **3** – Egypt (Baha El Din *pers. comm.*), **4** – Bahrain (Gallagher 1971),
5 – Sultanate of Oman (Gallagher & Arnold 1988), **6** – Khazakhstan (Gubin *pers. comm.*), **7** – Iran (Papenfuss *pers. comm.*), **8** – Kingdom of Saudi Arabia (Ross 1989).
All other references are for the UAE. * *B. tuberculatus* found to utilise sabkhat in the UAE during fieldwork conducted during this study.

Gallagher (1971) 4	Gallagher & Arnold (1988) 5	Gubin (*pers. comm.*) 6	Jongbloed (*pers. comm.*)	Jongbloed (2000)	Leviton et al. (1992)	Osborne (1994)	Papenfuss (*pers. comm.*) 7	Ross (1989) 8
A	A	A	A		A	A	A	A
S. khobarensis	*S. khobarensis*		*P. maculatus* *S. khobarensis*	*P. maculatus* S. khobarensis	T. flavimaculata	*S. khobarensis*	*P. maculatus*	*P. maculatus*
		B						
		A. sanguinolenta *A. horfieldi* *A. halys* *E. dione* *E. quatuorlineata* *E. arguta* *E. intermeida* *E. persica* *P. heliosopus*						

A - Sabkha dwellers **B** - Marginal sabkha dwellers **C** – Occasionally associated with sabkhat

1 – Iran (Anderson 1999), **2** – Sultanate of Oman (Arnold 1977), **3** – Egypt (Baha El Din *pers comm.*),
4 – Bahrain (Gallagher 1971), **5** – Sultanate of Oman (Gallagher & Arnold 1988), **6** – Khazakhstan (Gubin *pers comm.*),
7 – Iran (Papenfuss *pers. comm.*), **8** – Kingdom of Saudi Arabia (Ross 1989).
All other references are for the UAE. * *B. tuberculatus* found to utilise sabkhat in the UAE during fieldwork conducted during this study.

ECOLOGY OF SABKHA ARTHROPODS

PETER J. HOGARTH[1] & BARBARA J. TIGAR[2]

[1]Department of Biology, University of York, P.O. Box 373, York YO10 5YW, U.K.
[2]Department of Biological Sciences, University of Paisley, High Street, Paisley, PA1 2BE, Scotland, U.K.

Abstract

Sabkha is an extreme habitat for arthropods. There are a few resident and some vagrant terrestrial invertebrates: rapid colonizers of ephemeral aquatic habitats, however, can be very abundant. Terrestrial herbivorous insects such as Hemiptera, Homoptera, Lepidoptera and Coleoptera are confined to vegetated sabkha margins, with a few specialists feeding on algal mats. Pollen and nectar feeders such as Lepidoptera, Hymenoptera and Diptera occur where flowering plants are present. Granivores, particularly ants, may also occur. Detritivores are the most abundant terrestrial invertebrates, particularly tenebrionid beetles, which feed on allochthonous detritus. Opportunistic and vagrant predators include carabid and cicindelid beetles, scorpions, spiders, centipedes and camel spiders (Solifugae). These range across sabkha to hunt and scavenge invertebrate and vertebrate prey.

Semi-aquatic and aquatic insects (Diptera, Odonata, Hemiptera and Coleoptera) occur in seasonal pools of suitable salinity. The predominant Crustacea of sabkha pools are copepods, particularly Harpacticoidea; Ostracoda and Anostraca (particularly *Artemia*) may also intermittently be abundant. These are physiologically adapted to high and varying salinity, and encyst to survive desiccation. Encystment probably also facilitates dispersal between habitat patches. Virtually all Crustacea are filter-feeders, the principal exceptions being vagrant species (brachyuran crabs) from intertidal or other adjacent habitats.

Barth & Böer (eds.), SABKHA ECOSYSTEMS, 267-282

INTRODUCTION

Sabkha is an extreme environment even for arid zone invertebrates, which exhibit behavioral and physiological mechanisms to reduce heat stress and water loss (Cloudsley Thompson 1979, 1991).

Many invertebrates survive high temperatures and seasonal flooding by remaining deep inside burrows, and some species are limited by their ability to excavate burrows (Polis 1990). The hypersalinity of sabkha, its often impenetrable surface, and periodic inundation by salt water create further difficulties for arthropods. However, variations in salinity and water availability result in suitable microhabitats for both terrestrial and aquatic arthropods. Species with good dispersal abilities and rapid life cycles will colonise and exploit ephemeral habitats and resources.

Arabian arthropod studies are relatively new (Büttiker 1980) and there are few records from sabkha. This chapter is based largely on records of terrestrial invertebrates from the Arabian Peninsula, and aquatic and semi-aquatic species from Sinai. Invertebrates other than arthropods are not discussed, although it is likely that other groups are important in maintaining soil fertility (Zak & Freckman 1991). Further information on general invertebrate biology can be found in Polis (1990) and Walker & Pittaway (1987).

INSECTA, CHELICERATA AND MYRIAPODA

Insects contribute over half of the known animal species (Wheeler 1990). They are a major component of all terrestrial ecosystems and show diversity in life history patterns and food sources. The immature stages of holometabolous insects often utilize different foods and environments from adults. Some species require fresh or brackish water to complete their life cycle and may occur in sabkha pools. Arachnids are well-represented in arid regions and are exclusively predatory. General accounts of the terrestrial invertebrates in Arabia can be found in Dutton & Bray (1988), Gillett (1998), Shaw Reade et al. (1980), Tigar (1996a & b, 1998) and Walker & Pittaway (1987). Species recorded from sabkha are listed in Table 1.

Herbivorous insects

Phytophagous insects depend upon the distribution of their host plants and are confined to vegetated sabkha margins where shrubs from the families Chenopodiaceae and Zygophyllaceae grow (Satchell 1978, Roshier et al. 1996).

The orders Thysanoptera, Hemiptera and Homoptera are sap-feeders that occur widely in tropical Arabia and include many pests (Walker & Pittaway 1987). Those of native habitats are poorly studied, although Homoptera such as aphids, scale insects, mealy bugs and plant lice probably infest shrubs in sabkha.

Defoliating insects use mandibles to cut through leaves, and include the orders Orthoptera, Lepidoptera and Coleoptera (families Chrysomelidae, Curculionidae, Bruchidae and Meloidae). There are a few records for herbivorous Lepidoptera and Coleoptera (Tab. 1).

Other herbivores include those living inside plant tissues, as well as pollen and nectar feeders and seed-consuming granivores. Hymenoptera, particularly bees, wasps, flies (Syrphidae), beetles (especially Meloidea), and adult Lepidoptera feed on nectar and pollen, and are highly mobile. Invertebrate granivores have been widely studied in North American and Australian deserts where ants exert considerable effect on plant recruitment and community composition (MacKay 1991). Less is known in Arabia, although ants of the genus *Messor* harvest and store seeds in their nests (Collingwood 1985).

Along the shoreline of sabkha, mats of blue-green algae provide a food source for littoral invertebrates. These include female *Bledius* spp. beetles, which collect algae along the shore and feed it to their larva which inhabit burrows (Gerdes et al. 1985).

Detritivores

Primary productivity is low in deserts and even lower in sabkha therefore detrivores will be more abundant than herbivorous invertebrates. Most detritus consists of allochthonous material derived from wind-blown terrestrial and marine sources. Detritivores generally forage opportunistically and include many of the typical desert tenebrionid beetles, termites and ants (Tigar 1998). Some of the larger beetles, such as *Prionotheca coronata* and *Blaps kollari* move considerable distances in search of food. Some detritivores are more specialised, for example coprophages like the dung beetle *Scarabaeus christatus*, or carrion feeders including blowflies (Calliphoridae) and histerid or silphid beetles.

Predators

Omnivory and intraguild-predation are well documented in desert food webs (Polis & Yamashita 1991) and opportunistic species do not distinguish between live and dead prey. The main Arabian invertebrate predators are arachnids (scorpions, spiders and camel spiders), insects (beetles, ants and parasitoid wasps) and centipedes. Active, fast moving predators include two centipede species that are highly resistant to desiccation: *Scolopendrida mirabilis* and *S. valida* (Lewis & Gallagher 1993). Other active predators are large carabid and cicindelid beetles, hunting spiders (Gnaphosidae), wolf spiders (Sparassidae) and solifugids (Solifugae). Scorpions either sit-and-wait at the entrance of a burrow or actively search for prey. Arabian scorpions are listed in Vachon (1980) and Vachon & Kinzelbach (1987) and include many active hunters of the family Buthidae.

Other predators occur in proximity to plants: for example, web-building spiders require vegetation in which to build their webs and to provide cover from predators. Cryptically colored predators, such as crab spiders (Thomisidae) and preying mantids, are highly conspicuous if not on appropriate foliage or flowers.

Insect parasitoids (Diptera: Tachinidae, Hymenoptera: Braconidae and Mutillidae) are limited by the availability of their hosts, usually herbivorous insects. In addition, many require nectar sources during their adult stage. In a heterogeneous environment like sabkha their populations will be low and patchily distributed.

Aquatic and semi-aquatic insects

Aquatic insects spend their entire life cycle in water, while semi-aquatics live in water for only part of their life cycle, usually the immature stage. Aquatic herbivores feed on algae, detritivores and filter feeders usually rely on suspended detritus, and scavengers and predators feed on animal prey. Temporary and permanent pools in sabkha may be highly productive when conditions are suitable. Many aquatic insects are aerial colonists with rapid life cycles and some possess resistant stages that survive desiccation.

Odonata and Diptera (mosquitoes, Chironomidae, Chaoboridae and Stratiomyidae) have fully aquatic immature stages. Other taxa live above the meniscus including pond striders (Hemiptera: Gerromorpha) and whirligig beetles (Gyrinidae). Victor & Victor (1997) found several unspecified chironomids (Diptera), Odonata, Ephemeroptera, Coleoptera and Heteroptera from a coastal lagoon in Oman. In Sinai, Gerdes et al. (1985) recorded six aquatic insect species that tolerated up to 14% salinity and nine species that tolerated up to 30% salinity (Tab. 1). The surprising abundance of Coleoptera and lack of Hemiptera was attributed to the profusion of algal mats at their study sites.

Salt-water mosquito and chironomid larvae are osmoregulators that maintain constant internal osmotic pressure over a range of environmental salinities. Many species become osmoconformers at salinities above 1-2%. In general, mosquito larvae take up ions from the water by their anal papillae, whose size is reduced in species living in salt water. For other aquatic insects, such as adult beetles, water loss is restricted by cuticular impermeability (as in terrestrial species), while the malpighian tubules and rectum are generally responsible for the excretion of excess ions and the reduction of water loss (Mordue et al. 1980).

Crustacea

Crustacea, essentially aquatic animals, are largely restricted to wet sabkha habitats. The dominant crustacean groups found in sabkha pools are Copepoda (particularly Harpacticoidea), Ostracoda, and Branchiopoda. Species recorded from sabkha are listed in table 2.

Copepoda

The paucity of data on sabkha copepods is largely due to under-recording: copepods are small and inconspicuous species. Moreover, most if not all of the sabkha-associated species are benthic or meiofaunal in habit, hence easy to overlook. The relatively well studied Di Zahav pool in the eastern Sinai Peninsula is recorded as having 10 species (Por & Dor 1975), the Sirbonian Lagoon (Sabkhat el Bardawil), Egypt, either 16 (Por 1973) or 21 (Por 1971). In contrast, studies of hypersaline habitats of the Saudi coast of the Arabian Gulf list only four (undetermined) species (Basson et al. 1977) or two (Jones et al. 1978).

Copepods of sabkha are filter or deposit feeders, essentially primary consumers feeding on photosynthetic diatoms and cyanobacteria. There is specialisation within this feeding mode: analysis of the gut contents of harpacticoids in the sabkha of Sinai suggests that some species (e.g. *Robertsonia salsa*) feed primarily on diatoms and coccoid cyanobacteria, others (*Nitocra* sp.) on filamentous cyanobacteria (Gerdes et al. 1985).

An important unresolved question is the extent to which filter-feeding species of sabkha pools depend on allochthonous sources. Desert pool ecosystems in both South Africa and Nevada receive a large input of wind-borne organic particles (Hutchinson et al. 1932, Hutchinson 1937 in Cole 1968), and it seems likely that this may also apply to sabkha pools in windswept regions such as the Arabian Gulf coast or Sinai (Por & Dor 1975). Coastal sabkha may accumulate large amounts of algal and seagrass remains.

Sabkha pools are generally highly saline. Copepod species differ in salinity tolerance. Of the species found in the Sirbonian Lagoon, three occurred in areas where salinity was below 4.5%, four were limited to salinities below 6.5%, a further two species to 9% and one to 11-12%. Two species - *Cletocampus confluens* and *Nitocra lacustris* - occurred at salinities of 16% (more than 4 x sea water), and these did not occur at salinities as low as that of sea water (Por 1973).

Nothing is known directly about how sabkha copepods survive highly saline conditions. Some indications, however, come from a temperate harpacticoid species, *Tigriopus brevicornis*. This lives at the bottom of small supratidal pools exposed only to occasional splash from waves. Evaporation results in salinities of more than 10%, comparable to sabkha pools. Salinity and temperature tolerance interact, individuals acclimatised to high temperatures being less able to survive high salinities (Davenport et al. 1997).

Tigriopus is an osmoconformer over a wide range of salinities, an energetically efficient strategy for a small organism. As salinity of the external medium increases, so does its density: by increases in ionic concentration of its body fluids, and decrease in body volume, *Tigriopus* maintains a density greater than that of its surroundings. This means that, as a benthic forager, no energy is required to maintain position at the bottom of the water column (McAllen et al 1998). It seems likely that other harpacticoids, such as sabkha species, will be found to function similarly.

Sabkha copepods must also be adapted to the ephemeral nature of the environment. Many freshwater copepods produce thick-shelled dormant eggs, and in some species the adult aestivates by encysting within a secreted covering (Ruppert & Barnes 1994). Copepod cysts can remain viable for decades, or even centuries (Hairston et al. 1995). Although neither mechanism has been demonstrated specifically in sabkha species, encystment, as egg or adult, seems essential to survival. Encysted eggs would also provide a means of dispersal between pools when carried on the feet or bodies of birds or other animals.

Anostraca

Branchiopods (Order Anostraca) are a further significant group of filter-feeding primary consumers found in hypersaline sabkha pools, where they may be a principal food source for birds such as flamingos. *Branchinella spinosa* has been recorded from several locations in Saudi Arabia and Oman, and is probably widespread in the Arabian Peninsula (Thiéry 1996), while the brine shrimp (*Artemia* sp.) has been recorded from the Sinai Peninsula, Egypt (Por 1972), Iraq (Triantaphyllidis et al. 1998), Kuwait (Thiéry 1996) and Abu Dhabi (Aspinall & Hellyer 1999).

Currently, *Artemia* taxonomy is confused. '*Artemia salina*' is no longer a valid species, and is now known to comprise a number of species and varieties. These vary in salinity tolerance, life history strategy, and reproductive mode, some being parthenogenetic and others sexual. Competitive interactions between the different forms are complex and poorly understood (Browne et al. 1988, Browne & Halanych 1989, Browne & Wanigasekera 2000, Triantaphyllidis et al. 1998). The form recorded from Iraq is *A. parthenogenetica*, but otherwise it is not known which species of *Artemia* occur in the region (Triantaphyllidis et al. 1998). Following the collection of fresh material, the species reported from Abu Dhabi appears to reproduce sexually, hence is not *A. parthenogenetica* (G. Van Stappen, pers. comm.).

Artemia graze principally on bacteria and cyanobacteria. In contrast to the majority of harpacticoid copepods, they are primarily planktonic rather than benthic feeders, often diving down into the more saline and anoxic deeper layers to feed (Gerdes et al. 1985). An adult *Artemia* may take 6.4×10^6 cells daily, and total productivity can be as high as 62.6 kg ha^{-1} (Cole 1968).

Artemia is an osmoregulator, maintaining the ionic concentration of its body fluids higher than that of its surroundings at low environmental salinities and lower in hypersaline media (Beadle 1943). This disparity is facilitated by relative impermeability of the body surface, and the reduction in the ratio of surface area to volume implied by increased size: adult *Artemia* are more than an order of magnitude larger than harpacticoid copepods. Internal ionic constancy is achieved as the result of a balance between water and ion uptake and removal. Water uptake is principally through the gut lining, while salt elimination is chiefly through the branchiae of adults, and a specialised neck organ in naupliar larvae: NaCl is actively secreted in an energy-demanding process (Croghan 1958a, 1958b, 1958c; Conte et al. 1972).

A result of salt elimination in hypersaline media is that the density of *Artemia* becomes consistently less than that of the surrounding water, making it more feasible to remain planktonic. This contrasts with the benthic copepods (see above). Like copepods, branchiopods produce large numbers of dormant encysted eggs. There is uncertainty about the extent to which hatching is endogenously or environmentally determined (Brendonck 1996). It might seem obvious that cysts should respond directly to environmental conditions. However, branchiopods may play a more subtle game. The appearance and duration of ephemeral pools is highly variable with time. It might therefore be a more successful long-term strategy for an individual to spread hatching of its offspring across several successive appearances of standing water, rather than for them all to be committed to what might prove too transient for full development and effective reproduction. There is evidence for such bet-hedging in related species, although not so far in *Artemia* and *Branchinella* (Simovich & Hathaway 1997). Cysts, in addition to enabling survival of desiccation, therefore probably also allow dispersal in time (Brendonck 1996).

Encystment also facilitates dispersal in space, allowing survival in transit between pools. Although wind-borne dispersal of branchiopod eggs can occur (Brendonck & Riddoch 1999), it is more likely that birds serve as vectors over long distances.

Ostracoda

Several ostracod species have been recorded from sabkha pools of the Sinai Peninsula and Arabian Gulf region, including *Cyprideis torosa* and *C. littoralis*. These are primary consumers, particularly of diatoms and coccoid cyanobacteria, and are often present in enormous numbers, particularly at high salinities. Most are hyperstenohaline, flourishing in salinities of 5-10% (Gerdes et al. 1985, Jones et al. 1978, Por 1971, 1972, Por & Dor 1975).

Other sabkha crustaceans

Remaining crustacean groups, including amphipods, decapods and isopods, are represented in sabkha largely as temporary immigrants from adjacent habitats. Permanently wet sabkha with intermittent contact with the sea may have more or less permanent populations of essentially marine species. Among decapods, omnivores such as the brachyuran crab *Metopograpsus messor* have been observed by the principal author in an anthropogenic pool some distance from the open sea, but which floods at occasional high tides. Given the requirement for a planktonic stage in the life cycle of the Brachyura, such populations are unlikely to be self-sustaining. Predatory crabs such as *Portunus pelagicus*, and penaeid prawns (*Penaeus latisulcatus*, *Metapenaeus stebbingi*) have also been recorded from pools with intermittent contact with the sea (Jones et al. 1978). There are no records of terrestrial isopods from Arabian sabkha (S. Taiti, pers. comm.), but a few species may occur in moist sabkha habitats. *Halophiloscia* sp., an amphibious isopod that feeds on decaying organic matter, has been found in seasonally-exposed algal mats in littoral sabkha, where it appears to move into burrows vacated by the beetle *Bledius* as the moisture increases (Gerdes et al. 1985).

DISCUSSION

The terrestrial invertebrate fauna of Arabian sabkhat has generally been ignored on the assumption that it is a poor habitat for terrestrial species. While it is true that the drier sabkhat has generally low arthropod biomass, wet sabkhat, and in particular sabkha margins with some salt-bush vegetation or cyanobacterial growth, seem to be more favourable habitats.

However, even in dry and hypersaline environments some species can survive and flourish. Terrestrial arthropod collection in the future should use traps, baits and automatic methods of sampling in order to capture a full range of sabkha dwellers. Their dispersal ability, especially by flight, allows many insects rapidly to colonize ephemeral habitats and exploit food and other resources. In addition, the physiology of many species allows them to survive periods of food limitation and environmental stress, including extremes of salinity and temperature. Specialist sabkha species of all arthropod groups are probably widespread and have been under recorded, while terrestrial species probably use sabkhat when conditions are favourable.

Sabkha crustaceans are restricted to wet habitats, and are represented by relatively few species. With the exception of incursions by marine or maritime species in sabkha pools, virtually all sabkha Crustacea are filter-feeding primary consumers, ingesting small organic particles, particularly diatom, bacterial or cyanobacterial cells (Gerdes et al. 1985). This contrasts with the relatively higher trophic diversity of other sabkha arthropods.

Species diversity of sabkha Crustacea is also low, and decreases with increasing salinity (and decreasing marine influence) (Beadle, 1943, Gerdes et al. 1985). As ephemeral freshwater and brackish pools frequently show high faunal diversity (Lahr et al. 1999, Williams et al. 1998), low species diversity results from high salinity, rather than from the ephemeral nature of sabkha pools. High abundance of individual species often accompanies this low biodiversity (Beadle 1943, Cole 1968, Por & Dor 1975).

In general, the fauna of inland saline waters are predominantly freshwater, rather than marine, in origin (Bayly 1967, Cole 1968, Macan 1963). This is probably also true of sabkha Crustacea, although in many cases the affinities are uncertain. In contrast to those in marine habitats, inland aquatic organisms often need to survive in variable and ephemeral habitats. An inland aquatic species is therefore more likely than a marine one to extend into wet sabkha habitats, which often show extremes of temperature and salinity and are likely to dry out completely. *Artemia* is a classic example, showing a form of osmoregulation typical of freshwater organisms, and the ability to survive desiccation by encystment.

Although the lack of taxonomic information makes generalisation risky, sabkha crustacea tend to be cosmopolitan species, or representatives of cosmopolitan genera. Examples among the copepods are *Heterolaophonte quinquespinosa* and *Nitocra affinis*, circumtropical in their distribution, and *N. lacustris*, which occurs from Mexico to Iran (Cole 1968, Gerdes et al. 1985, Por 1973). *Artemia* species also occur naturally worldwide, except in Australia (Triantaphyllidis et al. 1998, Williams et al. 1998). The adaptations that enable sabkha species to survive desiccation and to disperse between patches of suitable habitat also facilitate dispersal and continuity over longer ranges.

FURTHER INVESTIGATIONS

There are many aspects of the ecology of sabkha invertebrates about which information is virtually non-existent, and on which future sabkha research should focus. Among the most important unresolved issues are the extent to which wet sabkha ecosystems depend on allochthonous organic inputs rather than autochthonous primary production; whether observed differences in species distribution and community composition between different sabkha locations are due to environmental differences, biogeographical differences, or represent different stages of successional sequence; the nature of interactions between sabkha invertebrate species; and the quantitative significance, in terms of energy flow, of the sabkha invertebrate community.

In area, sabkhat are often a major constituent of the regions in which they occur, and are of commensurate regional ecological and economic significance. In general, the arthropods of sabkhat are probably of relatively minor economic importance. However, in other parts of the world, rearing *Artemia* for the aquaculture and aquarium markets is a major, and lucrative, industry. The economic potential of *Artemia*-rearing in the Arabian Peninsula has so far not been assessed.

ACKNOWLEDGEMENTS

We should like to thank Mike Gillett for his unpublished records from Ain Al Faydah, Abu Dhabi, Benno Böer for discussions on sabkha plants, Stefano Taiti for information on terrestrial isopods, Simon Aspinall for information about *Artemia*, and for collecting specimens, and Mark Beech and members of the Abu Dhabi Islands Archaeological Survey (ADIAS) for assistance during a field visit to Abu Dhabi.

References

Aspinall, S. & Hellyer, P. 1999. The history and development of Al Wathba Lake, Abu Dhabi. *Tribulus* 9/2: 22-25.

Bayly, I.A.E. 1967. The general biological classification of aquatic environments with special reference to those of Australia. In: Weatherley, A.H. (ed.), Australian inland waters and their fauna. Australian National University Press, Canberra, pp. 78-104.

Basson, P.W., Burchardt, J.E., Hardy, J.T. & Price, A.R.G. 1977. Biotopes of the western Arabian Gulf. Aramco, Dhahran.

Beadle, L.C. 1943. Osmotic regulation and the faunas of inland waters. *Biological Reviews* 18: 172-183.

Brendonck, L. 1996. Diapause, quiescence, hatching requirements: what can we learn from large freshwater branchiopods (Crustacea: Branchiopoda: Anostraca, Notostraca, Conchostraca). *Hydrobiologia* 320: 85-97.

Brendonck, L. & Riddoch, B.J. 1999. Wind-borne short-range dispersal in anostracans (Crustacea: Branchiopoda). *Biological Journal of the Linnean Society* 67: 87-95.

Browne, R.A. & Halanych, K.M. 1989. Competition between sexual and parthenogenetic *Artemia*: a re-evaluation (Branchiopoda, Anostraca). *Crustaceana* 57: 57-71.

Browne, R.A. & Wanigasekera, G. 2000. Combined effects of salinity and temperature on survival and reproduction of five species of *Artemia*. *Journal of Experimental Marine Biology and Ecology* 244: 29-44.

Browne, R.A., Davis, L.E. & Sallee, S.E. 1988. Effects of temperature and relative fitness of sexual and asexual brine shrimp *Artemia*. *Journal of Experimental Marine Biology and Ecology*. 124: 1-20.

Büttiker, W. 1980. Zoological collections in Saudi Arabia. *Fauna of Saudi Arabia* 1: 1-22.

Cloudsley Thompson, J.L. 1979. Adaptations of Arthropoda to arid environments. *Annual Review of Entomology* 20: 261-283.

Cloudsley Thompson, J.L. 1991. Ecophysiology of desert arthropods (Adaptations of desert organisms series). Springer-Verlag, Berlin.

Cole, G.A. 1968. Desert limnology. In Brown, G.W. (ed.), Desert Biology. Special topics on the physical and biological aspects of arid regions. (Academic Press), pp. 424-486.

Collingwood, C.A. 1985. Hymenoptera: Fam. Formicidae of Saudi Arabia. *Fauna of Saudi Arabia* 7: 230-302.

Conte, F.P., Hootman, S.R. & Harris, P.J. 1972. Neck organ of *Artemia salina* nauplii - a larval salt gland. *Journal of Comparative Physiology* 80: 239-246.

Croghan, P.C. 1958a. The osmotic and ionic regulation of *Artemia salina* (L.). *Journal of Experimental Biology* 35: 219-233.

Croghan, P.C. 1958b. The mechanism of osmotic regulation in *Artemia salina* (L.): the physiology of the branchiae. *Journal of Experimental Biology* 35: 234-242.

Croghan, P.C. 1958c. The mechanism of osmotic regulation in *Artemia salina* (L.): the physiology of the gut. *Journal of Experimental Biology* 35: 243-249.

Davenport, J., Barnett, P.R.O. & McAllen, R.J. 1997. Environmental tolerances of three species of the harpacticoid genus *Tigriopus*. *Journal of the Marine Biological Association of the United Kingdom* 77: 3-16.

Dutton, R.W. & Bray, D.A. (eds.) 1988. The scientific results of the Royal Geographical Society's Oman Wahiba Sands Project 1985-1987. Journal of Oman Studies Special Report 3. Hindson Print Ltd., Newcastle Upon Tyne.

Gerdes, G., Spira, J. & Dimentman, Ch. 1985. The fauna of the Gavish Sabkha and the Solar Lake - a comparative study. In Friedman, G.M. & Krumbein, W.E. (eds.), Hypersaline Ecosystems. The Gavish Sabkha. *Ecological Studies* 53, Springer Verlag, pp. 322-345.

Gillett, M. 1998. Insects. In: Vine, P.J. (ed.), Natural Emirates. Trident Press, London, pp. 95-106.

Hairston, N.G., Van Brunt, R.A. & Kearns, C.M.1995. Age and survivorship of diapausing eggs in a sediment egg bank. *Ecology* 76: 1706-1711.

Hamer, I. 1988. Hymenoptera Highlights V. *Emirates Natural History Bulletin* 35: 6-12.

Jones, D.A., Price, A.R.G. & Hughes, R.N. 1978. Ecology of the high saline lagoons Dawhat as Sayh, Arabian Gulf, Saudi Arabia. *Estuarine and Coastal Marine Science* 6: 253-262.

Lahr, J.D., Ndour, A.O., Badji, K.B. & Diouf, A. 1999. Phenology of invertebrates living in a sahelian temporary pond. *Hydrobiologia* 405: 189-205.

Lewis, J.G.E. & Gallagher, M.D. 1993. Scolopendromorph and Geophilomorph centipedes from Oman and UAE. *Fauna of Saudi Arabia* 12: 171-199.

Macan, T.T. 1963. Freshwater Ecology. Longmans, London.

MacKay, W. P. 1991. The role of ants and termites in desert communities. In: Polis, G.A. (ed.), The ecology of desert communities. The University of Arizona Press, Tucson, pp. 99-112.

McAllen, R.J., Taylor, A.C. & Davenport, J. 1998. Osmotic and body density response in the harpacticoid copepod *Tigriopus brevicornis* in supralittoral rock pools. *Journal of the Marine Biological Association of the United Kingdom* 78: 1143-1153.

Mordue, W., Goldsworthy, G.J., Brady, J. & Blaney, W.M. 1980. Insect Physiology. Blackwell Scientific, Oxford.

Polis, G.A. & Yamashita, T. 1991. The ecology and importance of predaceous arthropods in desert communities. In: Polis, G.A. (ed.), The ecology of desert communities. The University of Arizona Press, Tucson, pp. 180-222.

Polis, G.A. (ed.) 1990. Biology of Scorpions. Stanford University Press, California.

Por, F.D. 1971. The zoobenthos of the Sirbonian lagoons. Rapports de la Commission Internationale de la Mer Méditerranée. 20: 247-249.

Por, F.D. 1972. Hydrobiological notes on the high-salinity waters of the Sinai Peninsula. *Marine Biology* 14: 111-119.

Por, F.D. 1973. The benthic Copepoda of the Sirbonian Lagoon (Sabkhat el Bardawil). *Cahiers de Biologie Marine* 14: 89-107.

Por, F.D. & Dor, I. 1975. Ecology of the metahaline pool of Di Zahav, Gulf of Elat, with notes on the Siphonocladacea and the typology of near-shore marine pools. *Marine Biology* 29: 37-44.

Roshier, D.A., Böer, B.B. & Osborne, P.E. 1996. Vegetation of Abu Dhabi and a preliminary classification of its plant associations. In: Osborne, P.E. (ed.), Desert ecology of Abu Dhabi – a review and recent studies. Pisces Publications, Newbury, pp. 50-65.

Ruppert, E.E. & Barnes, R.D. 1994. Invertebrate Zoology. Saunders, London, 6th Edition.

Satchell, J.E. 1978. Ecology and environment in the United Arab Emirates. *Journal of Arid Environments* 1: 201-285.

Shaw Reade, S.N., Sale, J.B., Gallagher, M.D. & Daly R.E. 1980. The Scientific Results of the Oman Flora and Fauna Survey 1977 (Dhofar). Journal of Oman Studies Special Report No. 2.

Simovich, M.A, & Hathaway, S.A. 1997. Diversified bet-hedging as a reproductive strategy of some ephemeral pool anostracans (Branchiopoda). *Journal of Crustacean Biology* 17: 38-44.

Thiéry, A. 1996. Large branchiopods (Crustacea: Anostraca, Notostraca, Spinicaudata, Laevicaudata) from temporary inland waters of the Arabian Peninsula. *Fauna of Saudi Arabia* 15: 37-98.

Tigar, B.J. 1996a. A preliminary assessment of the arthropods of Abu Dhabi. In: Osborne, P.E. (ed.), Desert ecology of Abu Dhabi – a review and recent studies. Pisces Publications, Newbury, pp. 172-195.

Tigar, B.J. 1996b. Terrestrial Arthropods (excluding insects). In: Vine, P.J. (ed.), Natural Emirates. Trident Press, London, pp. 107-120.

Tigar, B.J. 1998. The abundance and diversity of desert invertebrates in Abu Dhabi and their role in the diet of the houbara bustard. Ph.D. thesis, University of London.

Triantaphyllidis, G.V., Abatzopoulos, T.J. & Sorgeloos, P. 1998. Review of the biogeography of the genus *Artemia* (Crustacea, Anostraca). *Journal of Biogeography* 25: 213-226.

Vachon, M. & Kinzelbach, R. 1987. On the taxonomy and distribution of the scorpions of the Middle East. In: Krupp, F, Schneider, W. & Kinzelbach, R. (eds.), Proceedings of the symposium on the fauna and zoogeography of the Middle East, Mainz 1985. Reicher Verl., Wiesbaden, pp. 91-103.

Vachon, M. 1980. Arachnids of Saudi Arabia: Scorpions. *Fauna of Saudi Arabia* 1: 30-36.

Victor, R. & Victor, J.R. 1997. Some aspects of the ecology of littoral invertebrates in a coastal lagoon in southern Oman. *Journal of Arid Environments* 37: 33-39.

Walker, D.H. & Pittaway, A.R. 1987. Insects of Eastern Arabia. Macmillan, London.

Wheeler, W.C. 1990. Insect diversity and cladistic constraints. *Annals of the Entomological Society of America* 83: 91-97.

Williams, W.D., De Deccker, P. & Shiel, R.J. 1998. The limnology of Lake Torrens, an episodic salt lake of central Australia, with particular reference to unique events in 1989. *Hydrobiologia* 384: 101-110.

Zak, J.C. & Freckman, D.W. 1991. Soil communities in deserts: microarthropods and nematodes. In: Polis, G.A. (ed.), The ecology of desert communities The University of Arizona Press, Tucson. pp. 28-54.

Table 1. Insect and arachnid records from sabkha and related environments in the Middle East. Sources: (1) Gerdes et al. 1985, (2) Gillett, M., in litt., (3) Hamer 1988, (4) Walker & Pittaway 1987, (5) Wiltshire 1990.

Feeding classification **CLASS:** **Order** Family	**Species**	**Food plant and habitat**	**source**
Herbivores			
INSECTA			
Lepidoptera			
Noctuidae	*Anumeta asiatica*	*Cyperus conglomeratus*	4
	Cardepia sociabilis cineraria	Chenopodiaceae in salt deserts and coastal terrain	5
	Clytie benenotata	*Tamarix* spp. near salt flats	4
	Clytie sancta	near salt flats	4
	Heliothis nubigera	*Zygophyllum* spp.	4
Arctiidae	*Utetheisa pulchella*	*Heliotropium* spp.	4
	Streblote siva	*Tamarix* spp.	4
Lasiocampidae			
Geometridae	*Hyperythra muselmana*	*Tamarix* spp.	4
Coleoptera			
Curculionidae	*Ammocleonus aschabadensis*	*Zygopyllum* spp	4
	Bothynoderes anxius	*Salsola* spp.	4
	indet sp	salty terrain	2
Buprestidae	indet	salt bushes on salty terrain	2
Melolonthidae	indet sp	salty terrain	2
Staphylinidae	*Bledius angustus*	sabkha wetland, feed on algal mats	1
	Bledius capra	sabkha wetland, feed on algal mats	1
Detritivores			
INSECTA			
Coleoptera			
Tenebrionidae	8 species from tribes Erodiini, Tentyriini, Pimeliini	salty terrain	2

Predators and parasitoids			
INSECTA			
Mantodea	*Blepharosis mendica*	*Zygophyllum* spp., *Haloxylon* spp.	4
Hymenoptera			
Sphecidae	*Cerceris tricolorata*	adults feed on *Heliotropium* spp.	4
	Bembex dahlbomi	adults feed on *Heliotropium* spp.	4
	Chrysis ehrenbergi	adults feed on *Heliotropium* spp.	4
	Oxybelus lamellatus	coastal salt flats	3
Coleoptera			
Carabidae	*Anthia duodecimgutata*	adults hunt over wide area, including sabkha	2, pers. observ.
	Scarites subcylindricollus	salty terrain	2
	Pherosophus africana	salty terrain	2
Cicindelidae	*Myriochile melancholica*	adults hunt over seashores and salt-flats, may construct burrows near sabkha	4
	Megacephala euphratica	salty terrain	2
	Lophyra histrio	salty terrain	2
	Lophyridia aulica	sabkha wetland	1
ARACHNIDA			
Pseudoscorpiones	*Halominniza aegyptiaca litoralis*	sabkha wetland	1
Aranea	Clubionidae (indet.)	sabkha wetland	1
Aquatic or semi-aquatic			
INSECTA			
Odonata	*Hemianax ephippiger*	adults: salt marshes, salt flats	4
Coleoptera			
Dytiscidae	*Deronectes* sp.	hyperstenohaline pools	1
	Erectes stictus	hyperstenohaline pools	1
	Guignotus sp.	hypereuryhaline pools, predatory	1
	Scarodites sp.	hyperstenohaline pools	1
Hydrophilidae	*Anacaena* sp.	hypereuryhaline pools	1
	Enochrus sp.	hyperstenohaline pools	1
	Paraberosus melanocephalus	hypereuryhaline pools	1
Hydraenidae	*Ochthebius* cf. *auratus*	hypereuryhaline pools	1
Diptera	*Atissini* sp.	hypereuryhaline pools	1

Atylotus agrestis	hypereuryhaline pools	1
Bezzia sp.	hypereuryhaline pools	1
Ephydra flavipes	hypereuryhaline pools, feed on cyanobacteria and diatoms	1
Ephydra macellaria	hyperstenohaline pools, feed on cyanobacteria and diatoms	1
Hecamede grisescens	hypereuryhaline pools, feed on cyanobacteria and diatoms	1

Table 2. Sabkha Crustacea. With the exception of *Halophiloscia* which feeds on decaying organic matter, all of the species listed are filter-feeders (see text). Most records in the table come from saline pools in the Sinai Peninsula. Sources: (1) Por 1971, (2) Por 1972, (3) Por 1973, (4) Por & Dor 1975, (5) Gerdes et al. 1985, (6) Thiéry 1996.

CLASS		species	source
BRANCHIOPODA			
	Anostraca	*Artemia sp.*	2, 5, 6
		Branchinella spinosa	6
COPEPODA			
	Calanoidea	*Ridgewayia canalis*	4
	Harpacticoidea	*Cletocampus retrogressus*	2
		Robertsonia salsa	2, 3, 4, 5
		Nitocra sp	6
		Cletocampus confluens	2, 3
		Canuella perplexa	3
		Canuellina insignis	1, 3
		Enhydrosoma vicinum	3
		Harpacticus flexus	3
		Longipedia minor	3
		Longipedia sp.	1
		Mesochra rostrata	3
		Nitocra affinis	3
		Heterolaophonte sigmoides	1
		Laophonte inornata	4
		Nitocra lacustris	2, 3
		Orthopsyllus linearis	4
		Paramphiascella sirbonica	3, 4
		Robertgurneya similis	3
		Robertsonia knoxi	1, 3
		Stenhelia minuta	3
		Typhlamphiascus confusus	4
	Cyclopoidea	*Microcyclops dengizicus*	2
		Neocyclops salinarum	1, 2, 3
		Euryte sp	1

OSTRACODA	*Aglaiocypris* sp.	1
	Bairdia sp.	4
	Cyprideis littoralis	2
	Cyprideis torosa	1, 4, 5
	Paracyprideinae sp.	5
ISOPODA	*Halophiloscia* sp.	5

FISHES OF SABKHA-RELATED HABITATS

UWE ZAJONZ[1], MARK BEECH[2] & ANTHONY C. GILL[3]

[1]Section Ichthyology, Senckenberg Research Institute and Natural History Museum Senckenberganlage 25, 60325 Frankfurt am Main, Germany
[2]Department of Archaeology, The King's Manor, University of York, York Y01 7EP, U.K.
[3]Fish Research Group, Department of Zoology, The Natural History Museum Cromwell Road, London SW7 5BD, U.K.

Abstract

The taxonomy and ecology of fish species known to live in sabkha-related habitats is poorly known. An annotated account of seven species is provided and additional fish taxa that potentially occur are listed. There are no fish species known to date that are found exclusively in sabkha habitats. The species composition essentially depends on colonisation with marine species of the adjacent ocean. In addition amphidromous and very few freshwater fishes are potentially encountered in sabkhat. The fish species can be classified ecologically as transients represented by tidal and seasonal visitors or residents represented by true and partial residents. Only *Lebias dispar* seems to be a true resident. Migratory and vagrant species form the largest proportion of the sabkha fish community. To live under the environmental conditions of sabkhat presents unique problems to fishes. The disadvantages are apparently outweighed, at least for a limited suite of preadapted species, by a comparatively comfortable situation as far as competition for food and space, and predation are concerned. The present review suggests that sabkha fish are more significant in terms of number of species and individuals, and hence of greater ecological importance, than previously expected.

INTRODUCTION

Sabkha-related aquatic habitats are characterised by environmental factors which strongly fluctuate and temporarily reach extreme values, e.g. temperature, salinity, oxygen content, pH, nutrient content and several others. The taxonomy and ecology of fish species known to live under these environmental conditions around the

Barth & Böer (eds.), SABKHA ECOSYSTEMS, 283-298

Arabian peninsula is poorly known. In spite of the large area of sabkhat in the region and the apparently small number of species in question, ichthyological literature specifically on sabkhat is scarce. This paucity of data is, firstly, due to a lack of systematic collecting and survey work. Secondly, species recorded from related aquatic or semi-aquatic habitats have not been referred to as "sabkha fishes" but as inhabitants of e.g. tidal creeks or lagoons. The sabkha ecosystem definitions previously in use have not been 'aquatic' enough to promote the study of fishes since subsurface, interstitial water was considered the main water body of a sabkha environment in sensu strictu. The definitions set by this book, especially for coastal and supralittoral sabkhat shape the delimitations towards aquatic habitats linked to sabkhat that form permanent or ephemeral indentations of marine, or freshwater origin. On this assumption the aquatic realm merits its consideration as an integral part of certain sabkhat. Still, few species of fish have been hitherto recognised.

In essence this chapter represents a literature review, largely based upon studies carried out in the Gulf region. Due to the lack of studies specifically on sabkhat references have been used dealing with "restricted environments" which are spatially and ecologically linked to sabkhat. Beside a few true records from these habitats the species considered have been extrapolated from the suite of littoral and coastal migratory fishes which are known to thrive in tidal creeks, tide channels, mangrove swamps and mudflats in the area. Important contributions to the knowledge of restricted coastal environments of the Arabian Gulf were made by Evans et al. (1973), Hughes Clarke & Keij (1973), Basson et al. (1977) and MEPA/IUCN (1989). Only scattered data on fishes are included in these publications. A treatise of the intertidal fishes of Kuwait was provided by Clayton & Seedo (1986). A study of the fishes of the coastal lagoons of Dhofar, Oman, was presented by Fouda (1995). Comprehensive accounts of the fishes of the Gulf, or of certain parts of it, were published by Blegvad (1944), Al-Baharna (1986), Kuronuma & Abe (1986), and Carpenter et al. (1997), of Oman by Randall (1995), and of the Red Sea by Goren & Dor (1994).

As with faunistic studies, the ecological importance of sabkha habitats for fishes has also received little attention from researchers. First, owing to the paucity of original data, information about the ecology and habitat demands of the above considered species were compiled from Carpenter et al. (1997), Randall (1995) and FishBase (Froese & Pauly 1999). Second, ecological strategies of salt marsh fishes identified from other regions (Moyle & Cech 1982) were compared to the available data, then used to interpolate a picture of some order of probability.

The nomenclature used herein follows Eschmeyer (1998), in accordance with the International Code of Zoological Nomenclature (1999). The authors trust that the bulk of standard literature dealing with coastal wetland, mangal and estuarine ecology is being referred to elsewhere in this volume. The fossil record of fishes petrified during sedimentation processes of ancient sabkhat is not treated in this contribution. A general review of the fossil fishes of the intertidal zone can be found in Schultze (1999).

SPECIES ACCOUNT

The number of fish species present in a salt marsh is typically fewer than 15 (Moyle & Cech 1982). This figure refers to salt marshes in higher latitudes. The definite number that occurs at any time in our area of concern is not known. An annotated account of seven species is provided which have been so far reported from sabkha-related environments. In addition a list of fish species which potentially occur is given below. The data on habitat and biology, size and distribution ranges were compiled from Carpenter et al. (1997), Fischer et al. (1990), FishBase (Froese & Pauly 1999), Lieske & Myers (1994), Randall (1995), Whitehead (1985) and Wildekamp et al. (1986), unless stated otherwise.

Abbreviations: SL = standard length; TL = total length

Family **Clupeidae** - Herrings, sardinellas, sardines, and shads
Sardinella sindensis (Day, 1878)
Common name: Sind sardinella (English) - Abed or Uma (Arabic, Oman), Mashineh or Moomagh (Arabic: Bahrain, Iraq, Kuwait, Qatar, Saudi Arabia, UAE), 'Um (Arabic).
Habitat & Biology: Saltwater, pelagic. Schooling in coastal waters. Marketed fresh and dried-salted.
Size: Commonly to 14 cm SL, maximum to 17-20 cm SL.
Range: Western Indian Ocean, Arabian Sea, from Gulf of Aden to the Arabian Gulf and Bombay, also reported from Mauritius.
Remark: Additional clupeids with similar life habits occur in the area and may therefore be encountered in sabkhat.

Family **Cyprinodontidae** – Pupfishes, killifishes
Lebias dispar dispar Rüppell, 1829
Common name: Arabian killifish / pupfish (English) - Afty (Arabic, Qatar), Harsun (Arabic).
Habitat & Biology: Demersal, lives and reproduces under freshwater, brackish and marine conditions. Survives high salinities and temperatures. Occurs in coastal zones near the surface, usually found in estuaries, but also has been taken in shallow reef areas. Found in inland pools with hypersaline to fresh water. A schooling, omnivorous species.
Size: Commonly to 5 cm TL, males reported to 8 cm, females to 6 cm.
Range: Indian Ocean, Red Sea, Arabian Gulf, Dead Sea; landlocked populations in Saudi Arabia, Iran, Siwa Oasis, western Egypt. Immigrant via the Suez Canal into the Mediterranean.
Remark: The genus name *Lebias* has priority over its widely used synonym *Aphanius* (Lazara 1995). In the United Arab Emirates, according to Gary Feulner (pers. comm.), "it is probably the single most abundant fish in the Ras al-Khaimah and Umm al-Qaiwain mangrove areas discovered to date". Probably the species with the greatest abilities to cope with the harsh and fluctuating environmental conditions of sabkha-related habitats. Several congeners are recorded from the area and are expected to occur in sabkhat, e.g. *Lebias stiassnyae* Getahun & Lazara, 2001 has recently been described from the hypersaline Lake Afdera, Ethiopia.

Family **Sparidae** - Porgies and seabreams

Argyrops spinifer (Forsskål, 1775)

Common name: King soldierbream (English) - Andaq (Arabic, Qatar), Kofar or Kofer (Arabic, UAE), Cophra, Da'ara, Kawfar or Rababa (Arabic, Oman), Kufar (Arabic).

Habitat & Biology: Saltwater, demersal. Only the young fish occur in very shallow waters of sheltered bays and lagoons; larger individuals in deeper water, being most common between 5 and 100 m. Inhabits a wide range of bottoms, but especially over mud or sand substrata. Hunting macrofauna, mainly benthic invertebrates such as molluscs. Important as foodfish.

Size: Commonly to 30 cm TL (Bauchot & Smith 1984), maximum to 70 cm TL.

Range: Indo-West Pacific; from Red Sea, Arabian Gulf, South Africa and East Africa to the Indo-Malayan archipelago and northern Australia.

Remark: The species has frequently been confused with *Argyrops filamentosus* (Valenciennes, 1830).

Family **Serranidae** – Groupers and seabasses

Epinephelus coioides (Hamilton, 1822)

Common name: Orange-spotted grouper (English) - Hamoor (Arabic, UAE), Hamour (Arabic, Qatar), Balool = juvenile (Arabic).

Habitat & Biology: Marine and brackish water, reef-associated. Found along continental and large island shores; often occurs in brackish areas, also inhabits turbid coastal reefs down to depths of 100 m. Juveniles are common in shallow waters of estuaries and mangal biotopes over soft sediments. Hunting macrofauna, mostly small fishes, shrimps, and crabs. Important and highly prized food fish.

Size: Up to 100 cm SL (Kottelat et al. 1993).

Range: Indo-West Pacific; from the Arabian Gulf and the Red Sea south to Natal, South Africa and east to the western Pacific, including Australia. Has invaded the Mediterranean as Lessepsian migrant.

Remark: The species has previously been misidentified as *E. malabaricus* (Bloch & Schneider, 1801) or *E. tauvina* (Forsskål, 1775) in the Gulf, both of which do not occur there.

Family **Terapontidae** – Terapons, grunters, or tigerfishes

Terapon jarbua (Forsskål, 1775)

Common name: Jarbua terapon, thornfish or crescent perch (English) – Baam, Jimjam, Yanam (Arabic, Oman), Yamyam (Arabic, U.A.E.), Dheeb (Arabic).

Habitat & Biology: Saltwater, freshwater, demersal, catadromous. Found in schools in shallow waters above soft sediment bottoms. Enters estuaries and rivers. Spawns in the sea and juveniles migrate into fresh water. Omnivorous. Capable of producing sound.

Size: Commonly around 25 cm SL, maximum reported to 33 cm.

Range: Indo-Pacific; Red Sea to Southeast Asia, and western Pacific.

Family **Gerreidae** – Silver-biddies and mojarras

Gerres oyena (Forsskål, 1775)

Common name: Common silver-biddy or blacktip mojarra (English) - Badhah (Arabic, Qatar), Bedhah (Arabic, UAE), Badha or Mussalakh (Arabic, Oman), Badh ar-rayash (adult), Musallakh (juvenile) (Arabic).

Habitat & Biology: Brackish, saltwater, demersal. Commonly found in shallow water, lagoons, estuaries and mangrove sloughs. Occurs singly or in groups over shallow sandy areas (Sano et al. 1984). Feeds on small organisms, thriving in sandy substrates. Marketed fresh or used for fishmeal.

Size: Commonly to 15 cm SL, maximum to 30 cm (Roux 1986), or to about 20 cm respectively.

Range: Indo-Pacific; Red Sea, Arabian Gulf, to the south-eastern coasts of India and East Africa, extending to the western Pacific.

Family **Mugilidae** - Mullets
Liza subviridis (Valenciennes, 1836)
Common name: Greenback mullet (English) - Araby or Beyah (Arabic, Saudi Arabia), Anubah, Biyah, Gawafa or Gutarana (Arabic, Oman)
Habitat & Biology: Brackish, saltwater, demersal. Forms schools in shallow coastal waters and enters lagoons, estuaries, and freshwater to feed, juveniles may enter rice fields and mangroves. Feeds on small algae, diatoms and benthic detrital material taken in with sand and mud; fry take zooplankton, diatoms, detrital material and inorganic sediment. Spawning occurs at sea. Marketed fresh and salted.
Size: Commonly to 25 cm SL, maximum to 30 cm SL.
Range: Indo-Pacific, Red Sea, Arabian Gulf to western Pacific, south to Natal, South Africa.
Remark: Listed as *Chelon subviridis* in Randall (1995). Part of the type series originates from the River Ganges. A frequently used synonym is *Mugil dussumieri* Valenciennes. Several species of the family Mugilidae with similar life habits occur in the area and may therefore by recorded from sabkhat.

Additional fish taxa that potentially occur in sabkhat are listed below. The selection is based on physiological abilities, habits, life histories, and on literature records from habitats which occur in association with sabkhat, such as tropical coastal lagoons (Yáñez-Arancibia et al. 1994, Fouda 1995), and marsas (or mersas = shallow embayments) and mangrove swamps (Khalil 1994, Ormond & Edwards 1987, Por et al. 1977). Taxa marked by an asterisk are exclusively freshwater fishes.

Dasyatidae
Elopiidae
Elops machnata (Forsskål, 1775)
Megalopidae
Megalops cyprinoides (Broussonet, 1782)
Ophichthidae
Lamnostoma orientalis (McClelland, 1844)
Muraenichthys schultzei Bleeker, 1857
Clupeidae
Nematalosa nasus (Bloch, 1795)
Thryssa hamiltonii (Gray, 1835)
Engraulididae
Chanidae
Chanos chanos (Forsskål, 1775)
Cyprinidae*
various species
Heteropneustidae*
Heteropneustes fossilis (Bloch, 1801)
Ariidae
Arius tenuispinis Day, 1877
Plotosidae
Batrachoididae
Atherinidae
Belonidae
Tylosurus crocodilus (Peron & Lesueur, 1821)
Hemiramphidae
Hemiramphus far (Forsskål, 1775)
Cyprinodontidae
Lebias sophiae (Heckel in Russegger, 1846)
Lebias stiassnyae Getahun & Lazara, 2001
Scorpaenidae
Pseudosynanceia melanostigma Day, 1875
Platycephalidae
Platycephalus indicus (Linnaeus, 1758)
Ambassidae
Ambassis gymnocephalus (Lacepède, 1802)
Terapontidae
Pelates quadrilineatus (Bloch, 1790)
Terapon puta (Cuvier, 1829)
Terapon theraps (Cuvier, 1829)
Silliganidae
Sillago sihama (Forsskål, 1775)
Carangidae
Caranx spp.
Trachinotus spp.
Monodactylidae
Monodactylus argenteus (Linnaeus, 1758)
Leiognathidae
Leiognathus decorus De Vis, 1884
Gerreidae
Gerres spp.
Lutjanidae
Haemulidae
Sparidae
Petrus belayewi Hora & Misra, 1943

Nemipteridae
Polynemidae
Sciaenidae
Mullidae
Pomacanthidae
Pomacanthus maculosus (Forsskål, 1775)
Mugilidae
Liza abu (Heckel, 1843)
Liza klunzingeri (Day, 1888)
Pinguipedidae
Trichonotidae
Blenniidae
Omobranchus spp.
Gobiidae
Apocryptodon madurensis (Bleeker, 1849)
Aulopareia cyanomos (Bleeker, 1849)
Boleophthalmus spp.
Periophthalmus barbarus (Linnaeus, 1766)
Periophthalmus waltoni Koumans, 1941
Scartelaos histophorus (Valenciennes, 1837)
Scartelaos tenius (Day, 1876)
Valenciennea persica Hoese & Larson, 1994
Microdesmidae
Ephippidae
Platax spp.
Scatophagidae
Scatophagus argus (Linnaeus, 1766)
Siganidae
Siganus spp.
Soleidae
Euryglossa orientalis (Bloch & Schneider, 1801)
Paralichthyidae
Pseudorhombus sp.
Triacanthidae
Triacanthus biaculeatus (Bloch, 1786)
Tetraodontidae
Lagocephalus spp.
Torquigener spp.

ECOLOGY

At the interface between aquatic and terrestrial ecosystems, coastal and supralittoral sabkhat are among the most demanding habitats for higher organisms, in particular for fishes.

Water

Water is a critical requirement for fishes; the vast majority of fishes are fully aquatic and only a few have behavioural and physiological adaptations for drought or limited exposure to air. In the tropics and subtropics e.g. certain gobioids and blenniids are able to survive out of water for a limited time. For those species which are unable to do so the exploration of sabkha habitats poses the problem of how to not get accidentally trapped under unfavourable conditions, whether caused by the receding tides or by evaporation. There are no fish species known to date that are found exclusively in sabkha habitats. The species composition essentially depends on colonisation with marine species from coastal and littoral habitats of the adjacent ocean, or perhaps rarely with species of freshwater origin.

The tidal regimes of the Gulf are very complex, their periodicities vary greatly over the length of the area, and with the seasons. The surfaces of the lower salt marshes flood and drain, and coupled with the tidal regimes, this causes the littoral fringe periodically to expand far into the coastal wetlands. The range of the actual tidal amplitudes between the mean high and the mean low water can be very wide. For Kuwait, Jones (1986) reported a vertical amplitude of 3-4 m in the north to less than 2 m in the south. Horizontally the tidal amplitudes reach up to 2 km in certain areas (Wright et al. 1990). In addition, at times of maximum and spring tides also the supratidal sabkhat become temporarily flooded as well. This gives rise to a wealth of ephemeral aquatic habitats, as well as to permanent aquifers like drainage channels and tidal creeks (Gibson 1999).

In classifying the fish species ecologically, one needs to distinguish first between temporary immigrants, say transients, which are visitors or irregular "strays", compared to a very few species which are resident at least during a certain time of their life. Transients are species for which the intertidal provides nursery areas for juveniles or feedings grounds for subtidal species on foraging excursions. Residents are defined as species which spend their entire lives or at least a major part of it interacting with other sabkha organisms (Gibson 1969), and which show only limited horizontal movement with the tidal fluctuations. An elaborate classification has been devised by Moyle & Cech (1982) which divides the transients into tidal visitors, e.g. sciaenids, flounders, and halfbeaks, and seasonal visitors, e.g. drums, anchovies, mullets, and mojarras. The former move into the marshes at high tides in order to feed, the latter use the salt marshes as spawning or nursery areas, or as seasonal refuges from predators. Among the residents these authors distinguish between true residents e.g. killifishes, which complete their entire life cycle on the salt marshes and partial residents e.g. silversides, which occur only as juveniles. There is certainly no sharp delimitation between partial residents and seasonal visitors. Gundermann & Popper (1984) divided mangrove fishes into three categories: 1) true residents, 2) closely associated species, and 3) loosely associated species (Khalil 1994). Another grouping has been suggested by Peterson & Turner (1994), according to the extent the species use the marsh surface. Based upon Louisiana salt marsh fishes, they separate: 1) interior marsh residents that remain in flooded depressions on the marsh surface at low tide, 2) interior marsh users that move onto the marsh at high tide but return to the edge of the creeks at low tide, 3) edge mark users, resident and transient species that occupy only the margins of marshes at the edge of creeks, and 4) a group of subtidal species that do not move onto the flooded marsh at all. Several studies of salt marsh fishes and their patterns of tidal migrations are available, mainly from temperate and boreal regions, but nothing has been published yet from our area of concern.

Active emergence, terrestrial shuttling to and from the waterline, aestivation in burrows and wet pockets, as well as the ability to breathe air on occasion are responses to varying or adverse environmental conditions (Martin & Bridges 1999). The emerging species have been classified into skippers, tide pool emergers, and remainers by Martin according to the extent of a species' behavioural tendency to actively leave the water. Beside drought, emergence behaviour is primarily triggered by other physical factors like temperature, salinity, oxygen content or pH. If aquatic conditions become inhospitable the emergence strategies no doubt increase a species' fitness in the stressful sabkha environment. Such amphibious specialisations are generally found among less motile groups of fishes which are preadapted to upper littoral habitats e.g. the various species of rock and mudskippers, or toadfishes. At present only *Lebias dispar* seems to be a true resident in sabkhat. Among over 30 species recorded from Sudanese mangals (Khalil 1994) only *L. dispar*, *Gerres oyena* and some undetermined gobiids proved to complete their life cycles in the mangroves. Gobiid mudskippers of the genus *Periophthalmus* live in the mangroves, and on mud flats where they are active at low tides. This genus has been reported to occur sympatrically with other mudskippers of the genera *Boleophthalmus* and *Scartelaos* in the region (Clayton & Vaughan 1982, Zander et al. 1999). Intertidal catch studies (Wright et al. 1990) indicate that *Periophthalmus* species stay in their burrows during the high-tides. The genera *Boleophthalmus* and *Scartelaos* are more closely associated with the upper littoral below the *Periophthalmus* zone, but are also found on regularly inundated mudflats associated with sabkhat.

Still, any of these adaptations remains rather limited in terms of survival time gained under conditions of desiccation and exposure. The highly specialised species of the African lungfish family Protopteridae, which hibernate underground, enveloped in clay burrows or mucus cocoons, do actually occur in African Red Sea countries, but are confined to freshwater and therefore are unlikely to be found in sabkhat.

The prevalent suite of species suggests that migratory and vagrant species which are not restricted to certain benthic habitats form the superior proportion of the sabkha fish community. They are capable of rapidly colonising ephemeral aquatic habitats on occasion. They move in and out with the tides in search of food, and respond to unfavourable conditions by following the receding waterline or by leaving the zone completely. For several species it has been shown that they follow high flood-waters, e.g. *Liza klunzingeri* and *Leiognathus decorus* (Abou-Seedo et al. 1990). They belong to a suite of 76 species in 36 families which are frequently caught in intertidal fish traps in Kuwait (Abou-Seedo 1992, Gibson 1999).

Environment & habitat

Aquatic sabkha habitats produce extreme environments for its inhabitants which have to face the risk of dehydration during periods of emersion, and extreme values and severe changes of temperature, salinity, oxygen content, pH and other parameters caused by evaporation or precipitation. Thus, to maintain respiration, osmo and acid-base regulation, excretion and other metabolic processes presents unique problems to fishes and restrict sabkha environments to a small but select suite of preadapted species. However, it is not implicit that the adverse factors necessarily render the habitats unsuitable for the establishment of robust and productive communities. Restricted environments are often rich in number of individuals made up by relatively few species. Certain species of micro-algae, foraminifera, molluscs, annelid worms, crustaceans, echinoderms, and also fishes have adapted to hypersaline conditions and may happen to occur in large populations in sabkha-related aquatic habitats. Still, the diversity undoubtedly decreases as environmental factors like salinity, temperature, oxygen depletion and pH increase. This then usually coincides with a reduced interchange of the water body concerned and the adjacent ocean.

The restricted coastal environments of the Gulf were classified by several studies (Hughes Clarke & Keij 1973, Basson et al. 1977, MEPA/IUCN 1989) into ecological subdivisions, according to the prevalent salinities, general biological characteristics and biota recognised including species of fish.

The "normal marine environment" - salinities of up to 50 ppt: Metahaline biotopes with a similar or slightly higher salinity than the adjacent ocean, like lagoons, mangrove swamps and tidal channels; housing lower-diversity assemblages which generally depend on replenishment from the marine realm. A good example of such an environment is the Gulf of Salwah which, in comparison to the rest of the Gulf combines a higher salinity with a relatively lower biological diversity (Apel & Türkay 1999). A broad scale of littoral, vagile inshore and particularly immature fishes potentially occur. In discussing permanent tide pools of creek beds Basson et al. (1977) reported the young and partly grown individuals of the following species: *Acanthopagrus* sp., *Lebias dispar*, *Gerres oyena*, *Gerres* sp., *Hemirhamphus far*, *Liza subviridis*, Mugilidae gen. sp., *Nematalosa nasus*, Platycephalidae gen. sp., *Pseudorhombus* sp., *Sardinella* sp., *Terapon jarbua*, *Pomacanthus maculosus*. During a monitoring survey of nine khawrs (coastal lagoons) of varying salinity in the Dhofar province of Oman, Fouda (1995) reported 47 species, of which 18 were recognised as being immature. According to Evans et al. (1973), as with bony fishes, small sharks and rays are commonly observed in lagoons and are therefore found on inundated sabkha surfaces, brought in with the inflowing current. A stranded stingray was found by West (1997) at the seaward margin of the Umm Said Sabkha in Qatar.

The "restricted environment" - salinities of 50 to ca 70 ppt: Moderately hypersaline biotopes with a higher salinity than the adjacent ocean like tidal flats, tide pools, backwaters, mangrove swamps, or isolated lagoons. These happen to be periodically cut off from the sea, then experience severe evaporation; housing impoverished assemblages of largely marine origin. The fishes which are reported (MEPA/IUCN 1989) include *Lebias dispar*, *Argyrops spinifer*, *Gerres oyena*, *Sardinella sindensis*, *Epinephelus coioides*. Particularly the last species has been reported by BASSON et al. (1977) to be abundant even at salinities approaching 70 ppt.

The "highly restricted environment" - salinities greater than 70 ppt: Strongly hypersaline biotopes with salinities of up to more than 200 ppt commonly occur in isolated lagoons, ponds, salt lakes, and high tide pools. They are subject to frequent drought, housing virtually unknown communities of continental and marine origin. Although poor in species, sometimes abundant populations of photosynthetic cyanobacteria and other micro-algae are found. No species of fish are reported, except of undetermined healthy young (Basson et al. 1977), *Lebias dispar* and *Liza subviridis*.

The "freshwater/brackish environment" - salinities below ca 30 ppt: This category has not been considered by the above cited studies. Not all aquatic sabkha habitats are characterised by salinities above the level of the adjacent ocean. In certain cases freshwater of indigenous or anthropogenic origin may control the water chemistry (Fouda 1995). The young of several families of marine fishes are amphidromous; namely terapontids, lutjanids, sciaenids and gobiids are known to actively enter nearshore brackish and freshwater. In April 2000 the second author visited a freshwater/brackish outlet pool for the Ruwais Housing settlement in western Abu Dhabi region just east of Jebel Dhanna (N 24°7'42.4", E 52°40' 0.1"). The pool was situated about 3 km from the present coastline, had a storm drain on its seaward side and was surrounded by sabkha. The salinity was about 6 ppt where the waste freshwater entered the pool, but increased to about 20+ ppt at the seaward side close to the storm drain which connected to an area of inundated wet sabkha. It was found to be colonised by abundant *Terapon jarbua* of various sizes from 7 to ca 13 cm total length. This site is one of the few recorded locations on the Abu Dhabi coast where fishes have exploited a niche in what is largely a classic sabkha-like environment. Also the glassfish *Ambassis gymnocephalus* is known to typically inhabit brackish inlets and estuaries. It has been recorded from Qurm Creek near Muscat, Oman (Randall 1995). Introduced *Tilapia*, *Oreochromis* or *Poecilia* species have been observed to flourish under hypo and hypersaline conditions. It is possible that these species occur elsewhere in sabkha-related fresh or brackish water habitats.

If freshwater courses discharge through sabkha environments, especially after periods of strong precipitation, it is possible that even primary freshwater fishes occur. The largest marsh area in the southern part of Iraq, the Hor al Hammer, is a freshwater environment proper. The composition of its fish assemblages was known to be nearly similar to that of the rivers Euphrates and Tigris, which is dominated by the freshwater family Cyprinidae. Khor al Zubair is a Northwest extension of the Gulf, housing a fish community of marine origin. Owing to a recently built channel, the Schatt al Basrah, which connects it to Hor al Hammar the fish communities are now reported to mix (Al-Hassan & Muhsin 1986). Several freshwater fishes e.g. *Barbus luteus* (Heckel, 1843) and *Heteropneustes fossilis* have been recorded from Khor al Zubair, while in a reverse process marine species have also invaded the freshwater of the inland marsh area. Only in a few of Arabia's wadis does open water reach lower altitudes. Several populations of *Garra*, *Cyprinion* and *Lebias* are known to occur in streams of the Oman mountains which drain towards the opposite direction into the Rub'al Khali and an area of inland salt marshes. Likewise, these species are known to occur in streams that drain over relatively short distances towards coastal khawrs (Banister & Clarke 1977, Fouda 1995, Feulner 1998).

Food & interrelationships

Beside physical factors and habitat availability, determined by the mere presence or absence of water referred to above, biological factors like the availability of food and nutrients, inter and intraspecific competition, and prey-predator interrelationships may well exist to control the diversity and abundance of sabkha fish communities. The diets of intertidal fishes of the tropics are poorly known (Norton & Cook 1999). Even less is known about the diets of the fish assemblages associated with mangroves and salt marshes, in spite of the large coastal areas covered by these ecosystems. A notable exception are the mudskippers (Barak et al. 1994, Clayton 1993) and *Leiognathus decorus* (Wright 1989) from the Gulf.

The trophic capacity of a biotope is determined by its primary production, intrinsic food web and productivity imported from adjacent aquatic and terrestrial biotopes. The salt marshes e.g. of the Pacific and Atlantic coast of America are very productive (Moyle & Cech 1982). Although less is known about the productivity of the salt marshes of Arabia, it is evident that their productivity is high as well. Some seaward sabkhat are believed to be the most productive ecosystems in the Gulf. Microscopic Cyanophyta forming extensive blue-green algal mats and diatoms are a source of excess consumable organic matter. Beside the indigenous primary productivity, lagoonal, mangal or mudflat biotopes which connect to sabkhat are thought to export substantial productivity into wet sabkhat. Macro-algae and seagrass detritus is washed into sabkhat, markedly enriching the nutrient content. Mangal, in particular, represents a very productive habitat of coastal wetlands in the area (Por et al. 1977, Price et al. 1987). Beyond the primary productivity rendered by the mangrove trees themselves, their pneumato-phores and stems provide solid substrate for rich communities of epiphytic organisms like macro and microscopic algae, bryozoans and sponges. A rich mix of macroinvertebrates, especially of detrivores and herbivores, lives associated with these communities. The abundant food organisms and the shelter provided by the mangrove thickets support juveniles of various groups of fish (Khalil 1994). Mudflats support vast numbers of small, especially burrowing animals, including crabs, gastropods, ostracods, nematodes, flatworms, copepods and annelid worms, biomass which is partly exported to sabkhat. The limited habitability of the semi-aquatic habitats and the extreme environmental premises pose major constraints to the productivity of macroscopic aquatic plants e.g. turf or macro-algae of the various phyla, or seagrasses. Higher vascular plants occupy the low mudbanks and the lower marsh adjacent to semi-aquatic sabkha habitats. A distinct zonation exists recognising community types of emergent, halophile macrophytes like grasses, herbs, and shrubs (Halwagy 1986, Jones 1986). These are considered the major source of productivity in salt marshes.

The high productivity of restricted ecosystems is often markedly contrasted by the comparatively low biological diversity. The excess organic matter of sabkhat is not directly consumed by a diverse community of higher trophic categories like fishes, but by-passed through a detrital food chain (Moyle & Cech 1982, Norton & Cook 1999) to a rather limited number of species. Despite the general environmental disadvantages, at least for a small number of fishes, the search for food is successful enough to balance the high energy costs expended in the tidal movements and in maintaining a physiological equilibrium under the harsh conditions.

Herbivores, grazing on diatoms and cyanobacterial mats of salt marshes and mudflats (*Boleophthalmus*) and omnivores (*Lebias*, *Terapon*, *Leiognathus*, *Gerres*, *Liza*, *Scartelaos*), which opportunistically ingest detrital matter, diatoms, micro-algae, and invertebrates like annelid worms, copepods, ostracods, and small crabs (for sabkha arthropods see Hogarth & Tigar, this volume) represent the majority of the species in consideration. Because there is little hard substrate present, the food of this group of fishes are mainly small items found somehow suspended (plankton), and adherent (cyanobacterial mats) or buried into the sediment (infauna). Beyond indigenous planctonic production planctivorous species like herrings, atherinids, and halfbeaks certainly feed on oceanic plankton imported with the tides. Not only zooplankton but also algal and detrital drift make up substantial proportions of the food of these fishes (Parrish 1989, Phillips, this volume).

Few species are predators (*Epinephelus*, *Argyrops*, *Periophthalmus*) preying upon other fishes (piscivores), on epifaunal macroinvertebrates like gastropods and crabs, and on shrimps. Where aquatic or semi-aquatic insects like odonata, coleoptera, diptera, and hemiptera inhabit sabkhat they are undoubtedly preyed upon on occasion. The Arabian pupfish *Lebias dispar* and terapontids take insects e.g. those living above and below the meniscus like halotolerant mosquito and chironomid larvae, pond striders (Hemiptera: Gerromorpha), or whirligig beetles (Gyrinidae). It is possible that terrestrial arthropods which actively enter into amphibious habitats like arachnids, or tenebrionid, carabid, and cicindelid beetles have prey-predator encounters with amphibious sabkha fishes to some extent. Finally, accidentally dislocated e.g. wind-blown terrestrial insects no doubt represent another source of food for sabkha fishes. For comprehensive treatises of the various aspects of intertidal fish biology in general the reader is referred to Horn et al. (1999).

DISCUSSION

The fish communities of sabkha environments have attracted very little attention by researchers in the past. The present review suggests that sabkha fish are more in number of species and individuals, hence of greater ecological importance than previously expected. Considering the fruitful studies of salt marshes accomplished in temperate and boreal regions the marshes of the Arabian Peninsula Region deserve a serious research attempt. The suite of species that will be recognised in future field work will depend on the actual physiographic delimitations set to define what is a sabkha habitat at a given study site.

The fact that fish species exploit sabkhat has possibly to be referred to niche displacement and predator avoidance. The disadvantages exerted by the harsh environmental conditions are apparently outweighed, at least for a selection of preadapted species, by a comparatively comfortable situation as far as competition for food and space, and predation are concerned. It has to be expected that sabkha fishes represent a food source for top predators like coastal birds. No data, however, are available. Observations of e.g. cormorants and terns, including studies of stomach contents shall potentially reveal interesting insights. In essence the interrelationships of sabkha fishes with other organisms largely remain to be studied.

As to what extent sabkha aquifers are used by fishes for spawning and as nursery grounds is not yet clear. The important role of mangal, lagoonal and estuarine biotopes in the avoidance of brood predation and as feeding grounds, rather than simply as spawning areas proper, have been demonstrated in a wealth of literature. It is desirable to study whether wet sabkhat have a potential to retain and sustain the planctonic young of coastal species, thereby supporting the replenishment last but not least of economically important species.

RECOMMENDATIONS

As very little is presently known about sabkha fishes it is desirable to initiate a research program specifically on the community composition and ecology of sabkha fishes, and the biotopes and biota they are associated with. The following research topics are considered to be of pivotal interest:

1. Mapping and classifying the sabkha habitats and biotopes: The positions and the extent of sabkhat around the Arabian peninsula should be mapped using Satellite Imagery. Bioinventory and habitat surveys within selected representative salt marsh areas shall be conducted to assess the different habitat and biotope types present and to identify their associated biota. The spectral reflectance at these ground-truthing sites could then be used to conduct a habitat and biotope classification of the sabkhat of the Arabian region by Satellite Image Processing. The data need to be stored into a customised GIS database.
2. Fish community composition: In selected sabkha-related habitats detailed qualitative surveys shall be conducted to study the species composition of the fish assemblages. Fishing and visual estimation techniques will have to be employed to collect time-series data on species abundances and community composition. It is crucial that the sampling sites represent the different habitat and biotope types.
3. Migration and replenishment: The research program should address the question where the populations of the different species do come from, for how long they stay, and when they leave sabkha habitats. Size-class data collected in tidal creeks and drainage channels of sabkhat, as well as in neighbouring coastal habitats, will elucidate the extent of fish migrations and the population structure of the species involved, hence patterns of replenishment. Different tidal levels,

different seasons and different sites throughout the Arabian region should be covered to assess the tidal, temporal and spatial variation.

4. Fisheries: Intertidal fishing is common throughout the area. Fishing activities within sabkha-related habitats like ponds, creeks and lagoons, as well as those which target areas in close proximity to sabkhat, shall be assessed. A comparison of catch data with those of the fish community, as well as migration studies will contribute towards answering the question whether the populations of species which are targeted in artisanal fisheries are replenished from adjacent, inundated sabkhat.

ACKNOWLEDGMENTS

We should like to thank the editors, Hans-Jörg Barth and Benno Böer, for the invitation to contribute to this volume. Benno Böer kindly informed us of the location of the Ruwais outlet pool in Abu Dhabi. Finally, we are grateful to David A. Clayton for the loan of specimens and to the colleagues which data we compiled for this review.

References

Abou-Seedo, F.S. 1992. Abundance of fish caught by stake traps (hadra) in the intertidal zone in Doha, Kuwait Bay. *Journal of the University of Kuwait (Science)* 19: 91-99.

Abou-Seedo, F.S., Clayton, D.A. & Wright, J.M. 1990. Tidal and turbidity effects on the shallow-water fish assemblage of Kuwait Bay. *Marine Ecology Progress Series* 65: 213-223.

Al-Baharna, W.S. 1986. Fishes of Bahrain. Directorate of Fisheries, Bahrain.

Al-Hassan, L.A.J. & Mushin, K.A. 1986. The presence of *Barbus luteus* and *Heteropneustes fossilis* in the Khor al Zubair, in the North-West of the Arabian Gulf. *Zoology in the Middle East* 1: 116-118.

Apel, M. & Türkay, M. 1999. Taxonomic composition, distribution and zoogeographic relationships of the Grapsid and Ocypodid crab fauna of intertidal soft bottoms in the Arabian Gulf. *Estuarine, Coastal and Shelf Science* 49 (supplement A): 131-142.

Banister, K.E. & Clarke, M.A. 1977. The freshwater fishes of the Arabian Peninsula. The Journal of Oman Studies, Special Report 1. The Oman Flora and Fauna Survey: 111-154.

Basson, P.W., Burchard, J.E., Hardy, J.T. & Price, A.R.G. 1977. Biotopes of the western Arabian Gulf: Marine life and environments of Saudi Arabia. Aramco, Dhahran.

Bauchot, M.-L. & Smith, M.M. 1984. Sparidae. In: Fischer, W. & Bianchi, G. (eds.), FAO species identification sheets for fishery purposes. Western Indian Ocean (Fishing Area 51). volume 4. [var. pag.] FAO, Rome.

Blegvad, H. 1944. Fishes of the Iranian Gulf. Einar Munksgaard, Copenhagen.

Carpenter, K.E., Krupp, F., Jones, D.A. & Zajonz, U. 1997. FAO species identification field guide for fishery purposes. The living marine resources of Kuwait, eastern Saudi Arabia, Bahrain, Qatar, and the United Arab Emirates. FAO, Rome.

Clayton, D.A. & Seedo, F.A. 1986. Class Osteichthyes. In: Jones, D.A. (ed.), A field guide to the sea shores of Kuwait and the Arabian Gulf. Univ. of Kuwait, Blandford Press, Dorset, pp. 167-175.

Clayton, D.A. & Vaughan, T.C. 1982. Pentagonal territories of the mudskipper *Boleophthalmus boddarti* (Pisces: Gobiidae). Copeia (1): 232-234.

Clayton, D.A. 1993. Mudskippers. *Oceanography and Marine Biology an Annual Review* 31: 507-577.

Evans, G., Murray, J.W., Biggs, H.E.J., Bate, R. & Bush, P.R. 1973. The oceanography, ecology, sedimentology and geomorphology of parts of the Trucial Coast Barrier Island Complex, Persian Gulf. In: Purser, B.H. (ed.), The Persian Gulf: Holocene carbonate sedimentation and diagenesis in a shallow epicontinental Sea. Springer Verlag, Berlin, Heidelberg, New York, pp. 233-277.

Eschmeyer, W.N. (ed.) 1998. Catalog of fishes. Special Publication, California Academy of Sciences, San Francisco.

Feulner, G.R. 1998. Wadi fish of the U.A.E. *Tribulus* 8.2: 16-22.

Fischer, W., Sousa, I., Silva, C., de Freitas, A., Poutiers, J.M., Schneider, W., Borges, T.C., Feral, J.P. & Massinga, A. 1990. Fichas FAO de identificaçao de espécies para actividades de pesca. Guia de campo das espécies comerciais marinhas e de águas salobras de Moçambique. Publicaçao preparada em collaboraçao com o Instituto de Investigaçao Pesquiera de Moçambique, com financiamento do Projecto PNUD/FAO MOZ/86/030 e de NORAD. FAO, Roma.

Fouda, M.M. 1995. Fish resources of Dhofar khawrs (coastal lagoons) in the Sultanate of Oman. *Fisheries Management and Ecology* 2: 209-225.

Froese, R. & Pauly, D. (eds.) 1999. FishBase 99: concepts, design and data sources. ICLARM, Manila.

Getahun, A. & Lazara, K.L. 2001. *Lebias stiassnyae*: A new species of Killifish from Lake Afdera, Ethiopia (Teleostei: Cyprinodontidae). Copeia (1): 150-153.

Gibson, R.N. 1969. The biology and behaviour of littoral fish. *Oceanography and Marine Biology an Annual Review* 20: 363-414.

Gibson, R.N. 1999. Movement and homing in intertidal fishes. In: Horn, M.H., Martin, K.L.M. & Chotkowski, M.A. (eds.), Intertidal fishes. Life in two worlds. Acad.emic Press, San Diego, pp. 97-125.

Goren, M. & Dor, M. 1994. An updated checklist of the fishes of the Red Sea (CLOFRES II). The Israel Academy of Sciences and Humanities, Jerusalem.

Gundermann, N. & Popper, D.M. (1984). Notes on the Indo-Pacific mangal fishes and on mangrove related fisheries. In: Por, F.D. & Dor, I. (eds.), Hydrobiology of Mangal. W. Junk Publishers, Den Haag, pp. 201-206.

Halwagy, R. 1986. Subkingdom Angiospermae. In: Jones, D.A. (ed.), A field guide to the sea shores of Kuwait and the Arabian Gulf. University of Kuwait. Dist. by Blandford Press, Dorset. pp. 36-39.

Horn, M.H., Martin, K.L.M. & Chotkowski, M.A. (eds.) 1999. Intertidal fishes. Life in two worlds. Academic Press, San Diego.

Hughes Clarke, M.W. & Keij, A. 1973. Organisms as producers of carbonate sediment and indicators of environment in the southern Persian Gulf. In: Purser, B.H. (ed.), The Persian Gulf: Holocene carbonate sedimentation and diagenesis in a shallow epicontinental sea. Springer Verlag, Berlin, Heidelberg, New York, pp. 35-56.

Jones, D.A. 1986. A field guide to the sea shores of Kuwait and the Arabian Gulf. University of Kuwait. Distributed by Blandford Press, Dorset.

Khalil, A.S.M. 1994. An ecological study on fishes of the mangrove ecosystems of the Sudanese Red Sea. M.Sc. thesis, unpublished. Department of Zoology, Faculty of Science. University of Khartoum.

Kottelat, M. 1993. Technical report on the fishes from fresh and brackish waters of Leyte, Philippines. Technical Report prepared for the Deutsche Gesellschaft für Technische Zusammenarbeit (GTZ) GmbH and ViSCA-GTZ Ecology Program, Visayan State College of Agriculture, Philippines. Cornol, Switzerland.

Kuronuma, K. & Abe, Y. 1986. Fishes of the Arabian Gulf. Kuwait Institute for Scientific Research, Kuwait.

Lieske, E. & Myers, R. 1994. Collins pocket guide. Coral reef fishes. Indo-Pacific & Caribbean including the Red Sea. Harper Collins Publishers, London.

Martin, K.L.M. & Bridges, C.R. 1999. Respiration in water and air. In: Horn, M.H., Martin, K.L.M. & Chotkowski, M.A. (eds.), Intertidal fishes. Life in two worlds. Acad. Press, San Diego, pp. 54-72.

MEPA/IUCN 1989. Arabian Gulf: Saudi Arabia - An assessment of biotopes and coastal zone management requirements for the Arabian Gulf. Meteorology and Environmental Protection Administration, Ministry of Defence and Aviation, Kingdom of Saudi Arabia. International Union for Conservation of Nature and Natural Resources, Gland, Switzerland. MEPA Coastal and Marine Management Series - Technical Report 5.

Moyle, P.B. & Cech, J.J., Jr. 1982. Fishes: an introduction to ichthyology. Prentice Hall, New Jersey.

Norton, S.F. & Cook, A.E. 1999. Predation by fishes in the intertidal. In: Horn, M.H., Martin, K.L.M. & Chotkowski, M.A. (eds.): Intertidal fishes. Life in two worlds. Academic Press, San Diego, pp. 223-263.

Ormond, R.F.G. & Edwards, A.J. 1987. Red Sea Fishes. In: Edwards, A.J. & Head, S.M. (eds.), Red Sea. Pergamon Press, Oxford, pp. 251-287.

Parrish, J.D. 1989. Fish communities of interacting shallow-water habitats in tropical oceanic regions. *Marine Ecology Progress Series* 58: 143-160.

Peterson, G.W. & Turner, R.E. 1994. The value of salt marsh edge vs interior as a habitat for fish and decapod crustaceans in a Louisiana tidal marsh. *Estuaries* 17: 235-262.

Por, F.D., Dor, I. & Amir, A. 1977. The mangal of Sinai: Limits of an ecosystem. *Helgoländer wissenschaftliche Meeresuntersuchungen* 30: 295-314.

Price, A.R.G., Medley, P.A.H., McDowall, R.J., Dawson-Shepherd, A.R., Hogarth, P.J. & Ormond, R.F.G. 1987. Aspects of mangal ecology along the Red Sea coast of Saudi Arabia. *Journal of Natural History* 21: 449-464

Randall, J.E. 1995. Coastal Fishes of Oman. Crawford House, Bathurst.

Roux, C. 1986. Gerridae. In: Daget, J., Gosse, J.-P. & Thys van den Audenaerde, D.F.E. (eds.), Check-list of the freshwater fishes of Africa (CLOFFA). Vol. 2. ISNB, Brussels; MRAC, Tervuren; and ORSTOM, Paris, pp. 325-326.

Sano, M., Shimizu, M. & Nose, Y. 1984. Food habits of teleostean reef fishes in Okinawa Island, southern Japan. University of Tokyo Press, Tokyo.

Schultze, H.-P. 1999. The fossil record of the intertidal zone. In: Horn, M.H., Martin, K.L.M. & Chotkowski, M.A. (eds), Intertidal Fishes. Life in two worlds. Academic Press, San Diego, pp. 373-392.

The International Commission on Zoological Nomenclature 1999. International Code of Zoological Nomenclature. The International Trust for Zoological Nomenclature 1999 c/o The Natural History Museum, London.

West, I. 1997. Sabkha (Salt Flats) field trip - West's geology field trips (http://www.soton.ac.uk/~imw/sabkha.htm), Dr. Ian West, Geology Department, University of Southampton.

Wildekamp, R.H., Romand, R. & Scheel, J.J. 1986. Cyprinodontidae. In: Daget, J., Gosse, J.-P. & Thys van den Audenaerde, D.F.E. (eds.), Check-list of the freshwater fishes of Africa (CLOFFA). Vol. 2. ISNB, Brussels, MRAC; Tervuren; and ORSTOM, Paris, pp. 165-276.

Whitehead, P.J.P. 1985. FAO species catalogue. Vol. 7. Clupeoid fishes of the world. An annotated and illustrated catalogue of the herrings, sardines, pilchards, sprats, shads, anchovies and wolf-herrings. Part 1 - Chirocentridae, Clupeidae and Pristigasteridae. FAO Fish. Synop. 7 (125) Pt. 1. Rome.

Wright, J.M. 1989. Biology of *Leiognathus decorus* (Leiognathidae) in Sulaibikhat Bay, Kuwait. *Australian Journal of Marine and Freshwater Research* 40: 179-185.

Wright, J.M., Clayton, D.A. & Bishop, J.M. 1990. Tidal movements of shallow water fishes in Kuwait Bay. *Journal of Fish Biology* 37: 959-974.

Yáñez-Arancibia, A., Domínguez, A.L.L. & Pauly, D. 1994. Coastal Lagoons as Fish Habitats. In: Kjerfve, B. (ed.), Coastal lagoon processes. *Elsevier Oceanography Series*, 60. Elsevier Science Publisher B.V., pp. 363-376.

Zander, C.D., Nieder, J. & Martin, K. 1999. Vertical distribution patterns. In: Horn, M.H., Martin, K.L.M. & Chotkowski, M.A. (eds.), Intertidal Fishes. Life in two worlds. Academic Press, San Diego, pp. 26-53.

SHORT COMMUNICATION

SEAGRASSES OF THE ARABIAN PENINSULA REGION, WITH PARTICULAR REFERENCE TO MINERALISATION IN SABKHAT

RONALD C. PHILLIPS

Commission of Environmental Research, Emirates Heritage Club, Abu Dhabi, United Arab Emirates

INTRODUCTION

Seagrasses create stable habitats and are a major food source for many, if not most, of the recreationally and commercially important marine food animals around the world. In the Arabian Gulf, seagrasses are the main food item for dugongs and two marine turtle species (Carlos Hasbun, National Avian Research Center, pers. comm.). Seagrasses demonstrate a very high rate of primary production (McRoy & McMillan 1977, Zieman & Wetzel 1980). This function, coupled with the perennial habit of the species, and the unusually high leaf densities, are probable keys to the diversity of organisms and complex food chains associated with seagrass beds which lead to these important animals.

Seagrass leaves tend to grow quickly, e.g., typically 5 mm/day, but up to 10 mm/day (Zieman 1982, Brouns 1987). This leads to massive leaf litter as the leaves age and slough off. These leaves break down into ever decreasing-sized particles, releasing dissolved organic carbon and other dissolved materials which form the basis for important food chains (Zieman & Wetzel 1980).

Many of these sloughed-off leaves remain in the meadow, but, at times, rafts of these leaves are carried offshore where they sink or are carried inshore to the beach or become trapped within mangrove thickets or to the coastal sabkhat. Wherever they are finally deposited, they mineralise, leaving their inorganic and organic matter in that habitat.

Barth & Böer (eds.). SABKHA ECOSYSTEMS. 299-302

This brief review will cover what is known about seagrasses in the Arabian Peninsula Region, but also, more importantly, what is not known and needs to be known about seagrasses and their association with sabkhat of the region.

DISTRIBUTION WITHIN THE ARABIAN PENINSULA REGION

Within the Arabian Peninsula Region, virtually nothing is known regarding seagrass growth in Iran. Seagrasses have been reported from Kuwait, eastern Saudi Arabia, Bahrain, Qatar, and the United Arab Emirates (Basson et al. 1977, Carpenter et al. 1997, de Clerck & Coppejans 1996, den Hartog 1970, Durako et al. 1993, Kenworthy et al. 1993, IUCN 1984, Phillips & Menez 1988, Phillips et al., unpublished data, Price & Coles 1992, Richmond 1996, Sheppard et al. 1992). Only three species have been reported within the Arabian Gulf, e.g., *Halodule uninervis* (Forsk.) Aschers., *Halophila ovalis* (R. Br.). Hook., and *H. stipulacea* (Forsk.) Aschers. A fourth species, *Syringodium isoetifolium* (Aschers.) Dandy, was reported from Bahrain (Jufaer 1950, on a coral reef; den Hartog 1970, Phillips & Menez 1988, Sheppard et al. 1992). However, the collection of Jufaer constitutes the only record of this species within the Gulf.

Outside the Gulf, the number of seagrass species increases, e.g., five species in the Arabian Sea (Jupp et al. 1996, Sheppard et al. 1992), 11 species in the Red Sea (Jupp et al. 1996, Sheppard et al. 1992, Aleem 1979, Jacobs & Dicks 1985, Hulings 1979, Lipkin 1977), seven species in the Gulf of Aqaba (Sheppard et al. 1992, Hulings & Kirkman 1982, Hulings 1979) and eight in the Gulf of Suez (Sheppard et al. 1992).

SEAGRASS ASSOCIATION WITH SABKHAT IN THE ARABIAN PENINSULA

Undoubtedly, many people have observed the close proximity of seagrasses to the coastal sabkhat. Many have probably also observed the rafts of sloughed-off seagrass leaves that wash into the sabkha and mineralise there. Few of these people have realised that these degrading leaves are contributing dissolved and particulate inorganic and organic matter, energy, and substrate material to the sabkha ecosystem.

Surely this fact contributes to the development of storm-berms at the seaward edge of coastal sabkhat, which then allows the establishment of halophytic vegetation, and the associated vertebrate and invertebrate fauna in the sabkhat. The sabkha substrate is stabilised by the rooted vegetation which diminishes wind erosion and blowing dust in coastal regions, as well as providing water retention in the coastal soils. The vegetation probably also is responsible for attracting extremely large numbers of birds to the shoreline areas which nest and feed in the stabilised zone on the fauna which use the vegetation as a refuge, shelter, and nursery area for their own feeding.

Sheppard et al. (1992) observed that seagrass detritus may become widely distributed to intertidal and subtidal communities. These authors also noted that in many places in the Gulf, the seaward edge of the sabkha grades into seagrass beds. However, it appears that no one has documented any specific interactions of sabkhat and seagrasses in the Arabian Region.

STUDIES NEEDED

Studies are needed to specifically document the areal amount of sabkhat which are adjacent to seagrasses. Needed also are studies to determine the leaf turnover rate of seagrasses, and the rates and amounts of sloughed-off leaves. Studies are needed to determine the fates of the leaves, e.g., where they go (how much seagrass leaf material is washed into the sabkhat), the rates of mineralisation of these leaves in the sabkha, the contribution of dissolved and particulate inorganic and organic matter to the sabkha, and the energy contribution of these seagrass materials to the sabkha. Thus, it appears that the extensive but almost neglected sabkha ecosystem could receive an enormous nutrient and energy input from the adjacent offshore seagrass ecosystem. This possible interdependence would make an extremely valuable research effort.

References

Aleem, A.A. 1979. A contribution to the study of seagrasses along the Red Sea coast of Saudi Arabia. *Aquat. Bot.* 7. 71-78.

Basson, P.W., Burchard, J.E., Hardy, J.T., & Price, A.R.G. 1977. Biotopes of the Western Arabian Gulf. Aramco, Dharan. 284 pp.

Brouns, J.J.W.M. 1987. Aspects of production and biomass of four seagrass species (Cymodoceoideae) from Papua New Guinea. *Aquat. Bot.* 27: 333-362.

Carpener, K.E., Krupp, F., Jones, D.A., & Zajonz, U. 1997. Living marine resources of Kuwait, eastern Saudi Arabia, Bahrain, Qatar, and the United Arab Emirates. Rome, FAO. 293 pp.

DeClerck, O. & Coppejans, E. 1996. Marine algae of the Jubail Marine Wildlife Sanctuary, Saudi Arabia. In: Krupp, F., Abuzinada, A.H., & Nader, L.A. (eds.), A Marine Wildlife Sanctuary for the Arabian Gulf. Environmental Research and Conservation Following the Gulf War Oil Spill. NCWCD, Riyadh and Senckenberg Research Inst., Frankfurt a.M. pp. 199-289.

Den Hartog, C. 1970. Sea-grasses of the World. North Holland Publ. Co., Amsterdam.

Durako, M.J., Kenworthy, W.J., Fatey, S.M.R., Valavi, H. & Thayer, G.W. 1993. Assessment of the toxicity of Kuwait crude oil on the photosynthesis and respiration of seagrasses of the northern *Gulf. Mar. Poll. Bull.* 27: 223-227.

Hulings, N.C. 1979. The ecology, biometry and biomass of the seagrass *Halophila stipulacea* along the Jordanian coast of the Gulf of Aqaba. *Bot. Mar.* 22: 425-430.

Hulings, N.C. & H. Kirkman. 1982. Further observations and data on seagrasses along the Jordanian and Saudi Arabian coasts of the Gulf of Aqaba. *Tethys* 10: 218-220.

IUCN. 1984. Report on the distribution of habitats and species in the Saudi Arabian Red Sea, Saudi Arabian Marine Conservation Programme, Report No. 4, 1984 (Red Sea). International Union for Conservation of Nature and Natural Resources, Gland, Switzerland.

Jacobs, R.P.W.M. & Dicks, B. 1985. Seagrasses in the Zeit Bay area and at Ras Gharib (Egyptian Red Sea coast). *Aquat. Bot.* 23:137-147.

Jupp, B.P., Durako, M.J., Kenorthy, W.J., Thayer, G.W. & Schillak, L. 1996. Distribution, abundance and species composition of seagrasses at several sites in Oman. *Aquat. Bot.* 53: 19-213.

Kenworthy, W.J., Durako, M.J., Fatemy, S.M.R., Valavi, H. & Thayer, G.W. 1993. Ecology of seagrasses in northeastern Saudi Arabia one year after the Gulf War oil spill. *Mar. Poll. Bull.* 27: 213-222.

Lipkin, Y. 1977. Seagrass vegetation of Sinai and Israel. In: McRoy, C.P. & Helfferich, C. (eds.), Seagrass Ecosystems: A Scientific Perspective. Marcel Dekker, New York. pp. 263-293.

McRoy, C.P. & McMillan, C. 1977. Production ecology and physiology of seagrasses. In: McRoy, C.P. & Helfferich, C. (eds.), Seagrass Ecosystems: A Scientific Perspective. Marcel Dekker, New York. pp. 263-293.

Phillips, R.C., Loughland, R.A. & Youssef, A. In prep. Seagrasses of Abu Dhabi, United Arab Emirates, Arabian Gulf.

Phillips, R.C. & Menez E.G. 1988. Seagrasses. Smithsonian Contributions to the Marine Sciences, No. 34. 104 pp.

Price, A.R.G. & Coles, S.L. 1992. Aspects of seagrass ecology along the western Arabian Gulf coast. *Hydrobiologia* 23: 129-141.

Richmond, M.D. 1996. Status of subtidal biotopes of the Jubail Marine Wildlife Sanctuary with special reference to soft-substrate communities. In: Krupp, F., Abuzinada, A.H. & Nader, L.A. (eds.), A Marine Wildlife Sanctuary for the Arabian Gulf. Environmental Research and Conservation Following the 1991 Gulf War Oil Spill. NCWCD, Riyadh and Senckenberg Research Inst., Frankfurt a.M. pp. 159-176.

Sheppard, C., Price, A.R.G. & Roberts, C. 1992. Marine Ecology of the Arabian region. Academic Press, London.

Zieman, J.C. 1982. The ecology of the seagrass of south Florida: a community profile. U.S. Fish and Wildlife Service. Office of Biological Services, Washington, D.C. FWS/OBS-82/25. 158 pp.

Zieman, J.C. & Wetzel, R.G. 1980. Productivity in seagrasses; methods and rated. In: Phillips, R.C. & McRoy, C.P. (eds.), Handbook of Seagrass Biology: An Ecosystem Perspective. Garland STPM Press, New York. pp. 87-116.

SHORT COMMUNICATION

THE COASTAL AND SABKHA FLORA OF THE UNITED ARAB EMIRATES

BENNO BÖER

UNESCO Regional Office in the Arab States of the Gulf, P.O. Box 3945, Doha, Qatar

Abstract

An information survey, based on the existing scientific literature, dealing with the flora of the United Arab Emirates, resulted in a detailed picture of the flora of the country's coastal and sabkha ecosystems; the flora consists of seawater tolerant halophytes, halophytes, and semi-halophytes, subject to different degrees of salinity. The good knowledge of the nation's native halophyte resources is a precursor for nature conservation, and the development of sustainable biosaline agricultural systems.

INTRODUCTION

The most comprehensive introduction to sabkha and coastal flora in the Arabian peninsula is probably the recent paper of Deil (1998). This UAE contribution, next to other documents presented in this volume, provide an even deeper insight into the flora of the region's countries. Substantial knowledge on the halophyte flora of the UAE has been generated over the last decade (Böer & Al Ansari 1999). In particular the exact distribution of plant species along the Abu Dhabi Gulf Coast received special attention (Böer 1991, Embabi 1993, Böer & Gliddon 1997, 1998, Böer & Saenger in press.). The territory of the Emirates, 84,000 km^2 in size, and in particular

Barth & Böer (eds.), SABKHA ECOSYSTEMS, 303-309

the coastal zone, undergoes massive landuse development processes (Böer 1998), and is subject to a number of prevailing anthropogenic impacts on the environment. These impacts may include the loss of suitable habitat, and a subsequent decline in biological diversity of the indigenous halophyte resources. In order to sustainably utilise these resources, professional marine, coastal, and terrestrial land use management plans are required. The exact knowledge of the composition of the halophyte flora is a precursor for the development of environmental management plans, as well as for the availability of native halophyte resources, which may play important roles in saline habitat restoration, and biosaline agriculture. This paper lists and categorises the native halophyte resources of the United Arab Emirates.

THE COASTAL AND SABKHA FLORA

Based on the literature sources as cited in Böer & Al Ansari (1999), as well as on the recently released checklist of the UAE flora (Jonbloed et al. 2000), a list of all known halophytes in the country was compiled. The species were so categorised, that the degree of salinity tolerance can be evaluated, based on field experience of the author, and the categories as suggested in Böer & Gliddon (1997). The list is based on the above methods, and can not be regarded as absolute. More detailed botanical surveys are necessary, in order to provide an even more comprehensive picture of the flora of the UAE. This paper is merely a first guideline to those species, which are in the upper level of salt tolerance, and promising for saline habitat manipulation.

A total of 76 plant species were identified as halophytes of the UAE, two of which are classified as true mangrove species. 14 are classified as seawater tolerant halophytes (high salinity tolerant = HH), 29 as halophytes (tolerant to salinity = H), 31 as semi-halophytes (tolerant to salinity, but probably dependent on brackish to fresh water for normal development = LH), and two are parasitic plants thriving on the roots of true halophytes, such as chenopods, and members of the Zygophyllaceae family. Several hundred other plant species occur in the study area, and approximately 10% of these are halophytes. Halophytes occur mainly in the coastal zones, around inland sabkhat, in intermediate zones between inland sabkhat and gravel plains (for example in Umm al Zumul), as well as in irrigated man-made systems with insufficient drainage, and with salt accumulation in the upper soil layers. Table 1 shows the halophytes of the United Arab Emirates, and the major ecosystems in which the species occur. The nomenclature follows Jongbloed et al. (2000).

Table 1. Halophytes of the UAE, with their salinity tolerance classes (LH = salinity tolerant, but probably dependant on brackish water influence; H = tolerant to salinity; HH = high salinity tolerant), and their major ecosystems.

Species Family	Halophyte type	Major Ecosystem Remarks
MONOCOTYLEDONAE		
Arecaceae		
Phoenix dactylifera L.	LH	Brackish plantations
Cymodoceaceae		
Halodule uninervis (Forssk.) Asch.	HH	Seagrass beds
Cyperaceae		
Bolboschoenus maritima (L.) Palla	H	Brackish plantations
Cyperus arenarius Retz.	LH	Coastal dunes
Cyperus laevigatus L.	LH	Brackish plantations
Hydrocharitaceae		
Halophila ovlis R.Br.) Hook.	HH	Seagrass beds
Halophila stipulacea (Forssk.) Aschers	HH	Seagrass beds
Poaceae		
Aeluropus lagopoides (L.) Trin.	H	Brackish plantations
Aeluropus littoralis (Gouan) Parl.	H	Brackish plantations
Arundo donax L.	LH	Brackish plantations
Halopyrum mucronatum	H	Coastal dunes
Paspalum vaginatum Sw.	LH	Brackish plantations
Phragmites australis (Cav.) Trin. Ex Steud.	LH	Brackish plantations
Sporobolus iocladus (Nees ex Trin.) Nees	H	Salt marshes
Sporobolus spicatus (Vahl) Kunth	H	Brackish plantations
Potamogetonaceae		
Potamogeton pectinatus L.	LH	Brackish ponds
Ruppiaceae		
Ruppia maritima L.	LH	Brackish ponds
Zanichelliaceae		
Zanichellia palustris L.	H	Brackish ponds
DICOTYLEDONAE		
Aizoaceae		
Mesembryanthemum nodiflorum L.	LH	Brackish moist ground
Sesuvium portulacastrum L.	HH	Brackish plantations
Sesuvium sesuvioides (Fenzl) Verde	H	Brackish moist ground
Sesuvium verrucosum Raf.	H	Brackish moist ground
Asclepiadaceae		
Calotropis procera (Aiton) W.T.Aiton	LH	Brackish moist ground
Asteraceae		
Pluchea dioscorides (L.) DC.	LH	Brackish plantations
Avicenniaceae		
Avicennia marina (Forssk.) Vierh.	HH	Mangroves

Boraginaceae		
Heliotropium bacciferum	LH	Several
Caryophyllaceae		
Spergularia marina (L.) Bessler	LH	Coastal dunes
Sphaerocoma aucheri Boiss.	LH	Coastal sand sheets
Chenopodiaceae		
Agriophyllum minus Fisch. & C.A.Mey.	LH	On sand over sabkhat
Anabasis setifera Moq.	LH	Sabkha edges; landfill
Arthrocnemum macrostachyum (Moric.) Moris & Delponte	HH	Salt marshes; sabkha edges
Atriplex canescens James	H	Sand sheets
Atriplex dimorphostegia Kar. & Kir.	H	Sand sheets
Atriplex farinosa Forssk.	H	Sand sheets
Atriplex leucoclada Boiss.	H	Sand sheets
Bassia arabica (Boiss.) Maire & Weiler	LH	Ruderal sand sheets
Bassia eriophora (Schrad.) Asch.	LH	Ruderal sand sheets
Bassia muricata (L.) Asch.	LH	Ruderal sand sheets
Beta vulgaris L. subsp. *maritima* (L.) Arcang.	LH	saline flats
Bienertia cycloptera Bunge ex Boiss.	H	Coastal dunes
Halocnemum strobilaceum (Pall.) M. Bieb.	HH	Salt marshes; sabkha edges
Halopeplis perfoliata (Forssk.) Bunge ex Asch.	HH	Salt marshes; sabkha edges
Haloxylon persicum Bunge	H	Sabkha edges; sand sheets over saline ground
Haloxylon salicornicum (Moq.) Bunge ex Boiss.	LH	Gravel plains; sand sheets
Salicornia europaea L.	HH	Salt marshes, mangroves
Salsola drummondii Ulbr.	H	Sabkha edges; edges of saline ponds
Salsola imbricata Forssk.	H	Sabkha edges; landfill
Seidlitzia rosmarinus Ehrenb. ex Bunge	HH	Sabkha edges; coastal dunes
Suaeda aegyptiaca (Hassel.) Zohary	H	Sabkha edges; brackish water influence
Suaeda maritima (L.) Dum.	HH	Salt marshes; mangroves
Suaeda vermiculata Forssk. ex J.F.Gmel.	HH	Salt marshes, rocky headlands; saline moist ground
Convolvulaceae		
Cressa cretica L.	H	Brackish moist ground
Ipomea pes-caprae (L.) R.Br.	LH	Brackish moist ground

Cynomoriaceae		
Cynomorium coccineum L.	Glycophyte ?	Parasite on halophytes
Fabaceae		
Alhagi maurorum Medik.	LH	Brackish moist ground
Lotus garcinii DC.	LH	Brackish moist ground
Frankeniaceae		
Frankenia pulverulenta L.	H	Brackish moist ground
Mimosaceae		
Acacia ehrenbergiana Hayne	LH	Saline gravel plains
Prosopis cineraria (l.) Druce	LH	The natural distribution of this tree is possibly a good indication for the groundwater salinity; gravel plains and sand sheets
Prosopis juliflora (Sw.) DC.	H	A true halophyte germinates in brackish moist ground; due to its superior salinity and grazing tolerance, it outcompetes *P. cineraria* in many areas in the UAE
Orobanchaceae		
Cistanche tubulosa (Schenk) Wight	Glycophyte ?	Parasite on halophytes; occurs in two totally different ecosystems: in seawater inundated salt marshes, as well as in dry sand dune areas; possibly sub-species ?
Plumbaginaceae		
Limonium carnosum (Boiss.) O.Kuntze	H	Brackish moist ground, and adjacent to sabkha
Limonium stocksii (Boiss.) Kuntze	HH	Coastal areas and adjacent to sabkha
Portulacaceae		
Portulaca oleracea L.	LH	Brackish ground in plantations
Rhizophoraceae		
Rhizophora mucronata Lamk.	HH	Mangroves; on one location on Mubarraz Island, ca. 200 specimen, re-introduced in 1984; formerly extinct in the Gulf

Salvadoraceae		
Salvadora persica L.	LH	Plantations; wild stock in the mountains
Solanaceae		
Lycium shawii Roem. & Schult.	LH	Low salinity gravel plains
Lycopersicum esculentum L.	LH	Low salinity moist ground
Tamaricaceae		
Tamarix aphylla (L.) Karst.	H	Saline moist sub-surface
Tamarix arabica Bunge	H	Brackish moist ground, and adjacent to sabkha
Tamarix aucherana (Decne.) Baum	H	Brackish moist ground, and adjacent to sabkha
Verbenaceae		
Phyla nodiflora (L.) Greene	LH	Brackish moist ground
Zygophyllaceae		
Fagonia indica Burm.f.	LH	Dry saline gravel plains
Zygophyllum boulosii A.Hosny	H	Saline ground
Zygophyllum mandavillei Hadidi	H	Sand dunes and gravel plains
Zygophyllum qatarense Hadidi	H	Coastal zones
Zygophyllum simplex L.	H	Saline gravel plains, and sand adjacent to sabkha

Distribution maps of most of the above species are available with the Geographical Information System of Abu Dhabi's Environmental Research and Wildlife Development Agency (ERWDA).

RESEARCH, ECONOMIC VALUE AND CONSERVATION

The halophyte flora of the UAE has in parts been described by a number of authors (references within Böer & Al-Ansari 1999). However, more focused and detailed research into the halophyte resources, and phytogeography, as recently carried out for Abu Dhabi Emirate by (Böer & Saenger in press.) is required, in order to gather a comprehensive picture of the halophyte flora, not only in Abu Dhabi Emirate, but within the entire UAE territory and in particular along the coastline. Sporadic surveys over the past 14 years revealed, that a lot is yet to be learned about the flora in the UAE, and new species are frequently being discovered. The storage and analyses of this information is essential as a starting point towards a national flora and vegetation monitoring project, which needs to be carried out, in order to manage the biological resources of the UAE on a sustainable basis. It is also needed for any successful saline vegetation restoration or manipulation projects.

Considering the immense ecological and potential economic importance of the halophyte flora, it is suggested to immediately carry out more botanical surveys, in particular in the emirates of Ajman, Fujairah, Ras al Khaimah, and Umm al Quwain. The vegetation of Abu Dhabi, Dubai, and Sharjah is by now pretty well known. However, the beaches should be surveyed on a regular basis for changes in the vegetation. It is also recommended to protect these halophytes within nature reserves, and in a living genetic resources collection. With the results presented in this paper we have now a better understanding of the halophyte resources.

References

Böer, B. 1991. Vegetationskartierung auf den Inseln Abu Dhabi und Rafiq mit Erfassung wichtiger ökologischer Parameter. M.Sc. Thesis, University of Osnabrück.

Böer, B. 1998. Anthropogenic factors and their potential impacts on sustainable development of Abu Dhabi's terrestrial biological resources. *International Journal for Sustainable Development and World Ecology* 5: 125-135.

Böer, B. & Al-Ansari, F. 1999. The vegetation and flora of the United Arab Emirates - a review. In: Abuzinada, A.H. & Joubert, E. (eds.), Proceedings of the workshop on the conservation of the flora of the Arabian peninsula. NCWCD & IUCN, Riyadh, pp. 63-77.

Böer, B. & Gliddon, D. 1997. Abu Dhabi's coastal zone - conservation requirements and land use potentials. International Conference on Water Management, Salinity and Pollution Control towards Sustainable Irrigation in the Mediterranean Region. Instituto di Bari, Italy, 22-26 September, 1997.

Böer, B. & Gliddon, D. 1998. Mapping of coastal ecosystems and halophytes (case study of Abu Dhabi, United Arab Emirates). *Marine and Freshwater Research* 49: 297-302.

Böer, B. & Saenger, P. in press: The coastal vegetation of Abu Dhabi Emirate. *Marine Pollution Bulletin.*

Deil, U. 1998: Coastal and sabkha vegetation. In: Ghazanfar, S. & Fisher, M. (eds.), Vegetation of the Arabian Peninsula. *Geobotany* 25. Kluwer Academic Publishers, Dordrecht/Boston/London, pp. 209-228

Embabi, N.S. 1993. Environmental aspects of geographical distribution of mangrove in the United Arab Emirates. In: Lieth, H. & Al Masoom, A.A. (eds.), Towards the rational use of high salinity tolerant plants. *Tasks for Vegetation Science* 27. Kluwer Academic Publishers, Dordrecht/ Boston/London. pp. 45-48.

Jongbloed, M., Western, R.A. & Böer, B. 2000. Annotated check-list for plants in the U.A.E. Zodiac Publishing, Dubai, 91 pp.

SHORT COMMUNICATION

BIRDS IN SABKHA ENVIRONMENTS

SIMON ASPINALL

Abu Dhabi Islands Archaeological Survey, P.O. Box 45553, Abu Dhabi, United Arab Emirates

In Gulf districts of the Arabian peninsula there is essentially no widespread large-scale utilisation of true sabkha for nesting or feeding birds, although it is inevitable that there is some use, albeit rather limited, by at least a small number of species. No species show an obligacy, although most species which occur are characteristically those of either, or both, sparsely vegetated or bare saline habitats. Certain species found are mostly only opportunistically present, this including intercontinental migrants in particular. In that sabkha occurs along both the coast and inland areas of southern Gulf districts, it is possible to draw up a shortlist of those species of bird that one would, or could, typically observe, even if none is actually a "sabkha specialist".

All species concerned are typically found in a variety of other arid and semi-arid habitats, (including especially sandgrouse *Pteroclididae*; larks *Alaudidae*; wheatears *Turdidae*), and, moreover, it is clear that it is due to the ecophysiological constraints imposed by sabkha that most of these birds occur as they do, rather than to any rigid ecological necessity or physiological adaptation possessed of any species. Within the ecological tolerance of many of those bird species described, it is the physiological limitations imposed on, and that ultimately dictates, the community which thereafter determines the likelihood of the presence or absence of any potentially able to exploit such an area. Although a higher flora is invariably absent on sabkha, it is the latter's physical structure and, almost perversely, its exacting nature that may generate mainly feeding possibilities for opportunistic scavengers and/or dry country specialists, whether granivorous, insectivorous or predatory.

Barth & Böer (eds.), SABKHA ECOSYSTEMS, 311-314

Ironically, it may once have been a bird now extinct, the flightless Syrian ostrich (*Struthio camelus syriacus*), that was a typical inhabitant of inland sabkha of the Arabian peninsula, most probably feeding around their margins (and otherwise traversing any patches encountered). Today, stone curlew (*Burhinus oedicnemus*) and cream-coloured courser (*Cursorius cursor*), both substantially terrestrial species, also forage along sabkha edges, or at least follow the sabkha perimeter once encountering an expanse. This much is revealed by tracing the course followed by their tracks, these typically meandering along the edges where dune sands abut the barren sabkha. Whether feeding opportunities are greater here or not has yet to be established, but it does seem probable, with the likelihood of an 'edge effect' operating. Although it is true to say that tracks would be unlikely to show on the sabkha itself, it is implausible to suggest that the sabkha is anything other than an area to be crossed or otherwise bypassed to get to a richer feeding area.

Amongst nesting species, Kentish plover (*Charadrius alexandrinus*) is the species most closely associated with sabkha, particularly those more continuously saturated areas, whether inland or along the coast, although the latter is greatly preferred. The young of this nidifugous species generally feed along an adjacent or nearby shoreline, even if the waters themselves may be hypersaline and with or without tidal influence. Nesting commences in March and the species may nest solitarily or semi-colonially. Either strategy may prove effective in bringing young to fledging. This, perhaps anti-predator, adaptation is discussed further below. Other species nesting in this terrain include Saunders' little tern (*Sterna saundersi*) and black-crowned finch lark (*Eremopterix nigriceps*), the former in coastal settings. The latter species can occur throughout coastal and inland sabkha, but generally prefers at least sparse vegetation cover (regularly of *Halopeplis perfoliata*, which is itself frequent along sabkha margins). Saunders' little tern will nest solitarily, as will black-crowned finch lark, although the latter is to be found feeding in flocks in the same general areas after the breeding season.

In the winter months, chestnut-bellied sandgrouse (*Pterocles exustus*) have been found roosting communally on coastal sabkha (Osborne et al. 1996). This has been suggested to be an anti-predator behaviour. Individuals at roost, if encountered by a ground predator such as an Arabian red fox (*Vulpes vulpes*), would of course be more likely to captured that when two or more individuals are present. Statistically speaking, communal roosting is an altogether safer bet, especially in an open habitat. A predator would then encounter all or none, and even if it did cross paths with a roost (which itself would display a collectively greater vigilance than any single individual otherwise might), it may, other than being outnumbered, be sufficiently "confused" or "distracted" by a number of individuals departing simultaneously such that all birds ultimately escape successfully.

A number of diurnal avian predators may be observed actively hunting alongside, but only rarely over, sabkha areas. Kestrel (*Falco tinnunculus*) and southern grey shrike (*Lanius meridionalis aucheri*) and steppe grey shrikes (*Lanius (m.) pallidrostris*) are the commonest of such species, with pallid harrier (*C. macrourus*) and marsh harrier (*C. aeruginosus*) being observed rather more infrequently. However, none breeds in such a setting, unless other more diverse habitats are present within their territories. Amongst nocturnal or crepuscular raptors, only little owl (*Athene noctua*) is at all frequent, and then only when suitable rock outcrops occur nearby, with their prey mainly invertebrates, including flying beetles and moths in particular, even if none actually originates from the sabkha. Any vertebrate prey (reptile, mammal or other birds) immediately available, see Cunningham & Loughland this volume, would be limiting and insufficient to be able to provision nestlings, hence the low density of this owl. A supply of suitable nest sites (e.g. crags), is, coincidentally, also typically absent in most sabkha areas in the southern Gulf (Aspinall 1996).

Several species of lark are found in and around sabkhat, these most probably exploiting resources collecting dead or live at their fringes. This would include invertebrates, and the seeds of goosefoots (Chenopodiaceae) and grasses (Gramineae). Typical species include hoopoe lark (*Alaemon alaudipes*), black-crowned finch lark (*Eremopterix nigriceps*), short-toed lark (*Calandrella brachydactyla*) and lesser short-toed lark (*C. rufescens*). The last-named may, as with the black-crowned finch lark, breed in something approaching true sabkha. Physiologically similar environments or habitats, such as lagoonal saltflats, are favoured in areas possessing a Mediterranean type climate, southern Europe, for example. Wheatears (Turdidae) may behave similarly in this respect, although no species breeds in the region under discussion, all being either passage migrants or winter visitors only. Invertebrates struggling across sabkha on foot may provide easy picking for vigilant wheatears, shrikes, kestrels and scavenging brown-necked ravens (*Corvus ruficollis*).

In the absence of any alternative, grounded migrants may find temporary shelter from direct sunlight (and safety from avian predators) under "teepees" formed by sabkha salt crust cracking and lifting. Nocturnal migrants may thus find shade in which to rest during daylight hours, even if feeding or drinking opportunities are absent. Fat reserves (built up in advance of undertaking migration) in many small songbird migrants, many warblers (Sylvinae), for example, may serve a dual role in being both an energy reserve for onward migration as well as a repository for water itself. The cost-benefit of such a strategy is, however, beyond the scope of the present discussion to be elaborated on here. Nonetheless, this must form only a minor contribution toward the survival of most, if not any at all, long-distance migrant species. Shorebirds, other than typical dry country coursers and stone curlews, such as greater sand plover (*Charadrius leschenaultii*) and lesser sand plover (*C. mongolus*), may make temporary migratory halts (stopovers) in sabkha areas en route to wintering areas. These two species, together with Kentish plovers, may also use sabkha areas as safe high-tide roosts, presumably, once again, as an anti-predator strategy, since their main avian predator in this region, the peregrine falcon (*Falco peregrinus*) mostly hunts along the immediate shoreline at high tide or over mudflats during low water. Other shorebirds may also join these roosts on occasion, dunlin (*Calidris alpina*) in particular.

Only a few species of bird exploit sabkha areas, and in all instances, as described above, it is the intrinsic nature of sabkha that may make its exploitation a possibility. No vertebrate can survive on sabkha alone, while even invertebrates may succumb in such an inhospitable environment. This may itself provide feeding opportunities for predators and scavengers. Locally there may also be a concentrated food supply, of winnowed seeds collecting along the sabkha fringe on which arid zone specialists may then take advantage, for example. Also, the security offered, certainly by inner parts of a sabkha, to nesting or roosting birds, in what to all appearances must present an unpromising and dangerous environment for diurnal predators or scavengers, must outweigh the problems associated with high temperatures and water conservation. It might be stated here that since the sabkha is still biologically very young, species may be ill-adapted to exploit it, but it is possible, given time, that certain birds at least may develop behavioural adaptations which will permit its fuller exploitation.

References

Aspinall, S.J. 1996. Status & Conservation of the Breeding Birds of the United Arab Emirates. Hobby, Dubai.

Osborne, P.E, Norton, J.A. & Aspinall, S.J. 1996. Desert Birds of Abu Dhabi. In: Osborne, P.E. (ed.), Desert Ecology of Abu Dhabi. pp. 98-123. Pisces Publications, Newbury.

SHORT COMMUNICATION

SABKHA AND COASTAL PLAIN DEVELOPMENT AND EVOLUTION ON THE NORTHEASTERN COAST OF THE U.A.E. GULF COAST

ASMA AL-FARRAJ

Geography Department, UAE University, P.O. Box 17771, Al Ain, United Arab Emirates

Abstract

The coastal plain in the north-eastern part of the UAE has developed in response to sea-level changes during the Holocene as a result of long- and short-term interaction between the landward terrain and winds and waves, longshore and tidal currents, in addition to wadi sediments input. Long-term interaction has developed the coast progressively from a series of khors (inlets) as spits build up in front of embayments. The sediment input from different sources (wave, wind, wadis) into the khors turned them into shallow lagoons. More sediment input into the lagoons turned them into series of sabkhat. As this series of sabkhat is linked together by removal of the sand bars between them, mainly by wind action, they become a coastal plain. This development is divided into four stages which are shortly described by this contribution.

INTRODUCTION

Many studies on sabkha (supratidal flat) evolution and development have been carried out on the west coast of the UAE (Evans et al. 1964, Kendall et al. 1998, Kirkham 1997, Patterson & Kinsman 1981, Wood & Wolf 1969). However, there are fewer and less detailed studies relating to sabkhat and coastal evolution on the northeast coast (Bernier et al. 1995, Goudie et al. 2000, Kirkham 1998, Purser &

Barth & Böer (eds.), SABKHA ECOSYSTEMS, 315-321

Evans 1973). Moreover, relatively minor or no studies consider the development of coastal plains as a stage in sabkha processes and development.

The northeastern coast of the UAE resulted from a combination of elements: the shamal (north) wind which is presently the prevailing wind in the area, sea-level rise, and longshore currents. The Holocene transgression between 1 and 2 m above the present sea-level filled the depressions between sand dunes (Kirkham 1997, 1998, Lumbeck 1996, Purser 1973). Shamal wind activity creates long shore drift. The southwest to northeast trend of the coast has been created as an interaction between the landward terrain, waves, wind and longshore and tidal current activities. The land terrain includes: extensive carbonate dunes near Um Al Qiwen otherwise quartz-rich dunes, wadi sediments including wadi deltas near Al Jazierah Al Hamrah and Um Al Qiwen, and alluvial fans near Ras Al Khaimah and Ar Rams. In the extreme northeast of the area there are also mountain cliffs. These different phenomena have resulted in different geomorphic interactions along the coast, and in return have resulted in several types of tidal flats. These different types of tidal flat have been seen as different stages of sabkha and coastal development (Al-Farraj, in prep).

The study area is located on the northeast coast of the UAE. It has a length of 82 km and the maximum width of the central zone is about 15 km near Um Al Qiwen. It has an arid climate. Temperatures may reach 50°C in summer and January minima are rarely below 14°C. The water salinity in the lagoons near the sabkhat range between 36.4 parts per thousand in Khor Ras Al Khaimah and 51 parts per thousand in Al Mudfag and Jazierah Al Hamrah.

NORTHERN UAE COASTAL MORPHOLOGY

Table 1 summarises the main coastal morphometry measurements made along the present coast. There are six spit systems in the study area at the present day protecting eight zones of tidal flats. Two of them are inactive; both located in the Shaam zone, the rest are active. The spit systems develop from south to north following the longshore currents and wind directions. The first three spit systems exhibit the interaction between the dunes and longshore currents and wave action. The rest show wave and longshore current interaction with alluvial fans. There is no interaction between previous process and the mountain cliffs.

The largest spit system is that in Um Al Qiwen (Fig. 1). Within this spit system there are small islands created by shoaling. Larger islands, around sand dune cores, were created as the sea-level rose.

The second spit system is that which starts at Al Mudfag and ends at north Khor Al Qarm. This spit system protects two tidal flat zones in the study area; Al Jazierah Al Hamrah and Khor Al Qarm (Fig. 1). Several channels creating several barrier spits separate the major spit system.

The third, spit system starts at Ras Al Khaimah. The spit has grown very long from Ras Al Khaimah to southern Jazirat Al Hilelah. However, this spit system is highly dissected by wave action, creating a series of barrier spits. This spit system protects three tidal flats; Ras Al Khaimah and Ar Rams, in addition to Khor Al Muireed which is located between the two of them.

Fourth, Jazirat Al Hilelah has a different style of tidal flat development. It started as an elongate dune island. Then, in response to the prevailing wind direction the sand has been swept northward creating a deflation sabkha over most of the island.

The final spit systems are those in the Shaam and Ghalilah. These two systems may have been created with availability of sediments from sand dunes in the Jazirat Al Hilelah at the south and silt and sand, which washes out from the wadi systems of Wadi Shaam and Wadi Ghalilah. Both are located at the northern end of fan toes.

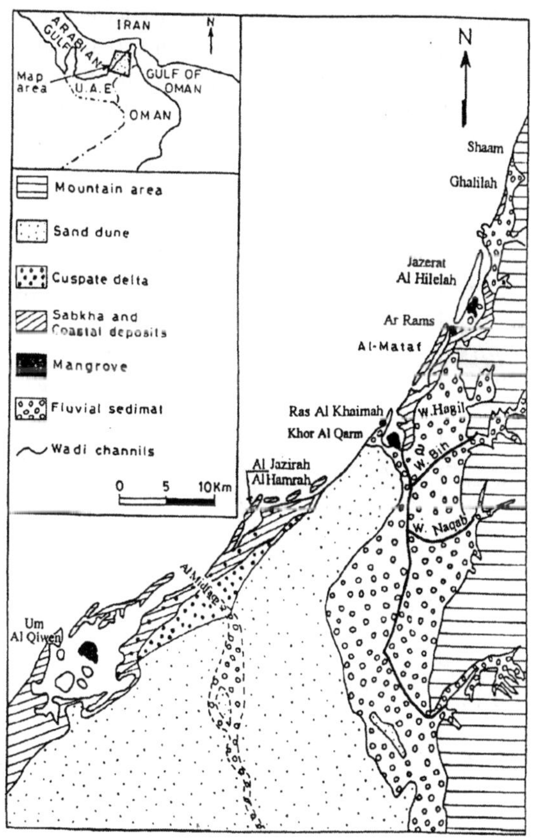

Figure 1. Location of the different spit systems.

COASTAL EVOLUTION AND SABKHA STYLE

The geomorphology of the northeast coast of the UAE exhibits long-term and short-term interaction between tidal currents, waves, and wind together with the landward terrain. The long-term interaction developed a series of khors and lagoons. Short-term interaction is evident only the spit system.

None of the khors (subtidal flat) of this coast was created as the result of the flooding of a wadi channel by seawater. Thus, all of the present khors in the study area may have been developed as embayments. Then long-term filling occurred and the sabkha environment was developed (Al-Farraj, in prep).

The evolution of the northeast coast of the UAE from the early Holocene until the present day can be divided into four main stages (Fig. 2). Pre-stage I, where embayments may have occurred as the result of sea level rise during the early Holocene occupying depressions between sand dunes (Fairbridge 1961, Kassler 1973). Stage I, The availability of sediments, especially from sand dunes, allows the development of spits where khors occurred. The availability of sediment controls spit size and khor size and/or khor number.

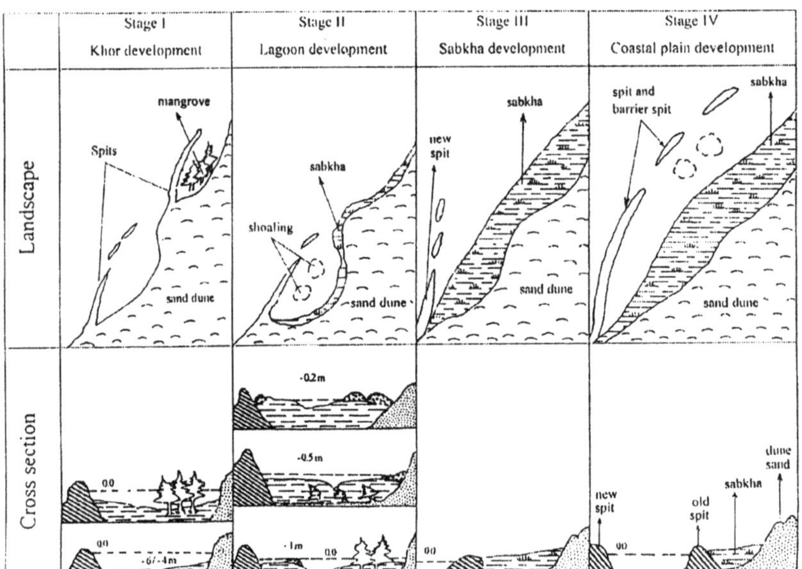

Figure 2. Main stages of coastal plain evolution.

There are two types of khors developed in the study area: khors with mangrove (e.g. Khor Ras Al Khaimah and Khor Um Al Qiwen) and khors without mangrove (e.g. Khor Al Muireed). Following to the khor formation, sedimentation processes started. There are two sources of sediment accumulating in the khors; 1) marine sediments accumulated by tidal currents and wave action, 2) terrestrial sediments accumulated by wind action and by fluvial action supplying wadi sediments into the khors. Within khors at the termination of wadis, such as Khor Ras Al Khaimah, mangroves developed. The mangroves trap sediments in the khor and transform it into a lagoon, such as Jazirat Al Hilelah. Khors, which do not have wadis draining into them like Khor Al Muireed, may still have some sediment input by currents and shoaling.

Stage II, sedimentation processes may produce a shallow lagoon such as Jazirat Al Hilelah, which represents the first stage of lagoon development. Water depth and water salinity allow mangroves to grow at Al-Hilelah. The second stage of lagoon development is the continuing sediment supply causing the lagoon to become shallower. If the lagoon shows mangrove development, it may be buried by sand and silt sediments which can be observed at Al Jazierah Al Hamrah (Al-Farraj & Goudie, in prep). The lagoon is now shallow with its surface exposed at low tide and its channels becoming shallower. Khor Al Qarm is an example of this stage of development

Stage III, the earliest phase of sabkha development starts when the lagoon and the channels are filled by sediment and flooding occurs only at the high spring tides once or twice a month. These are the sabkhat located adjacent to the present lagoons. With the availability of sediments, a situation we have at Um Al Qiwen, southward a new spit system may be built up in front of the lagoon.

Stage IV, the last stage in sabkha evolution is the coastal plain development. This occurs when wide sabkhat resulting from large tidal flats link together and create a coastal plain, such as those at Ar Rams and Um Al Qiwen. This type of sabkha coastal plain is flooded only after rainstorms. The sabkhat at Wadi Shaam and Wadi Ghalilah fan toes may be seen as representing this last stage of sabkha development on the basis of their elevation and flooding time. But because they are small they do not form a coastal plain. At this stage a new khor is being built up in front of the coastal plain which is happening at Um Al Qiwen as a result of sediment availability.

At the present there are short-term interaction controlling the modern process. This interaction is evident only on cutting and repairing the spit system after wind/wave storm (Goudie et al. 2000). In addition, it may be evident on sudden sediment input into the system after wind-, sand- and rainstorms.

Table 1. summery of tidal flat types and stage morphology according to their appearances from north to south.

Zone	Tidal flat type	Spit morphology			Sabkha morphology				Lagoon depth\m	Channel	
		Length \km	Max Width\ m	Max Height\m	Length \km	Width\ km	\m PSL	Flooding time		Width\m	Depth\ m
Shaam	Supratidal flat	1	200	4-5	1	0.5	1	After rain storm	-	-	-
Ghalilah	Supratidal flat	1	300-500	6	1	0.25	0.5	After rain storm	-	-	-
Jazirat Al Halilah	Intertidal flat	8.5 island length	Island width 1000	6m sand dune	round the lagoon	0.2-0.25	0	High tide	0.5	200	2
Ar Rams	Intertidal flat	7	250	2	9	3-5	1	After rain storm and spring tide	0.2	200-500	2-2.50
Al Muireed	Subtidal flat	3	250	2	5	6	1	After rain storm and spring tide	3	?	?
Ras Al Khaimah	Subtidal flat	7	750	4	Round the khor	0.2-0.5	0-1	High tide and spring tide	1?	Dendritic	8
Khor Al Qarm	Intertidal flat	9	250	0-2	9	0.25	0-0.5	High tide and spring tide	0.2	200	0.5
Al Jazierah Al Hamrah	Intertidal flat	8	300	1-2	Round the lagoon	0.25	0-0.5	High tide and spring tide	0.5	50-200	1-1.5
Um Al Qiwen	Subtidal flat	21	500	1-5	25	15	1-4	Spring tide and after rain storm	1?	200-1000	8

CONCLUSION

The coastal plain in the northern UAE has developed as a result of the interaction between landward terrain, wind and wave processes, and longshore and tidal currents. This coastal plain developed in a sequence of embayment, spit, khor (subtidal flat), lagoon (intertidal flat), sabkha and coastal plain (supratidal flat). In this coastal area a direct relationship between the sediment availability and spit size can be observed.

ACKNOWLEDGEMENTS

The author wishes to acknowledge the assistance of Prof. Adrian Harvey and Dr. Benno Böer. Many thanks to Mr. Zohdi Saleem for his help in drawing the figures.

References

Al-Farraj, A.M., in prep, Model of Sabkha Development and Processes: a case study of the UAE. northeastern coast.

Bernier, P., Dalongeville, R., Dupuis, B. & de Medvecki, V. 1995. Holocene shoreline variations in the Persian Gulf: example of Um Al Qiwen lagoon (UAE). *Quaternary international* 29/30: 95-103.

Evans, G., Kendall, C.G.St.C., Skipwith, D. & Patrick, A. 1964. Origin of the coastal flats, the sabkha, of the Trucial coast. Persian Gulf. *Nature* 202: 759-761.

Fairbridge, R.W. 1961. Eustatatic changes in sea-level. *Physics and Chemistry of the Earth* 4: 99-185.

Goudie, A.S., Parker, A.G. & Al-Farraj, A. 2000. Coastal change in Ras Al Khaimah (United Arab Emirates): a cartographic Analysis. The Geographical Journal *166: 14-25.*

Kassler, P. 1973. The structural and geomorphic evolution of the Persian Gulf. In: Purser, B.H. (ed.), The Persian Gulf. Berlin, Springer-Verlag, pp. 11-32.

Kendall, C.G.St.C., Alsharhan, A.S. & Whittle, G.L. 1998. The flood re-charge sabkha model supported by recent inversions of anhydrite to gypsum in the UAE sabkhas. In: (eds.), Alsharhan, A.S., Glennie, K.W., Whittle, G.L. & Kendall, C.G.St.C. (eds.), Quaternary desert and climatic change. Rotterdam, pp. 29-42.

Kirkham, A. 1997. Shoreline evolution, Aeolian deflation and anhydrite distribution of the Holocene, Abu Dhabi. *GeoArabia* 2: 403-415.

Kirkham, A. 1998. Pleistocene carbonate seif dunes and their role in the development of complex past and present coastlines of the UAE. *GeoArabia* 3: 19-31.

Lambeck, K. 1996. Shoreline reconstruction for the Persian Gulf since the last glacial maximum. *Earth and planetary science letters* 142: 43-57.

Patterson, R.J. & Kinsman, D.J.J. 1981. Hydrologic framework of a sabkha along Arabian Gulf. *American Association of Petroleum geologists Bulletin* 65: 1457-1475.

Purser, B.H. (ed.) 1973. The Persian Gulf, Holocene carbonate sedimentation and diagenesis in a shallow epicontinental sea. Springer-Verlag, New York.

Purser, B.H. & Evans, G. 1973. Regional sedimentation along the Trucial coast, SE Persian Gulf. In: Purser, B.H. (ed.), The Persian Gulf, Holocene carbonate sedimentation in shallow Epi-continental Sea. Springer-Verlag, New York, pp. 211-323.

Wood, G.V. & Wolfe M.J. 1969. Sabkha cycle in the Arab/Darb formation of the Trucial coast of Arabia. *Sedimentology* 12: 165-191.

PART III

SABKHA LAND USE AND DEVELOPMENT

BENNO BÖER & HANS-JÖRG BARTH

INTRODUCTION

The third section of this volume tries to provide some relevant information to the question, how were sabkhat historically utilised by man, and how has the situation changed in modern times. It will also try to provide prospects on future sabkha development and conservation. We know that sabkhat functioned as camel-caravan-routes along the coast. They also functioned for the production of edible salts, for example in the emirate of Umm al Quwain, and beside that, because of their extreme harsh environmental conditions, were largely avoided by humans. Pot-water was limited, bitter in taste, or non-existent, and the rangeland vegetation was too sparse to allow for grazing. Livestock had to find their feed either on the seaward halophytic storm-berm vegetation, or in the landward desert rangelands. If humans had to traverse sabkhat, then only after careful planning, well equipped with large resources of drinking water and protective clothing.

Today, the situation has changed drastically. Where we found sabkhat in the past, are now modern cities, build on land-filled areas, coastal harbours and dredged canals, as well as desalination and power plants. Long parallel highways are extending all the way through coastal sabkhat, and causing major impacts on sabkha hydrology. However, even today, with the advancement of the agricultural sector in the region's deserts, sabkhat are still not much used for plant production.

For the future, sabkhat offer great potential for seawater irrigation based halophyte systems. This can be both, farms for the production of salt tolerant cash crops, as well as man-made ecosystems for nature conservation purposes, both of which can generate jobs and income as well as contribute to the long-term process of desert greenification.

In the same time, the sabkhat in the region as part of the world, regional and national heritage, are unique landscapes, which are declining in size, due to human activities. Therefore, sabkhat deserve some kind of protection on the landscape level. The third section of this volume tries to address not only the historic importance of sabkhat, and the possibilities for future development, but also points out the importance to conserve these landforms, as part of our global landscape heritage.

Barth & Böer (eds.). SABKHA ECOSYSTEMS, 323-324

This section covers aspects on sabkha in relation to archaeology (Hellyer), the establishment of man-made seawater irrigation systems (de Soyza et al.), the establishment of man-made nature reserves (Aspinall & Hellyer), as well as landscape conservation requirements (Clusener-Godt).

THE ARCHAEOLOGY OF THE SABKHAT OF ABU DHABI

PETER HELLYER

Abu Dhabi Islands Archaeological Survey, P.O. Box 45553, Abu Dhabi, United Arab Emirates

Abstract

Archaeological investigations of areas of sabkha on the coast of the United Arab Emirates, in particular in the Emirate of Abu Dhabi, have failed to find sites of any significance on the surface of the sabkhat themselves. In many areas, however, there are raised areas within the sabkhat or in their vicinity, on or close to former shorelines, where archaeological sites have been identified. In Abu Dhabi, these date from the Late Stone Age to the Late Islamic period. While sabkha formation has been a continuing process since ca 4000 BP, affected in part, at least, by small-scale changes in sea level, the sabkhat provide an important indication of the position of older shorelines. Archaeological sites have been identified on these old shorelines. Finally, little or no investigation has been undertaken of the formerly shallow tidal or inter-tidal areas now concealed by the sabkhat. Archaeological sites may lie beneath the sabkha surface in some places.

INTRODUCTION

The presence of sabkha salt-flats in the United Arab Emirates has been well documented (e.g. Evans et al. 1969). These are of two main types. One are the coastal sabkhat, that stretch from the Sila'a peninsula in the west to the Ras Ghanadha area, in the north east, (all in the Emirate of Abu Dhabi), with intermittent sabkha also present on the northern Arabian Gulf coastline of the United Arab Emirates from Dubai to Ra's al-Khaimah, and in the Qurayyah and Kalba areas on the UAE's Gulf of Oman coast. The second type are the inland sabkhat to be found in the vicinity of the Liwa Oasis in the south-east of the UAE, just north of the mega-dunes of the Rub al-Khali. This paper seeks to examine briefly the nature of

Barth & Böer (eds.), SABKHA ECOSYSTEMS, 325-333

archaeological and palaeontological sites identified on or adjacent to the coastal sabkhat, with particular reference to Abu Dhabi.

It is widely recognised that travel across the sabkhat themselves was far from easy prior to the recent development of a network of tracks and roads. Even the hard crust that forms in summer is often thin, and easily broken through, while after rainfall, or at times of unusually high tides, they can become treacherous, now for vehicles and, in the past, for other forms of transport, such as camels. It has been noted (King 1996). that the preferred methods of travel in the pre-development era would either have been by boat, in the shallow waters offshore, or along the raised aeolianite dunes inland of the sabkhat. Thus it can be deduced that the presence of archaeological sites on the sabkhat themselves is unlikely. Extensive surveys by the Abu Dhabi Islands Archaeological Survey, ADIAS, on the coastal sabkhat of Abu Dhabi since 1992 have failed to find a single archaeological (or palaeontological) site on the sabkha surface, although, as will be seen later, sites have been found within the sabkhat on areas of raised relief.

The dating of the initial sabkha formation has been set at around 4000 BP, (Evans et al. 1969), and is discussed elsewhere in this volume. One result of their formation was a major change in the coastline of the UAE. Thus, prior to recent dredging and landfill activity, the island of Umm al-Nar, a major port-settlement site during the latter part of the Third Millennium BC (Frifelt 1991, 1995), was divided only by a narrow channel of no more than a few hundred metres in width from an area of sabkha and supra-tidal flats extending several kilometres inland. There is a distance of approximately 12 kilometres between Umm al-Nar and the low range of Pleistocene (largely aeolianite) hills inland on which Abu Dhabi International Airport now stands. This area is now sabkha, much of which has recently been infilled and developed. Before its formation, around 4000 BP, at a time when sea level was an estimated 1 to 3 metres higher than at present, it would have been shallow water or, possibly, inter-tidal flats, with the aeolianite hills forming the old coastline. The width of the coastal sabkhat varies, but can extend in excess of 15 km inland from the current shoreline in places, while the Sabkhat Matti, in the far west of Abu Dhabi, extends inland from the coast for almost 150 km., much of it being within Saudi Arabia.

Thus prior to the formation of the sabkha, the coastline of the Emirates was significantly different from that of today. As discussed below, there is also substantial evidence for the continuing evolution of shorelines since that date, as a result of small, but in some areas significant, changes in sea level. Within the sabkha zone, survey for archaeological sites should, therefore, concentrate on areas adjacent to the sabkha edge. Survey work in such areas undertaken in the United Arab Emirates, in particular in the Emirate of Abu Dhabi, has identified numerous sites, dating both from before the beginning of the formation of the sabkha and from later periods.

DISCUSSION

The first evidence of a human presence in the United Arab Emirates has been dated to the mid-Holocene, to around 7500 BP (5500 BC), in the Late Stone Age, thus pre-dating by at least 3500 years the suggested date for the beginning of the formation of the sabkha. Archaeological surveys have thus far failed to find any conclusive evidence of a human presence prior to this period, in the Lower Palaeolithic, for example, although earlier sites have been identified in Saudi Arabia (Nayeem 1990). It has been suggested that this may be due partly to the fact that prior to the mid-Holocene there was a lengthy period during which an inhospitable climate prevailed. Thus it has been proposed that "adverse climatic conditions during the last glacial era, with harsh weather and winds strong enough to transport sand, would have made living in the region at that time comparable to picknicking in a sandstorm for much of the year." (Potts 1998). It remains possible that there was Palaeolithic occupation of the Emirates, but evidence of it, if it exists, lies concealed beneath the major dune formations that cover much of the country.

Populations in the area prior to around 7500BP may have been concentrated in the more fertile and lower-lying plains of the extended Tigris-Euphrates valley, which, prior to the ice-melt at the end of the Last Glacial Maximum (LGM), around 21,000 to 14,000 years BP, extended to the Straits of Hormuz. Infilling of the Gulf, and the drowning of its floor, commenced around 14,000 BP, with the most rapid rise occurring between ca 12,000 - 8000 BP (Lambeck 1996). Populations living on the floor of the Gulf would have been obliged to move higher to escape the rising waters, leaving behind them the submerged evidence of their presence. Indeed, it has been suggested that the memories of the rising waters may have been the origins of the flood event recorded in the Sumerian Epic of Gilgamesh and of the Biblical Noah's Flood (Teller et al. 2000).

Rising sea levels appear to have reached their maximum of 1 to 3 m. above the modern level around 6000 - 4000 BP, following which there was a slow retreat (Uchupi et al. 1996 & Lambeck 1996, cited in Teller et al. 2000). Formation of the coastal sabkhat then commenced.

If, as appears to be the case, the inner edges of the coastal sabkhat roughly represent the shoreline that existed at this period (6000 - 4000 BP), it is reasonable to suppose that there should be some evidence of Late Stone Age occupation of the Emirates on, or adjacent to, this former shoreline. It should, of course, be noted that the movement of aeolian sands may have obscured the archaeological evidence in some areas.

It is perhaps significant that the oldest archaeological sites thus far identified, and securely dated by C14 dating of charcoal, in the Emirates date to between 7500 - 7000 BP, not long after this period of a rapid rise in the waters of the Gulf. The two oldest sites are on offshore islands in the waters of the Emirate of Abu Dhabi. The oldest, on the island of Marawah, has produced a date of a minimum of 7570 +/- 50

BP, while the second oldest, on Dalma, also in Abu Dhabi, has produced a date of 7115 +/- 65 BP. Dalma was probably already an island at this time, although it has been suggested that the chain of islands including Marawah may have still been a peninsula projecting from the mainland northwest to enclose the already-flooded Khor al-Bazm.

Both in the northern Emirates and in Abu Dhabi, sites from the Late Stone Age period have been found in the vicinity of the sabkha edge, although secure dating is rare. In the northern Emirates, one typical site is that of JH-57, a 5th - 4th Millennium BC shell midden at Jazirat al-Hamra in Ra's al-Khaimah (site 5 in figure 1), approximately a kilometre inland from the current coastline (Beech & Kallweit, H, in press). In Abu Dhabi, examination of the sabkha edge has produced few middens, all, in fact, on islands rather than on the mainland, where the movement of sand has more effectively obscured the shape of much of the former shoreline. Other types of archaeological sites have, however, also been identified both on islands and the mainland. At Abu Dhabi International Airport, for example, a site has been identified on a Pleistocene aeolianite dune immediately adjacent to the sabkha and representing the former shoreline. This has yielded evidence of occupation from at least five distinct periods. Three of these pre-date the formation of the sabkha around 4000 BP. The oldest, represented by lithic implements, can be dated to the Late Stone Age, ca 7000 BP (Hellyer 1998). A second phase is represented by the presence of ceramic material from the Jebel Hafit period (ca 5200 - 4500 BP) and a third by the presence of ceramic material from the Umm an-Nar period (4500 - 4000 BP) (De Cardi 1997).

Two later periods are also represented, also by ceramics, the Late Pre-Islamic period, ca 1st C. BC to 2nd C. AD and the Late Islamic period (16th - 19th C. AD) (De Cardi 1997).

Approximately 40 km south west of the city of Abu Dhabi, other Late Stone Age sites have been identified by ADIAS in the Rumaitha area (Fig. 1). In this area, the interface between the inner edge of the sabkha and the dune-covered aeolianite outcrops that represent the pre-sabkha shoreline is more complex than at Abu Dhabi Airport. Tongues' of sabkha extend in some places for a considerable distance into the dune-field, sometimes obscured by the presence of Holocene dune formations that sit on a sabkha surface so that pockets of sabkha exist that are today completely surrounded by dunes. In parts, the surface of these inter-dunal pockets is comprised of eroded and flattened Pleistocene aeolianite that merges almost imperceptibly into sabkha. Scatters of worked flint, including tools, have been found on the aeolianite on the edge of some of these sabkhat, as, for example, at the site of RM-5 in the Rumaitha area, providing evidence of a human presence during the Late Stone Age on and adjacent to the pre-sabkha shoreline (Hellyer & Aspinall, in prep.).

The majority of the archaeological sites that thus far have been recognised adjacent to the inner edge of the sabkha in coastal Abu Dhabi are of a Late Stone Age date. There is, however, some evidence that suggests that areas adjacent to the inner edge of the sabkha were also used during subsequent periods, such is as provided by a site north-east of the Bab oilfield. It lies on an interdunal aeolianite plain divided only by a narrow ridge of mobile sand from an adjacent sabkha "embayment". Worked flint, probably of a Late Stone Age date, has been collected from this plain, now traversed by a highway leading southwards from the coastal highway into the desert, but material from later periods has also been found, including imported glass ascribed a date in the 12th/13th Centuries AD and ceramic material of a Late Islamic date, or 16th - 19th C. AD (King et al. 1999). The presence of the later material suggests that the plain was used in the mid- and late Islamic periods by travellers making use of the easier going immediately inland of the sabkha edge, and to the north of the substantial dune field that lies to the south.

The inner edges of the sabkhat of the Emirates, therefore, represent areas where the presence of archaeological sites may be expected, even though, as has been mentioned, the mobility of the aeolian sand has undoubtedly obscured some sites.

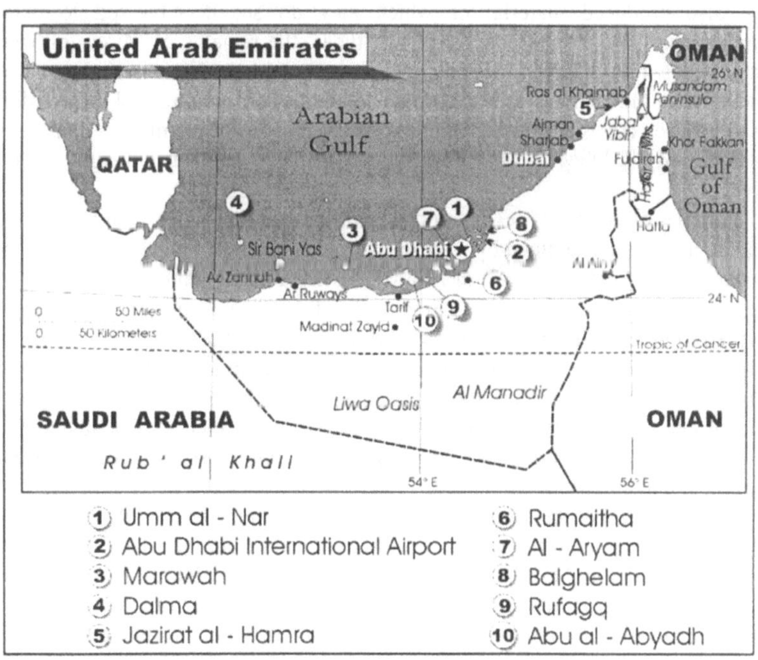

Figure 1. Location of archaeological sites in the UAE.

Other areas within the coastal sabkhat themselves have also produced archaeological, and palaeontological material. Rising out of the coastal sabkhat of

Abu Dhabi are numerous flat-topped mesas that represent the eroded remnants of a Miocene landscape, often capped with later geological strata, including limestone and flint. Prior to the formation of the sabkhat, these mesas would have been small islands, and, not surprisingly, several have produced archaeological artefacts and palaeontological material. The mesa of Barqat Rashid, (Site RM-12) to the south of Tarif, has extensive exposures of tabular flint, of a quality suitable for tool-making, while another, un-named mesa in the Rumaitha area (Site RM-2) has both tabular and nodular flint, and has also produced worked flint material. (Hellyer & Aspinall, in prep). Other mesa formations further west, such as those of Jebel Barakah and the island of Shuweihat, have produced vertebrate fossils of a Late Miocene date (Whybrow & Hill 1999).

Such formations, rising in some cases to over 60 metres above the sabkha, are sufficiently distinct to be distinguished easily from the sabkha, and are outwith the scope of this paper.

There are, however, less easily distinguishable formations within the sabkha that have also produced material or which may do so, if subjected to detailed examination. These are essentially of two types, the remains of Miocene outcrops that have lost their harder limestone or flint caps and have been eroded down to an elevation that may be only a metre or two above the surrounding sabkha and slightly raised beaches deposited during the peak of the mid-Holocene marine transgression that followed the LGM. This is now believed to have reached its height ca 6000 - 4000 BP.

Examples of the first type, the eroded Miocene outcrops in the Abu Dhabi coastal sabkhat, have been examined by ADIAS. One such outcrop, Muqatara-1, close to the coastal highway just south-west of Abu Dhabi, has a maximum elevation of no more than 3 metres above the sabkha. Fragments of Late Miocene fossilised eggshell, believed to be of a Struthionid ancestor of the ostrich, *Struthio camelus*, have been collected from the slopes of the outcrop. Another outcrop in the Rumaitha area, Site RM-1, a few hundred metres into the sabkha from the pre-sabkha shoreline, has yielded both Late Miocene fossilised eggshell and small pieces of tabular flint, the remains of the former hard cap, as well as worked flint, presumably of a Late Stone Age date (Hellyer & Aspinall, in prep.).

Good examples of the second type, the old and deflated beach ridges, are present in the coastal sabkhat to the south-west of Abu Dhabi from the island of Al-Aryam, in the east, to Abu al-Abyadh, in the west (site 7 and 10 in figure 1), and are visible on satellite imagery as linear features. On the surface, they can be identified by the sparse colonisation by halophytic vegetation, which is not present on the sabkha itself (Kirkham 1997). It has been suggested (see above), that the marine transgression that followed the LGM climaxed at around 6000 - 4000 BP at a level approximately 1 to 3 metres above present-day sea levels, at or above the current levels of these beach ridges. Subsequent deflation, however, would have reduced the ridges to their current levels. This climax of the post-LGM marine transgression would have occurred during the latter part of the Late Stone Age (dated in the UAE to ca 7500 BP to 5200 BP) or the early Bronze Age (5200 BP to 3300 BP). It is certainly possible, although as yet unproven due to a lack of survey work on such deflated beach ridges, that some may contain artefacts or other material relating to a Late Stone Age or Bronze Age presence, such as lithic implements or shell middens.

It is necessary to recall that the formation of the coastal sabkhat of the United Arab Emirates cannot be ascribed to a particular, finite period of time, even if their formation appears to have commenced around 4000 BP. Thus Late Holocene anhydrite is still forming on the seaward side of the storm beach system within the supra-tidal zone of the Khor Al Bazm and Khor Qirqishan (Kirkham 1997).

Investigation of the surface geology of a number of Abu Dhabi's offshore islands, carried out for the Abu Dhabi Islands Archaeological Survey, has produced evidence of changing sea levels during the Late Holocene period. Thus, on the island of Marawah, the dating of material from three beach ridges of varying levels has suggested they were formed at widely differing dates, *viz.* 4510 100 BP, 2230 50 BP and 1610 60 BP (Evans et al., in prep).

Also on Marawah (site 3 in figure 1), west of the village of Ghubbah, and adjacent to the archaeological sites MR-9, MR-11 and MR-12, there is an area of sabkha that appears to represent a former embayment, now infilled by natural sedimentation. While some of the sites overlooking this embayment date to the Late Stone Age or early Bronze Age, i.e. pre-dating the beginning of the formation of the sabkha, others have been dated to the late First Millennium AD (Elders 2000). Although further investigation of this embayment is required, it appears at least probable that it was still a tidal inlet during the late First Millennium AD. Elsewhere on Marawah, overlooking another area of inland sabkha, a number of hearths have been identified. Although these have not yet been examined in detail, they are of types that have produced C14 dates, from both Marawah and the islands of Balghelam and Rufayq, that range from approximately 4000 BP to 1800 BP, or after the date conventionally used for the beginning of sabkha formation. Two hearths on Rufayq dated to the early 1st Millennium BC and mid 1st Millennium BC lie close to the inland edge of an area of sabkha extending inland from the current shoreline.

On another island, Abu al-Abyadh, another archaeological site has been identified on a low raised area within what also appears to be an infilled embayment. The site, ABY-21, on the south-west of the island, has produced ceramics of Late Islamic date, but no earlier material (Hellyer et al. 2000). It is tentatively suggested that this site was a supra-tidal outcrop in a tidal inlet until relatively recently, although the whole inlet is today now silted up and supra-tidal.

Further study of the evolution of sea levels in the Arabian Gulf is required before a clear picture can be proposed, but studies by the Abu Dhabi Islands Archaeological Survey on the offshore islands of Abu Dhabi certainly suggest that at least some of the infilled embayments that are now sabkha were at least inter-tidal in the recent past.

Thus a review of the sabkhat of Abu Dhabi with relation to the archaeological evidence that may be present requires not only an examination of the inner edges of the coastal sabkhat and of any areas of raised relief within them, but also a study of areas close to the present-day coastline that may represent embayments that have been infilled during the 4000 years or so that have passed since the main process of sabkha formation commenced.

For this, as in so much else, a multi-disciplinary approach will be required if a full understanding of the coastal archaeology of Abu Dhabi is to be obtained. Over the last 4000 years, changes in sea level and, therefore, in shorelines, have been well attested in the Emirates. It is probable, therefore, that the formation of at least part of some of the coastal sabkhat is also relatively recent, postdating 4000 BP. Certainly a number of archaeological sites are in locations that, despite their dating, for example in the First Millennium BC and the First Millennium AD, suggest that they were deliberately sited adjacent to a now-disappeared coastline.

Finally, it is possible that the sabkhat themselves may conceal archaeological evidence that pre-dates their formation. Thus, at the site at Abu Dhabi International Airport, mentioned above, it is reasonable to assume that, prior to the formation of the sabkha, there may have been simple jetties to facilitate the shipping of goods offshore to the island of Umm al-Nar. No evidence of such jetties has thus far been identified, but they may lie beneath the sabkha that, as it formed and the sea retreated, rendered the jetties themselves redundant. The utilisation of techniques such as ground-penetrating radar may yet show that archaeological sites lie not only on the edges of the sabkha, but beneath its surface.

References

Beech, M. & Kallweit, H. 2001. A note on the archaeological and environmental remains from Site JH57, a 5th-4th Millennium BC shell midden in Jazirat al-Hamra, Ra's al-Khaimah. *Tribulus* 11 (1): 17-20.

Elders, J. 2000. Marawah, Spring 2000: Preliminary Survey and Excavation Report MR11 and MR12. Unpublished report of the Abu Dhabi Islands Archaeological Survey. 8 pp.

De Cardi, B. 1997. Third-Millennium and Later Pottery from Abu Dhabi Airport. Arabian Archaeology and Epigraphy 8,2: 161-173.

Evans, G., Kirkham, A., Hellyer, P. & Carter, R.A. in prep. The Geology of Marawah Island.

Evans, G., Schmidt, V., Bush, P. & Nelson, H. 1969. Stratigraphic and geologic history of the sabkha, Abu Dhabi, Persian Gulf. *Sedimentology* 12: 145-159.

Frifelt, K. 1991. The Island of Umm an Nar: The Third Millennium Graves. Jutland Archaeological Society Publications XXVI.1, Aarhus U.P. 188 pp.

Frifelt, K. 1995. The Island of Umm an Nar: The Third Millennium Settlement. Jutland Archaeological Society Publications, XXVI.11, Aarhus U.P. 260 pp.

Hellyer, P. 1998. Hidden Riches: An Archaeological Introduction to the United Arab Emirates. Union National Bank, Abu Dhabi. 226 pp.

Hellyer, P., Aspinall, S.J. & Verhage, M. 2000. Further Archaeological Sites on the island of Abu Al Abyadh. A report to the Private Department of His Highness Sheikh Khalifa bin Zayed Al Nahyan. 6pp. Unpublished report of the Abu Dhabi Islands Archaeological Survey.

Hellyer, P. & Aspinall, S.J. in prep.. A Gazetteer of Archaeological Sites in North-eastern Abu Dhabi).

King, G.R.D. The Arabian Gulf, the trade infrastructure and Nestorian fieldwork in Abu Dhabi and Ra's al-Khaimah. Unpublished lecture to the 5th Workshop on Late Antiquity and Early Islam. London, Nov. 1996

King, G.R.D., Hellyer, P. & Aspinall, S.J. 1999. An Archaeological Baseline Study of the ADCO oilfields. Part One: Coast and Islands. Unpublished report to the Abu Dhabi Company for Onshore Oil Operations, ADCO. 88 pp.

Kirkham, A. 1997. Shoreline Evolution, Aeolian Deflation and Anhydrite Distribution of the Holocene, Abu Dhabi. *GEOArabia* 2,4: December 1997. 403-416.

Lambeck, K. 1996. Shoreline reconstructions for the Persian Gulf since the last glacial maximum. *Earth and Planetary Science Letters* 142: 43-57.

Nayeem, M.A. 1990. Prehistory and Protohistory of the Arabian Peninsula, Vol. 1 (Saudi Arabia). Hyderabad. 36-51.

Potts, D.T. 1998. Maritime beginnings, in waves of time. In: Hellyer, P. (ed.), The maritime heritage of the United Arab Emirates. Trident Press, London, 1998. pp. 8-43.

Teller, J.T., Glennie, K.W., Lancaster, N. & Singhvi, A.K. 2000. Calcareous Dunes of the United Arab Emirates and Noah's Flood: the postglacial reflooding of the Persian (Arabian) Gulf. *Quaternary International* 68-71: 297-308.

Uchupi, E., Swift, S.A. & Ross, D.A. 1996. Gas venting and late Quaternary sedimentation in the Persian (Arabian) Gulf. *Marine Geology* 129: 237-269.

Whybrow, P. & Hill, A. (eds.) 1999. Fossil Vertebrates of Arabia. Yale University Press. 523 pp.

SALINE WETLAND RESERVE MANAGEMENT: A CASE STUDY FROM THE UNITED ARAB EMIRATES

SIMON ASPINALL & PETER HELLYER

Abu Dhabi Islands Archaeological Survey, P.O. Box 45553, Abu Dhabi, United Arab Emirates

INTRODUCTION

There is no broad international consensus on the perceived value to mankind of sabkhat (as defined elsewhere in this book), and of intermittently inundating and flooding coastal saltflats. Perceptions may range from being afforded a high value, because of their fragile ecology and importance to wildlife, to a low value because of their unsuitability for development or exploitation.

In the United Arab Emirates, one area of sabkha close to the Arabian Gulf coast is now the subject of a managed hydrological regime that makes opportunistic re-use of treated wastewater. While such an approach is likely to be feasible in only a limited number of instances, the UAE site, close to the country's capital, Abu Dhabi, has proved to be a success both in terms of its value for wildlife and in terms of human enjoyment. The site, Al Wathba Lakes, formerly an area of intermittently flooding sabkha, has been substantially enhanced by the inexpensive management of an available wastewater supply, together with minor earth-moving operations. As a result, an area once deemed locally of relatively minor value has now been accorded official recognition and receives protection as an important wildlife reserve.

Barth & Böer (eds.), SABKHA ECOSYSTEMS, 335-340

LANDUSE OPTIONS FAVOURING WILDLIFE

Amongst the many landuse options open to decision-makers, planners and developers when considering coastal sabkhat and other intermittently inundating saltflats is the possibility of what is popularly termed 'environmental enhancement'.

In the southern Arabian Gulf the choice of a 'leave-alone' management policy is, in most cases, improbable, given the importance of coastal and near-coastal development. This is particularly so in the United Arab Emirates, where the population has risen from less than 250,000 to just over 3 million inhabitants in the last thirty years. Historically, the coastal and near-coastal sabkha fringe of Abu Dhabi emirate, which, of the seven component members of the UAE has by far the longest coastline, have been of little value to the country's people (see Hellyer, this volume), and at present they are often perceived as being valueless except as real estate, which, if infilled, can be used for profitable urban or industrial development.

It is difficult to build a case on ecological grounds alone for *per se* preservation of sabkha in an unmodified state, and certainly when socio-economic argument favours its physical alteration to one of productive use by an individual or by society in general.

It should be noted that although this paper does not examine the intertidal edges of the sabkha of Abu Dhabi, it is conceivable that cyanobacterial mats growing on lower level intertidal coastal flats may be of major importance to energy flow and nutrient cycling both within and outwith the immediate ecosystem and thus worthy of scientific scrutiny. This hypothesis seems to have gone unheeded to date, and as a consequence the potential importance of this habitat remains of unknown value. Nonetheless it remains a possibility that a major contribution is made to energy and nutrient inputs in ecologically, and thus socioeconomically, productive coastal waters. In the absence of detailed study, this possibility should not be overlooked.

One possible alternative landuse which to date has gained support in single sites in each of Abu Dhabi and Dubai emirates has been the manipulation and management of saltflats for wildlife. In the latter emirate, a tidal area, Khor Dubai, with an adjacent sabkha or near-sabkha fringe, already known as an internationally important waterfowl reserve (Evans 1994, Scott 1995) has been enhanced. In the former, the Al Wathba Lakes had already been noted as being of value for wildlife prior to the introduction of water management around the site (Scott 1995). With this management, the lakes have now been transformed into a year-round nationally important waterfowl refuge. The Lakes are the subject of the case study described in detail below. It should be noted at the outset, however, that the emphasis in both instances was first and foremost to encourage greater flamingos *Phoenicopterus ruber* to nest. This is an indication of the cultural value attached to an essentially 'glamorous' subject - the flamingo.

In this regard, success was achieved in Abu Dhabi emirate in the winter of 1998/99 when breeding took place, with ten young flamingos fledging in the spring of 1999. This constituted the first documented breeding in the Arabian peninsula since that in Kuwait in 1922 (Ticehurst 1926), and first occasion actually to be recorded on the mainland of Arabia. This itself served, at a political level, to heighten both awareness and interest in birds and their management and conservation. It is this breeding event that is documented here. Only the physical management and related ecological aspects are described below, and not the socio-political context within which the extending of protection to the Lakes, and their subsequent emergence as an important refuge, took place.

AL WATHBA LAKES, ABU DHABI

In June 1993 a colony of nesting flamingos was discovered at an accidentally ponded area of saltflat adjacent to the truck road from Abu Dhabi to Al Ain at Al Wathba, 40 km east of Abu Dhabi and approximately eight kilometres inland (Aspinall & Hirschfeld 1993). Although this attempt was 'spontaneous' (i.e. opportunistic) on the part of the birds, it ultimately failed due to human interference. It was clear from this attempt, however, that disturbance and dropping water levels during the summer months would always ultimately be likely to prevent success locally, unless an effective management programme was put into place.

Annually thereafter in the summer from 1995 through to 1998 inclusive, small numbers of flamingos displayed at the northern of the two lakes in the area, also accidentally created due to bunding by the same truck road. Between nine and eighteen nests were built in at least three of the years. Rapidly dropping water levels apparently prevented each attempt getting far, and no eggs are known to have been laid on any occasion. The scheme was still blighted by the inability to regulate the water-level in the lake. Except in 1995/96, the lack of winter rain meant that by summer the area of the lake was so greatly reduced as to prove unsuitable for nesting by most waterbird species, flamingos included, while it also became hypersaline and thickly encrusted with salt deposits.

At the nearby Al Wathba Camel Race Track, treated water from the Al Mafraq sewerage treatment plant, used for irrigation at the fodder fields, was under-drained and pumped back up to a holding tank at the treatment plant in order to prevent salination of the fields. From the tank it was discharged gravitationally to the sea at Mussafah. At the fodder fields this water had, by percolating downwards, become saline and was clearly of no possible further use in irrigation. Early in 1996, once the perceived threat of mosquitoes breeding here had been allayed (the hypersalinity precluding breeding), permission was received for treated water to be released into the lake. Water was then let into the lake to regulate the level according to the season, being allowed to draw down naturally (through progressive evaporation) thus becoming suitable for shorebirds during autumn migration, for example. ERWDA (Environmental Research and Wildlife Development Agency), the Abu Dhabi

Government agency, has since commenced a programme of water quality monitoring of inflows, outflows and the lake waters themselves.

Three years later in the winter of 1998/99 (not, as might have been suspected, in the summer) breeding of flamingos took place. Ten young fledged successfully in April 1999. Although over one hundred nests were initially built, and at least forty contained an egg at some stage, many nests were destroyed or had their egg lifted out by wave action due to several days of strong winds in January when the lake level was at its highest. Regulation of the water level clearly required the possibility of it being lowered, as well as being raised. A subsequent breeding attempt in late 1999, failed for unknown reasons, perhaps as a result of human disturbance at a critical time. The planned construction of islands for nesting may alleviate the problem of wave action, as well as providing predator-free nest sites that are also isolated from human disturbance.

Where available, brine shrimps *Artemia sp.* form a large component of the diet of flamingos. In 1996 preliminary arrangements had been made to stock Al Wathba Lakes, should the need arise. Inspection of the lake revealed, however, the existence of already high populations of *Artemia sp.* aff. *fransiscana* and stocking was not, therefore, deemed necessary or, in the event, necessarily ecologically prudent.

The success of other species of bird has also been remarkable. Breeding by avocet *Recurvirostra avosetta* is now annual in the Al Wathba area, still the only such UAE site, while the black-winged stilt *Himantopus himantopus* population has also been restored to its former high level of over 150 pairs. Between 5000 and 10000 waterfowl may be present at any one time in winter, and while the turnover of passage wader populations can only be guessed at presently, many tens of thousand birds must use the lake through the course of spring and autumn migration. The highest national counts are now regularly obtained of several species of waterfowl, black-necked grebe *Podiceps nigricollis*, shoveler *Anas clypeata* and little stint *Calidris minuta* for example. On account of its importance to waterfowl, the site has been documented in the *Directory of Wetlands in the Middle East* (Scott 1995) and would also qualify nationally as an 'Important Bird Area' under the criteria outlined in *Important Bird Areas in the Middle East*, Evans (1994).

A supply of treated freshwater, or at worst, brackish water, tolerable to *Phragmites australis*, was used to saturate the existing reedbeds that had begun to form naturally in feeder and overflow channels. The high nutrient loading of the treated water permitted an accelerated growth rate of the reeds and probably, very usefully, catalysed brine shrimps *Artemia sp.* into mass-producing in the lake itself. In the early stages, reed growth was encouraged by saturating the incipient beds as necessary. The situation has now reached the point where reed cutting or other method of control needs to be introduced.

Management to create and maintain a mosaic of riparian and open wetland habitats has resulted in over 200 species of bird being recorded visiting the lake and surrounding area in the five years up to late 1999.

On the instructions of UAE President H.H. Sheikh Zayed bin Sultan Al Nahyan, a ban on hunting at the Al Wathba Lakes has been in force since the early 1990s. At the end of 1998, the lake was formally designated as a protected reserve. It was inaugurated in November 1999 by H.R.H. Prince Charles, Britain's Prince of Wales.

Management of the lake is now the responsibility of Abu Dhabi's Environmental Research and Wildlife Development Agency, ERWDA, on the instructions of UAE Minister of State for Foreign Affairs, Sheikh Hamdan bin Zayed Al Nahyan, Deputy Chairman of that Agency, while Sheikh Mohammed bin Butti Al Hamed, Chairman of the Abu Dhabi Municipality & Town Planning Department, has guaranteed the essential services of the Municipality in securing access and landscaping the site boundary. Aspinall & Hellyer (1999) documents the full history and development of the Al Wathba lake up to the summer of 1999.

As a corollary to the immediately preceding account, it should be noted that the original intention in Abu Dhabi was always to create a wetland close to the capital, suitable for flamingos to breed at, but which would serve primarily as an attraction for residents of the UAE and to visitors to the country, quite apart from its value as an educational resource for children. With the help of corporate sponsorship from the major international oil company, bp (formerly BP), ERWDA is now developing the area as a showpiece reserve, with hides, interpretive materials and guided nature trails. Research and management requirements for the site are adapted as an ongoing process, within the framework of stated conservation, research and educational goals, in parallel with the progressive ecological development of the area.

CONCLUSION

In countries such as the United Arab Emirates, with their rapid pace of development and associated population growth, it is difficult to justify the large-scale preservation of sabkha and related wetland systems in an unaltered state. Indeed, the construction of a main road across the sabkha flats at Al Wathba had already irrevocably altered the surface and sub-surface hydrology. The management regime now in place at Al Wathba Lakes, however, offers an environmentally and socially acceptable and, indeed, ecologically attractive alternative landuse option.

References

Aspinall, S.J. & Hellyer, P. 1999. The History and Development of Al Wathba lake. *Tribulus* 9.2: 22-25. ENHG, Abu Dhabi.

Aspinall, S.J. & Hirschfeld, E. 1993. The first breeding of the Greater Flamingo (Phoenicopterus ruber) in the UAE. *Tribulus* 3.2: 5-6. ENHG, Abu Dhabi.

Evans, M.I. (comp.) 1994. Important Bird Areas in the Middle East. BirdLife International, UK.

Scott, D.A. (ed.) 1995. A Directory of Wetlands in the Middle East. IUCN/IWRB, Slimbridge, UK

Ticehurst, C.B. 1926. Additional notes of the avifauna of Iraq. *Journal of the Bombay Natural History Society* 31: 110.

SUSTAINABLE DEVELOPMENT OF MANGROVES FOR COASTAL SABKHA ENVIRONMENTS IN ABU DHABI, UAE

AMRITA G. DE SOYZA[1], NAJAMUDDIN B. VISTRO[1] & BENNO BÖER[2]

[1]*Environmental Research and Wildlife Development Agency, PO Box 45553, Abu Dhabi, United Arab Emirates*
[2]*UNESCO Regional Office in the Arab States of the Gulf, P.O. Box 3945, Doha, Qatar*

Abstract

With an increasing demand for freshwater, and with freshwater resources becoming increasingly limited throughout the world, there is an urgent need to identify plant species that can grow with saline water. Mangroves, many of which can grow in the coastal inter-tidal areas and tolerate seawater concentrations of salts, have been identified as a group of plant species with high potential for expanded introduction to many areas of Abu Dhabi Emirate's coastal environment. With this in mind we have initiated projects to identify potential areas for expanding mangrove ecosystems. Projects include development of planting techniques, research on effects of various environmental factors on mangrove survival and growth, and the potential for introducing non-native species of mangroves. We are particularly interested in increasing mangrove (and other halophyte) cover in coastal areas including sabkhat (3550 km^2 of the Emirate's coastline) and the diversity of mangroves and associated plant and animal species along these coastal areas. Although these development activities have a high priority, the overriding priority remains that all developments must be sustainable and they must not have deleterious effects on existing ecosystems.

Barth & Böer (eds.). SABKHA ECOSYSTEMS, 341-346

INTRODUCTION

Abu Dhabi Emirate has a long-established and successful program of "greenification", where roadsides and other desert areas are planted with trees to reduce sand accumulation on roads, and to improve aesthetics and the general environment of the Emirate. Unfortunately, with insufficient precipitation to support these plantations naturally, these efforts require large volumes of fresh water for irrigation in order to maintain the plantations. Some of this water is recycled sewerage, but the vast majority of water used for agriculture, horticulture, and forestry in the Emirate is extracted from limited groundwater reserves or is produced by desalinisation of seawater. While freshwater supplies are limited or expensive to produce, there is no such limitation on seawater. As such, economically viable and ecologically sustainable halophytic agro-ecosystems are becoming increasingly popular.

In nature, salt tolerant plants occur in several distinct ecosystems among which are the tidal swamps and mangrove forests of the inter-tidal areas. Mangroves, which refer to a taxonomically diverse group of about 60 plant species that can grow in the coastal inter-tidal areas and tolerate seawater concentrations of salts, have been identified as one such group of species with high potential for expanded sustainable cultivation. Currently, only one species of mangrove is known to occur naturally in the Emirate - *Avicennia marina* (Forssk.) Vierh. - the grey mangrove.

In the Emirate of Abu Dhabi, the largest of the 7 Emirates that make up the United Arab Emirates, a large proportion, 3550 km^2, of the coastal zone is composed of sabkha. These flat hyper-saline plains do not naturally support any significant cover of higher plants. Relatively simple engineering can create a system that is naturally irrigated with seawater using tidal movements. Such systems can be used to both reduce the salinity of the sabkha and to provide saltwater irrigation for halophytic plants. However, not all sabkha areas are suitable for establishing halophyte plantations. Although, as previously stated, there is a dearth of higher plants, some coastal sabkhat do support lower organisms such as cyanobacteria sometimes in the form of large cyanobacterial mats which are very important components in the functioning of coastal ecosystems. In choosing areas to be planted with mangroves we feel it is important not to displace other functioning biological systems, and thus site selection for mangrove expansion must take into account the current functional condition of an area (such as the presence of living cyanobacteria mats).

Another question of concern is: *are all mangroves the same?* This question is relevant both within a species (e.g. *A. marina*) and among different species. Current (and past) practice in Abu Dhabi is to collect seed from large mature stands of *A. marina* and use these seeds to establish plantations elsewhere on the coast. The environments of these new locations are often extremely different from those of the seed source, and biological adaptation to specific environments may play a key role in the success or failure of plantation efforts. In the case of non-native species this issue is much clearer and possibly much more important. Mangrove species (like most other plants) are able to tolerate specific ranges of environmental conditions such as temperature and salinity. Introducing a species from a low salinity or cooler environment to the hot, highly saline environments of Abu Dhabi without considering its ability to tolerate the new environment could be a waste of time and effort.

Although mangroves are halophytes, a considerable amount of fresh water is often used during the early emergence and growth stages in plant propagation nurseries throughout the Emirate. In some cases mangroves are even grown supplemented with varying amounts of freshwater. While this may be acceptable at a small scale, at a larger plantation scale this will severely compromise the role of mangroves as a sustainable alternative to growing plants with freshwater. Other aspects of planting technology that may affect the success of mangrove plantations include the use of direct-seeding versus "polybagged" or pot-reared plants, the number of seeds used, and the season of planting, composition and depth of soil, average tide height and shelter from high energy wave action. Research to identify and/or develop appropriate propagation and planting techniques, taking into account success-rates, sustainability, and cost is therefore another vital aspect in developing the Emirate's mangrove ecosystems.

Another area of concern is that once time and effort have been spent to establish mangrove ecosystems, we should be able to protect these from adverse anthropogenic stressors. In Abu Dhabi Emirate these stressors are known to include direct activities such as physical changes to the environment due to uncontrolled development and indirect effects such as pollution from oil spills. No matter what the origin, anthropogenic stressors can cause widespread disruption to or destruction of the ecosystem. Given the importance of mangrove ecosystems to the Emirate, and the efforts taken to expand these ecosystems, it is appropriate to give due consideration to the question of how these ecosystems can be protected and preserved. In this paper we outline some of the research and development activities related to the development of mangrove ecosystems currently underway in the Emirate of Abu Dhabi, UAE.

RESEARCH AND DEVELOPMENT ACTIVITIES

Environmental conditions and genotypes

A. marina, the only mangrove in the UAE is probably the most widely distributed mangrove species in the world and occurs as far away as Australia. This wide distribution suggests that it also has high genotypic variability. We are in the process of quantifying environmental conditions in several large (natural) stands of mangroves to determine whether there are significant differences among these areas.

If so, the genotypes of mangroves in these areas may similarly be different and show adaptations to particular combinations of environmental factors. An understanding of these perhaps unique factor combinations and adapted genotypes would allow us to carefully select appropriate genotypes for mangrove expansion into particular areas. This is particularly important for sabkha areas where mangrove genotypes with adaptations to tolerate, for example, high salt concentrations would probably enjoy significant advantages over plants that are less able to tolerate high salt concentrations. This research is being carried out in collaboration with the International Center for Biosaline Agriculture (ICBA) based in Dubai, UAE.

Non-native mangroves

It has long been assumed that mangrove species other than *A. marina* do not occur naturally in the UAE because of the high prevailing seawater salinity and high air temperatures. However, other factors such as geographic isolation may be the actual cause. The introduction of exotic species without proper consideration of the consequences on the ecosystem at large can lead to environmental problems. Therefore we are undertaking studies (with ICBA) on the ability of several non-native species to tolerate prevailing environmental conditions and undertaking small-scale controlled planting of non-native mangroves in order to first assess their potential and their potential impacts. We hope that some non-native mangrove species may provide alternate genotypic material for establishment on sabkha areas and add to the currently low diversity.

Propagation techniques

Two general types of mangrove nursery are currently used to propagate mangroves in the UAE. The first type is the inter-tidal nursery where the nursery is constructed in an inter-tidal area and natural tidal fluctuations in sea level are employed to irrigate container-planted mangrove propagules. The second type is the land-based nursery where the nursery is constructed in supra-tidal areas and irrigated artificially. Irrigation is sometimes with seawater but most often initially with fresh water and followed with a brief period of acclimation to seawater before planting. Propagation success with the first (in-sea) type of nursery is variable and can sometimes be very low (>10%)%) - protection from wave-energy seems to be the crucial factor for high success rates. Supra-tidal nurseries are generally less variable and have higher success rates (75-90%). Our studies have also shown that using a 1:1 mixture of seawater/freshwater results in better growth of *A. marina* seedlings than does 100% freshwater or 100% seawater (Vistro & de Soyza 2001). This work is now being repeated for non-native mangroves such as *Ceriops tagal* and *Rhizophora mucronata.*

Research is also in progress to assess the suitability of various container types and sizes for mangrove propagation and subsequent planting. Studies are in progress to evaluate a range of propagation media including sand and various quantities and types of organic supplements to achieve the greatest propagule sprouting success and the subsequent success in establishment. These experiments are factorial in nature and will we hope enable us to maximise establishment success in various locations including sabkhat.

Planting techniques

Research projects are also underway to improve the establishment success of mangrove plantations. This includes studies to investigate the optimal number of seedlings (1 to 4) per container needed to achieve greatest transplantation success. In other ongoing research we are investigating the growth performance of *A. marina* saplings grown in full sunlight versus under shade.

Using information gathered we have also established medium-scale mangrove plantations in several locations around the city of Abu Dhabi. Initial analysis of survival and growth data together with continued monitoring of growth have enabled us to develop an understanding of some of the factors that must be taken into consideration for successful plantation of mangroves.

Mangroves in sabkhat

A current research project is also underway to investigate the potential of establishing mangrove plantations on some of the Emirate of Abu Dhabi's vast areas of coastal sabkha. In a collaborative project with the Japan Oil Development Company (JODCO) and scientists from the International Society for Mangrove Ecosystems (ISME) an area of sabkha has been identified, a channel system designed, and initial preparatory work underway to plant mangroves in an area of coastal sabkha. This project will draw on the combined understanding of mangrove ecology generated from the research outlined above to further understand and implement the potential of mangroves to grow in coastal sabkhat.

Protection

While historically mangroves of the UAE coast were probably most severely affected by direct human intervention - harvesting for livestock forage and for charcoal - the current threats are indirect. Locally, all trees including mangroves receive a high degree of protection and their destruction is banned. However, being in the inter-tidal zone of a seaway carrying large amounts of petroleum products mangroves are very prone to being adversely affected by marine pollution including oil spills. While studies are underway to evaluate the effects of marine oil spills on mangrove function and survival (de Soyza, in press), an environmental sensitivity atlas of Abu Dhabi's coastline (2001) has identified the locations of mangrove stands (and other sensitive areas). This will enable oil spill event coordinators to potentially prevent oil spills from reaching mangrove areas.

CONCLUSION

Mangroves hold high promise for use to increase plant cover in coastal areas. Of the Emirate's coastal areas the coastal sabkhat have the most potential for use to expand mangrove cover, and are a major emphasis of ongoing research. Although forestation of coastal sabkhat is considered to have high potential, it is being approached in an ecologically sensitive manner so that mangroves are not used as replacements for other functioning biological systems. In order to maximise the potential for success, and to minimise the potential for adverse environmental effects we are approaching the issue of successful propagation, planting, establishment, growth and protection of mangroves in coastal sabkhat in a logical and considered way. Research on individual issues relating to these issues will be followed by small-scale field trials is being used prior to an eventual large-scale planting effort. In addition, we are committed to collaborating with other interested parties such as the International Center for Biosaline Agriculture, the Japan Oil Development Company, Crown Prince (Pvt.) Management, Abu Dhabi Municipality, and the International Society for Mangrove Ecosystems to maximise the potential for success in this endeavor.

ACKNOWLEDGEMENTS

We thank the Crown Prince of Abu Dhabi, H.H. Sheikh Khalifa bin Zayed Al Nahyan, Chairman of ERWDA, and H.H. Sheikh Hamdan bin Zayed Al Nahyan, Deputy Chairman of ERWDA. We also wish to thank the many organisations and individuals within the UAE (Abu Dhabi Municipality, Crown Prince (Pvt.) Management, Ministry of Agriculture, International Center for Biosaline Agriculture, and ZADCO) and internationally (Japan Oil Development Company and the International Society for Mangrove Ecosystems) who have and are contributing toward the development of mangrove ecosystems in the UAE.

References

Böer, B. & Gliddon, G. 1998. Mapping of coastal ecosystems and halophytes (case study of Abu Dhabi, United Arab Emirates). *Marine and Freshwater Research* 49: 297-301.

de Soyza, A.G. (in press). Development of Quantitative Tools for Improved Environmental Decision-making in Arid Environments. *Environmetrics*

ERWDA 2001. Abu Dhabi Coastline Oil Spill Protection Priorities. ERWDA, Abu Dhabi, UAE, 120p.

Vistro, N. & de Soyza, A.G. (in press). Effect of water salinigy on seedling survival, growth, and physiology of grey mangrove (Avicennia marina) in the UAE. Proceedings of the conference on Prospects of Saline Agriculture in the GCC Countries, Dubai, UAE.

UNESCO's WORK FOR THE PROTECTION OF ARID ZONES WITH PARTICULAR EMPHASIS TO SABKHAT

MIGUEL CLÜSENER-GODT

UNESCO - Division of Ecological Sciences, Man and the Biosphere Programme (MAB) 1, rue Miollis, 75732 PARIS, Cedex 15, France

"Much is expected of the sciences to remedy the problems the world is experiencing in terms of health, nutrition, environment, resource management, preservation of the conditions of life on Earth, but also in terms of the knowledge of societies, people, civilisations, and of the understanding of the changes that are affecting humankind."

Koïchiro Matsuura

(Speech of the Director-General of UNESCO at the closing ceremony of the 30th session of UNESCO's General Conference, 17 November 1999)

INTRODUCTION

Arid and semi-arid zones account for more than one-third of the earth's land area and are inhabited by about 16 per cent of the world's population. Until recently, these lands had relatively low human and animal pressure and natural ecosystems kept in balance. However, population increase, global climate change, settling of a large number of the nomadic populations as well as over-use of renewable resources such as water has disturbed this sensitive balance. The results were weakening of ecosystems, decrease in potential and real production, which led to desertification and almost irreversible forms of soil degradation.

In this context, UNESCO was asked by its Member States concerned to elaborate strategies and to assist via its numerous programmes to overcome or alleviate part of

Barth & Böer (eds.), SABKHA ECOSYSTEMS, 347-352

these problems. With its broad mandate, covering all the sciences, including the social and human sciences, education, culture and communication, UNESCO is ideally placed to tackle environmental problems and their interaction with development factors. Among the main activities that UNESCO has been implementing over the past 40 years, one programme gives strong emphasis to sabkhat and similar ecosystems which high salinity: The Programme on Man and the Biosphere (MAB). Under this programme, many initiatives have been launched and are shortly described in the following paragraphs (UNESCO 1996).

While nationally implemented projects and activities will not be neglected, UNESCO is focusing mainly on projects of a regional and sub-regional character, as well as on world-wide initiatives, such as the World Network of Biosphere Reserves (UNESCO 2000).

Arid and semi-arid regions are vulnerable areas and merit special attention of the world scientific community. By preserving the natural renewable resources and the traditional knowledge of a region and by well determining its given carrying capacity, these areas will keep their productivity and will keep their unique living conditions for the benefit of their inhabitants.

MAN AND THE BIOSPHERE PROGRAMME

The Man and the Biosphere Programme (MAB) has been launched in 1971, and deals mainly with terrestrial and coastal resources management, terrestrial and coastal ecology. The main goal of this programme is to harmonise the relationship between human beings and nature through conservation and sustainable development. Main topics of this programme cover arid and semi arid zones with special emphasis on network building, technology transfer and improvement of management know-how for protected areas, such as biosphere reserves.

The main issues of this programme and its ways of operating aim at:

- mobilising the scientific community working in a specific field, such as arid or semi arid zones,
- extending available knowledge and distributing it,
- improving the technological and scientific standards valid in that region,
- promoting or introducing new approaches, methods and techniques,
- exchanging information, expertise, ideas (UNESCO et al., 1998),
- strengthening the scientific and technological capacities through training of junior and senior scientists,
- acting directly carrying out pilot projects which aim at proving on the ground the different techniques and making them available at the regional level and understood by the local decision-makers.

Biosphere Reserves are areas of terrestrial and coastal ecosystems, which are internationally recognised within the framework of UNESCO's Man and the Biosphere Programme. Collectively, they constitute a World Network. They are nominated by national governments and must meet a minimal set of criteria and adhere to a minimal set of conditions before being admitted into the World Network. Each Biosphere Reserve is intended to fulfil three basic functions, which are complementary and mutually reinforcing:

- a conservation function - to contribute to the conservation of landscapes, ecosystems, species and genetic variation;
- a development function - to foster economic and human development which is socio-culturally and ecologically sustainable;
- a logistic function - to provide support for research, monitoring, education and information exchange related to local, national and global issues of conservation and development.

Individual Biosphere Reserves remain under the sovereign jurisdiction of the countries in which they are situated. Some countries have enacted legislation specifically to establish Biosphere Reserves. However, in many cases, advantage is taken of the existence of areas already protected under national law to establish Biosphere Reserves. Biosphere Reserves consist of a core area, a buffer zone and a transition area and only the core area requires legal protection. A number of Biosphere Reserves simultaneously encompass areas protected under other systems (such as National Parks or Nature Reserves) and other internationally recognised sites (such as World Heritage or Ramsar wetland sites).

Ownership arrangements may vary too. The core areas of Biosphere Reserves are mostly public land, but may also be privately owned or belong to non-governmental organisations. In many cases, the buffer zone is in community or private ownership, and this is generally the case for the transition area.

Establishing a Biosphere Reserve obviously poses an enormous challenge, namely to set up an appropriate mechanism, for instance a committee, to plan and co-ordinate all the activities that will take place there. This human dimension of Biosphere Reserves makes them special, since the management of a Biosphere Reserve essentially becomes a "pact" between the local community and society as a whole. Management needs to be open, evolving and adaptive. Such an approach requires perseverance, patience and imagination. But it will allow the local community to be better placed to respond to external political, economic and social pressures, which would affect the ecological and cultural values of the area.

PROJECT ON INTEGRATED BIODIVERSITY STRATEGIES FOR ISLANDS AND COASTAL AREAS (IBSICA)

UNESCO also establishes projects which have strong field experiment components, such as the Project on Integrated Biodiversity Strategies for Islands and Coastal Areas (IBSICA). This project deals mainly with the protection of coastal areas and the protection, rational use and development of their living resources. Following programme areas include as well arid and semi arid zones:

- Integrated management and sustainable development of coastal areas, including exclusive economic zones,
- Marine environmental protection,
- Sustainable use and conservation of marine living resources under national jurisdiction.

Within the framework of this project, which is sponsored by Germany through the Federal Ministry for Economic Co-operation (BMZ), particular attention is given to those countries having or planning biosphere reserves on islands and/or coastal zones and to those countries whose governments have already signed and ratified the Convention on Biological Diversity. These countries represent nearly 30% of the ratifying countries.

Major actions of this were the "Mobile demonstration module to show the operation of the sustainable halophyte utilisation with sea water irrigation", which was started in early 1997. It was executed in close co-operation with the Concerted Action Programme "Sustainable Halophyte Utilisation in the Mediterranean and Subtropical Dry Region" of the European Commission aimed at the testing and demonstration of successful application of seawater irrigation during the course of the project. During the development of the programme it became clear that crucial aspects of our system could be shown at relevant fairs and international meetings. UNESCO has provided a module for this undertaking as a special international support to the programme.

HALOPHYTIC PLANTS FOR SUSTAINABLE DEVELOPMENT OF COASTAL AND SALT AFFECTED AREAS IN MOROCCO

Another project deals with "Halophytic plants for sustainable development of coastal and salt affected areas in Morocco". The aim of this project is to investigate Moroccan and acclimated halophytic plant species in order to estimate their potential as economically useful plants under saline irrigation. The use can be either in terms of greenification, dune stabilisation or biomass production. The importance of such species resides in the fact that they can tolerate irrigation with highly saline water. As it was stated in the introduction water scarcity is becoming a structural problem and there is increasing competition between different uses. The possibility to use saline water for plant irrigation will have two positive consequences. First it will relieve pressure on fresh water and second it will enable the creation of improved environments in terms of green areas and stabilised dunes in addition to the possibility of useful biomass supply to local populations (fodder, fuel, vegetables).

The work was conducted at the Institut Agronomique et Vétérinaire Hassan II, Agadir (30° 20' Northern latitude, 9° 30' Western longitude, 32 m above sea level) in a greenhouse designed especially for the purpose of investigation on halophytes.

In addition to the closely monitored experiments under greenhouse conditions the establishment of a demonstration site was considered as an important issue with the aim of showing in reality to the potential users the feasibility of the utilisation of halophytes under saline irrigation. This was the reason for the contraction of a tripartite protocol for the establishment of such a demonstration site between the I.A.V. Hassan II, the Municipality of Agadir and the Beach Club Hotel in Agadir.

CONCLUSION

UNESCO is giving support both logistic and financial to all kind of actions concerning these activities and would encourage interested agencies and institutions to participate in these challenging ventures. UNESCO invites all concerned countries to become active members in the conventions and in the scientific programmes, such as the Man and the Biosphere Programme (MAB). As these programmes aim to harmonise the impact of human action with nature for the benefit of the local population, they are the ideal tools for sustainable development and conservation of renewable natural resources.

Finally, it is important to mention that the World Science Conference, held in Budapest in June 1999, included an agenda for action in arid and semi-arid regions in the list of priority follow-up activities. Only the engaged participation of all governmental, non-governmental and intergovernmental agencies and operators can guarantee it's success (UNESCO & ICSU 2000).

In conclusion, and recalling the role of Biosphere Reserves in arid zones for nature conservation and sustainable development, the establishment of such reserves in the Arabian Peninsula is highly recommend by UNESCO and its international partners. The sabkhat of this region, and in particular those in Abu Dhabi, which are the largest of the world, are perfectly situated for conservation of dry ecosystems and for the development of sustainable land-use techniques with seawater irrigation. This is particularly relevant for the interaction of sabkhat with adjacent terrestrial or marine ecosystems, such as active sanddunes, mangroves, seagrass beds and other coastal systems.

References

UNESCO, 1996. Arid Zones in UNESCO's programmes. UNESCO, Paris (in English and French), 33 pp.

UNESCO, 2000. Map of the World Network of Biosphere Reserves.

UNESCO, BMZ, Institut du Sahel, and OSS, 1998. Bilingual CD-ROM (English and French) on Arid Zones and Desertification.

UNESCO and ICSU, 2000. Science for the Twenty-First Century: A new Commitment, 48 pp.

Index

MIX
Papier aus verantwortungsvollen Quellen
Paper from responsible sources
FSC® C105338

If you have any concerns about our products,
you can contact us on
ProductSafety@springernature.com

In case Publisher is established outside the EU,
the EU authorized representative is:
Springer Nature Customer Service Center GmbH
Europaplatz 3, 69115 Heidelberg, Germany

Printed by Libri Plureos GmbH
in Hamburg, Germany